LAND-USE SURVEY IN TARAI TRACT

LAND-USE SURVEY IN TARAI TRACT
A Study of Eastern Uttar Pradesh

S.C. SHARMA

Foreword by
Prof. MOHAMMAD SHAFI

CONCEPT PUBLISHING COMPANY, NEW DELHI 110059

The Indian Council of Social Science Research gave the financial support towards the publication of this work. However, the responsibility for the contents of the book — facts, opinions, conclusions, etc. — is entirely that of the author and not of the ICSSR.

A.K. Tiwari
ICSSR Consultant

ISBN 81-7022-337-7

First Published 1991

Published and Printed by
Ashok Kumar Mittal
Concept Publishing Company
A/15-16, Commercial Block, Mohan Garden
NEW DELHI 110059 (India)

Lasertypeset by
Microtech Advance Printing Systems (Pvt.) Ltd.
H-13, Bali Nagar
NEW DELHI 110015

DEDICATED
TO
MY REVEREND PARENTS

FOREWORD

THE BOOK *Land-use Survey in Tarai Tract (A Study of Eastern Uttar Pradesh)* is a new addition in the land-use planning of an area which has hitherto been backward and underdeveloped. The independence of the country opened new vista for the development of this region. Large scale colonisation has taken place. Population has increased substantially making heavy demands on land resources for greater production of food. More and more land has been cleared of the forests, which is affecting the ecological balance. The need of the hour is a scientific land-use planning of the region.

Dr. Sharma in this book has attempted to map out the land-use of the region on the basis of field-work and has analysed the spatial and temporal aspects of land-use. The book is divided into four parts. After discussing the general pattern of land-use, he has examined intensively the land-use and associated features and problems of each sample village selected for the study on the basis of field-work and makes concrete suggestions for the improvements.

The book gives a mine information on the basis of field-work and should be useful to researchers, planners and all those interested in the uplift of the Tarai Region.

MOHAMMAD SHAFI
Emeritus Professor

Aligarh

ACKNOWLEDGEMENTS

AT THE outset, I wish to express my sincere gratitude to the University Grants Commission, New Delhi for making available liberal financial support and sponsoring this study and to the Indian Council of Social Science Research, New Delhi which have come up with a publication grant making it possible to bring out this investigation in a book form.

I owe a debt of gratitude to all those who extended their assistance, cooperation and help in the completion of this geographical investigation. To mention some of them, I am thankful to all the district, tahsil and block authorities who permitted us access to the published as well as unpublished official records and helped us in the collection of data and other relevant information. I owe a special debt of gratitude to the authorities of the Gandak Command Area and Regional Soil Testing Laboratory, Kurhaghat, Gorakhpur, for not only providing the technical guidance for systematic detailed soil survey and analysis, but also for taking up the physical and chemical analysis of the soil samples. All the staff always welcomed to share with their technical expertise and rich experience. I would like to acknowledge my thanks to the officials of the Survey Department of India, Dehradun for supplying the spot heights of the required stations of the area, and to the officials of the Regional Meteorological Department, New Delhi for making available the climatic data required by us.

I take this opportunity of placing on record my sincere gratitude to Prof. M. Shafi, Aligarh Muslim University, Aligarh for writing the foreword. It is my pleasant duty to express my deep gratitude to my reverend teachers, Dr. A.N. Bhattacharya, Retd. Professor of Geography, Udaipur University, Udaipur, Dr Indra Pal, Rtd. Professor of Geography, Rajasthan University, Jaipur, and Dr. D.P. Saxena, Retd. Reader and Head of the Geography Department, D.A.V. College, Kanpur for the stimulus and encouragement they have given for the valuable suggestions they have poured so liberally.

I find no words to express my gratitude to Dr. R.S. Murthy,

Director, National Bureau of Soil Survey and Land-use Planning, Nagpur, Professor S.B. Despandey, G.B. Pant University of Agriculture and Technology, Pantnagar, Dr. S.K. Roy, Assistant Registrar General, India, Mr. G.K. Dutt, Joint Director, National Atlas Organisation, Calcutta, Professor A. Ramesh, Madras University, Madras, Professor E. Ahmad (Retd.) and late Professor P. Pandey, Ranchi University, Ranchi, Professor (Retd.) R.L. Dwivedi, Allahabad University, Allahabad and Dr. V.K. Srivastawa, Gorakhpur University, Gorakhpur for the valuable suggestions I received from them on matters of their specialisation.

I take this opportunity of placing on record my appreciation and thankfulness to the villagers with whom we came in contact and who provided us with valuable information.

I would be failing in my duty if I do not express my thankfulness to my research assistants, particularly Mr. Jagannath Tiwari, who worked hard in collection, tabulation, compilation and cartographic representation of data, and to Messrs Puttu Lal Tripathi and Ram Kewal Gupta who took great pains in helping me in revising and up to dating the data and maps of this study.

Last but not the least, I am thankful to my wife for providing great encouragement and enthusing inspiration in me when they were needed most.

Mr. Ashok Kumar Mittal, Concept Publishing Company, New Delhi, have taken up the challenging task of printing and publishing this book in such a short span of time and they have come up to my expectations. I am thankful to them.

S.C. SHARMA

Department of Geography
M.L.K. (P.G.) College,
Balrampur, Gonda, U.P.
January 26, 1990

CONTENTS

LIST OF FIGURES

LIST OF TABLES

INTRODUCTION

LAND IS the most significant of all the natural resources of our country, as the larger proportion of its inhabitants depends on the exploitation of this resource. It largely contributes to the well-being of human society. Through various ways, it provides material for human needs. Today, when we are faced with the crucial problem of incessantly growing pressure of population on the insufficiently exploited limited land resources, attended by widespread poverty, malnutrition and low economic level, it seems imperative to undertake a factual objective study of these resources, i.e. use and misuse of land, its actual and potential resources and their proper exploitation and conservation, for the proper use of land is not only directly linked with the welfare of the rural population but also with the general well-being of the country as a whole. The need of the hour is that keeping in view the basic principles of optimum utilisation of land resources, every acre of country's land must contribute to the maximum of its capacity and must serve some purpose useful to the society. This necessitates the proper planning of the land resources, which must be preceded by a thorough and careful factual survey of the present position and its scientific interpretation.

Such a detailed and intensive land-use survey is needed for the entire country, but there are some 'Problem Areas' which although endowed with rich land resources, are not contributing their due share to the national economy and well-being of the community, where very little work has been done by various agencies, which have remained forbidden and underdeveloped lands for centuries and which throw a great challenge before the land-use planners. These 'Pioneer Fringes' must receive the immediate attention of the National Research and Development (R&D) schemes and be taken up for study on the priority basis.

The east-west narrow, elongated strip of the *Tarai* of Eastern Uttar Pradesh, south of the Indo-Nepal boundary, bounded by the river

Ghaghra in the west and the river Gandak in the east, is such a 'Pioneer Fringe' which calls for our immediate attention. It is a low, in some parts marshy, stretch of land with low gradient and is infested even today, although in only a few parts, with tall grasses or forests under natural conditions. Its micro-pattern reveals a surface that is entirely criss-crossed by innumerable streams and *nalas*. It is generally marked by high water-table and fertile though highly leached soils with excessive moisture in the sub-soil and high clay factor with extensive water-logging and flood hazards in many parts. It has been notorious for its unhealthy malarial climate which is characterised by excessive heat and humidity and moderately heavy rainfall. Essentially an agricultural region, the *tarai* is marked by a comparatively lower density of population, poor means of transport and communication reducing its accessibility, by stagnation and under development. It is backward to all intents and purposes and is still the home of some aboriginal tribes such as Tharus.

Objective

The objective of the study is to undertake a detailed and intensive land-use survey of this 'Pioneer Fringe' and to analyse and interpret it scientifically with a view to exploring and unfurling the potential land resources of the region and suggesting an optimal use for the future in the national interest and thus converting this region of backward economy into a region of sound and flourishing economy and also suggesting the suitable methods and techniques of land-use analysis for such other areas in the country.

The actual plan of the work is as follows:

(i) To undertake a comprehensive study of the physical and cultural factors affecting land utilisation,
(ii) To work out a clear picture of the existing and past land-use on the micro pattern on the basis of the detailed field survey, collection of data and cartographic representation,
(iii) To undertake a systematic examination and scientific analysis and interpretation of the existing and past land-use and thus to evaluate the various factors which have determined the very complex and intricate pattern of land-use,
(iv) To adjudge the potential land resources and work out land capability classification keeping in view the degree of

limitations in the use of different categories of land, fertility status of soils, their slope, drainage and other properties, present and past land-use etc. and find out whether the land resources are properly utilised,

(v) To suggest the actual land-use planning for the future, i.e. the optimal use of the various categories of land according to their capability and suitability in the national interest and to suggest proper steps to eradicate the handicaps and problems that hamper such an utilisation, and

(vi) On the basis of the above study, to suggest the methods and techniques of land-use analysis, most suited to our conditions.

Overview of Literature

The principle that a factual survey of the present position and its scientific interpretation must precede any future planning of land-use received the attention of the geographers in the early part of the present century. It was first emphasized in some papers published by the professional geographers in the early part of this century.[1] This was followed by numerous land-use studies in various parts of the world by eminent geographers. The works of Stamp and Buck are really pioneer in this field.[2] This subject was given international recognition in 1949 when a Commission was set up by the IGU to institute and direct world land-use survey.

Inspired by Prof. Stamp, Chatterjee emphasised the necessity of land-use survey in India in his presidential address to the Section of Geography and Geology in 1940 Session of the Indian Science Congress. Since then the topic has widely been discussed and the problems of scientific utilisation of land have attracted the attention of the Indian geographers greatly. Various land-use studies ranging from inventories of land-use surveys to topical or regional descriptive accounts and applying recently quantitative techniques in the analysis of various land-use components, have been carried out. Chatterjee conducted the land-use survey of two districts of West Bengal[3] and discussed the problem of covering the whole of India by land-use surveys after the British Model, as such reliance on sample surveys in the vast country such as India with immense regional variations would not be very useful.[4] The pioneer work on land-use was done in South India by V.L.S. Prakasa Rao.[5] He emphasised the techniques of the soil survey that he adopted in the study of Godavari region. He advocated that the field to field survey, though essential, would be difficult to undertake in such a large country and hence sample technique should be adopted. Shafi whose work is pioneer in Northern India also suggested this technique.[6] He made a strong plea to carry out a land utilisation survey combining with its land capability survey in order to determine the measures for maximum utilisation of land in relation to its intricate quantities.[7] Some of the other early works covering regional surveys are those of Ameen, M., A.K. Sen, C.B. Joshi, O.P. Bharadwaj, A.N. Raina, A.T.A. Learmonth, M.S. Honras, N.P. Ayyar and Srivastava and S.C. Bose etc.[8] A number of studies have been made in the recent past applying the modern techniques of land-use analysis in various parts of India.[9] It is not possible to enumerate all them here. In Uttar Pradesh where the work of land-use survey and analysis on the lines of the model of Prof. Shafi has continuously been done by many students, a number of studies have been carried out,[10] but no systematic land-use survey has been conducted in this 'Pioneer Fringe'. This fact together with the familiarity of the author with the region and the special obligation on account of his institution being situated in this area has prompted him to undertake this important study.

Coverage and Methodology

The study area comprises of 4127 revenue villages grouped into 31 Development Blocks which are the basic territorial units for geographical analysis at the micro-level. The data relating to land-use, which were initially collected at the village level from the village land records, viz., Area statement, Cropwise Statement and Assessment Reports etc., have been compiled at these block levels. Triennial averages have been used to allow for the annual erratic fluctuations. Thus for exhibiting the existing pattern of land-use, the averages of three years, viz., 1985, 1986 and 1987 have been worked out, and for evaluating the changes in the past, averages for 1955, 1956 and 1957; 1965, 1966 and 1967; and 1975, 1976 and 1977 have been calculated. Since there have been great variations in the block boundaries and that in the previous years, the units for recording the data of land-use after the villages were Revenue Circles or *Tappa* etc., all these data had to be taken at the village level involving a tremendous work.

The ideal method of a quantitative study is by a systematic field work.[11] It plays a significant role in land-use studies by supplying an adequate and systematised body of relevant data for generalisations and for better future planning. Such a field to field survey based on 'British Model' was recommended by the International Geographical Union (IGU) for India. It got the favour of the eminent geographers like Prof. Chatterjee, as it has already been mentioned. But due to the large number of the villages of the area, the immensity of the data to be handled, small number of the trained workers available for the job and the financial limitations, it could not be possible to undertake such a detailed survey for the entire study area and the data regarding these villages were collected, as mentioned earlier, from secondary sources. For detailed field to field study of the land-use and for other relevant information, 14 sample villages were selected from different physiographic regions employing the stratified random sampling technique.

The author does admit that the number of the villages is inadequate as regards the principle of sampling is concerned and it can also not be claimed with certainty that they do represent their respective regions, but under the limitations already emphasised upon, greater number could not be taken.

These villages were visited a number of times, first for gathering the general information and then thrice in a year (in kharif, rabi and zaid seasons). The land-use data were recorded on the cadastral maps and the other relevant information was collected through questionnaires and village schedules. It is in the study of these villages that the land capability classification has been attempted and a pragmatic scheme for land-use planning has been suggested.

No systematic soil survey has been undertaken in the study area, except in the portion coming under the Gandak Command Area in Deoria and Gorakhpur districts. This required a systematic soil survey at the micro-level. Since it again involved enormous time and funds, 58 sample villages were selected by employing the same stratified random sampling technique. They were selected from different topographical sub-regions. About 400 soil samples were taken from these villages and they were analysed in the laboratory for determining their physical and chemical properties. On the basis of these results together with the information obtained from the District Gazetteers, Settlement Reports and other revenue records, the soils of the study area have been classified into different groups.

The population data were complied from the *District Census Handbooks* for the years 1951, 1961, 1971 and 1981. As regards the population data of the sample villages for the year 1987, they were collected personally during the field study. All the climatic data were obtained from the secondary sources.

Various statistical techniques have been adopted for representation of the data. Absolute figures pose vexing problems because the spatial comparisons and relationships between various associated phenomena cannot be ascertained and these data do not fit into a standardised statistical test and cartographic representation unless changed into ratio-converts.[12] Hence ratios have been extracted by relating one category to the other and quantitative maps, especially those employing choropleth technique have been prepared. Averages, standard variations, percentages, co-efficient of variation, correlation etc. have been worked out. Doi's method has been used for finding out crop-combinations and for the quantitative measurement of the overall volume of change in the general and crop land-use Weaver's index has been employed.

Since most of the data have been obtained from secondary sources, the statistical approach for representing these data is limited by their quality and degree of accuracy, although every care has been taken to do away with the discrepancies and anomalies. The study involves large calculation and computation work, all of which has been done manually. In this tedious work again, great care has been taken to avoid errors, but even then the author cannot claim of the flawless perfection.

Design of the Investigation

The whole study is divided into the following four parts:

Part one deals with the physical and cultural factors affecting the land-use, namely, location, geology and relief, drainage, climate, soils, population and irrigation.

Part two discusses the patterns and dynamics of general land-use, intensity of cropping, crop land-use and their changing patterns together with the crop-combinations of the region.

Part three is devoted to the field to field land use survey of the sample villages and discusses the data pertaining land classification, irrigation, manuring, general land-use, crop land-use, population and land-use and carrying capacity of land etc. together with the recommendations for the optimum use of the land according to its capability in the national interest.

Part four gives the summary of the main findings and recommendations, i.e., the problems and prospects of the land-use of the region. It suggests the methods and techniques of land-use analysis and planning, most suited to our conditions.

NOTES

1. Saucor, C.O., (1919), 'Mapping the Utilization of Land', *Geographical Review*, Vol. 4.
   Jones, W.D. and Finch, V.C. (1925) 'Detailed Field Mapping of the Agricultural Area', *Annals of Association of American Geographers*, Vol. 15, pp.19-25.
2. Stamp, L.D. (1947), *Land of Britain: Its Use and Misuse*. Buck, J.L., *Land Utilization in China*.
3. Chatterjee, S.P. (i) (1945), 'Land Utilization in the District of 24 Pargana', *Bengal, B.C. Law*, Part 2, Calcutta, (ii) (1952), Land Utilization Survey of Howrah District, *Geo. Rev. of Ind.*, 14(3).
4. Chatterjee, S.P., (1953), 'Land Utilization Survey in India', *Observer*, 1, (1956), 'Preliminary Remarks on Land-use Survey in India', *Proceedings of the International Geographical Seminar*, Aligarh; (1962), 'Land-use Survey in India', *Proceedings of the Summer School in Geography*, Simla.
5. Prakasa Rao, V.L.S., (1956), 'Land-use Survey in India: Its Scope and Some Problems', *Proceedings of the International Geographical Seminar*, Aligarh; (1947), 'Soil Survey and Land-use Analysis', *Indian Geog. Journal*, 22(3).
6. Shafi, M., (1960), 'Land Utilization in Eastern Uttar Pradesh', Aligarh.
7. Shafi, M., (1961), 'Fieldwork in Land-use Research under Indian Conditions', *Bombay Geographical Magazine*, 8 & 9(1)
   (1951), 'A Plea for Land Utilization Survey', *Geographer*, 4(2).
   (1966), 'Techniques of Rural Land-use Planning with Reference to India', *Geographer*, 13.
8. Ameen, M., (1956), 'Land-use Survey of Delhi State', *Proceedings of the International Geographical Seminar*, Aligarh.
   Sen, A.K., (1957), 'Land Utilization and Agricultural Planning in Matpalsa, Birbhum', *Geogr. Rev. of Ind.*, 19(4).
   Joshi, C.B., (1961), 'The Ratnagiri Taluka', *Bombay Geographical Magazine*, 8 and 9 (1).
   Bharadwaj, O.P., (1961), 'Land-use in the Lowlands of the Beas in the Dist-Jullundur Doab', *National Geographical Journal of India*, 7(4).
   (1964), 'Land-use in the Lowlands of the Sutlej in the Dist-Jullundur Doab, Sample Studies', *N.G.J.I.*, 10(1).
   Raina, A.N., (1962), 'Floating Gardens — Land-use Study', *Proc. S.S. G.*, Simla.
   (1962), 'Scope of Land-use in Rural Kashmir', *National Geographer*, 5.
   (1963), 'Observations on Soils and Land Utilization in Kashmir,', *Geographical Rev. of India*, 25(3).
   Learmonth, A.T.A., (1962), '*Sample Villages in Mysore State in India*, Liverpool, United Kingdom.

Honrao, M.S., (1962), 'Land-use Pattern in the Lower Kali Basin', *Proc. S.S.G.*, Simla.

(1953), 'Halge; A study in Land-use in North Kanara Coast Lands', *Bom. Geo. M.* 1(1).

Ayyar, N.P. and Srivastawa, S., (1968), 'Land-use and Nutrition Beas Basin', *Geographer*, 15.

Bose, S.C., (1965), 'Land-use in the Malnad Region', *Proceedings of the Summer School in Geography*, Nainital.

(1969), 'Environmental Control on Land-use in the Himalayas and Its Scope of Development', *Essays on Agricultural Geography*, in B. Banerjee (ed.) Calcutta, pp. 52-59.

Sinha, R.N.P., (1965), 'Present Land-use Pattern in the Canal Irrigated Area of Patna District, Bihar', *Geographical Outlook*, Vol. IV.

9. Singh, J., (1974-75), *An Agricultural Atlas of India: A Geographical Analysis*, Vishal Publication, Kurukshetra.

(1976), *An Agricultural Geography of Haryana*, Vishal Publications, Kurukshetra.

Shinde, S.D., (1980), *Agriculture in an Underdeveloped Region: A Geographical Survey*, Himalayan Publishing House, Bombay.

Jha, B.N., (1980), *Problems of Land Utilization: A Case Study of Kosi Region.*

Das, M.M., (1981), 'Land-use Patterns in Assam', *Geographical Review of India*, Vol. 43(3).

Basu, S. (1984), 'Land-use and Cropping Pattern of the Lower Damodar Valley', *Geographical Review of India*, Vol. 46(4).

Ramanaiah, Y.V., and Charles, N.K., (1987), 'Agricultural Geography of Anantapur District, Andhra Pradesh', *The Indian Geographical Journal*, Vol. 62(2).

10. Ganguli, B.N., (1953), 'Land-use Survey and Agricultural Planning in Uttar Pradesh', *Geographical Review of India*, Vol. 15(2).

Singh, V.R., (1958), 'Land Utilization in the Sonpar (Mirzapur)', *N.G.J.I.*, Vol. 4(3).

Amani, K.Z., (1966), 'Land Utilization in Aligarh District', *Geographia*, 5.

(1976), '*Agricultural Land-use in Aligarh District*, Aligarh.

Roy, B.K., (1968), 'Measurement of Land-use in Azamgarh, Middle Ganga Valley', *Geographer*, Vol.15.

Roy, B.K., (1967), 'Land-use in the Riverine Region of the Ghaghra', *Geographical Thought*, Vol. 3(2).

Hussain, M., (1968), 'Land Utilization in the Upper Ganga-Yamuna Doab', *Geographical Review of India*, Vol. 30(3).

Sharma, S.C., (1968), 'Land Utilization in Etawah District' An unpublished thesis.

11. Singh J., (1976), *op. cit.*, p. 6.
12. *Ibid.*, p. 12.

PART I

THE PHYSICAL AND CULTURAL FACTORS AFFECTING LAND-USE

ONE

PHYSICAL FACTORS AFFECTING LAND-USE

LAND-USE is a function of four variables – land, water, air and man. Each has its own role to compose its life history. Land constitutes its body, water runs through its veins such as blood, air gives its life and man acts as dynamic actor to reflect its types, pattern and distribution. In fact, man – the user of land, is himself the product of atmospheric behaviour, hydrological action and lithospheric expression.[1] Land-use pattern in a region is thus governed, in a large measure, by physical controls and thereafter modified by socio-economic and technical organisational variants. It reflects man's reaction to the advantages and shortcomings of the natural environment, in that he adjusts his uses of the land to what he perceives to be its economic use capability.[2] It is thus clear that the proper understanding of the effects of these physical factors is very essential for any land-use analysis. An effort has been made here to discuss the important facets of the physical environment of the region under the following sections.

LOCATION AND EXTENT

Immediately below the Siwalik outerhills and skirting them, there is a dry, boulder strewn tract of *Bhabar* (literally meaning porous land) where the bulk of the moisture contributed by the rainfall and the small hill streams is absorbed in the loose talus. This tract is followed by a low marshy stretch of land where the water absorbed by the porous soils of the Bhabar reappears by symphonic influence and which is conspicuous by the 'ubiquitous' presence of water. It is infested at places with reeds, tall grasses and scrub forests, marked by high water-table and fertile though highly leached soils with excessive moisture and high clay factor.[3] This is known as the *Tarai*, literally which means 'moist or wet land'. In Nepal, it is known as the *Taryani* and with some specific additions as Morang. Atkinson,[4] J.H. Batten[5] and Madden[6] have called it '*Turaee*', while C.S. Thomason has christened it as the '*Terraie*', and Hodgson, Wadis, Spate and some others have resorted to the use of the word '*Tarai*' or *Terai*'.[7]

This Indo-Nepalese Tarai belt lies between Phike or the Ramganga in the west and Tista in the east.[8] The Tarai tract of eastern Uttar Pradesh, forms a part of this narrow belt and extends between 26°55′ and 28°19′ N. latitudes and 81°4′ and 84°2′ E. longitudes. The maximum length from east to west is about 325 km and the width from north to south about 40 km. It covers approximately an area of 9035.6 sq. km (Fig. 1.1)

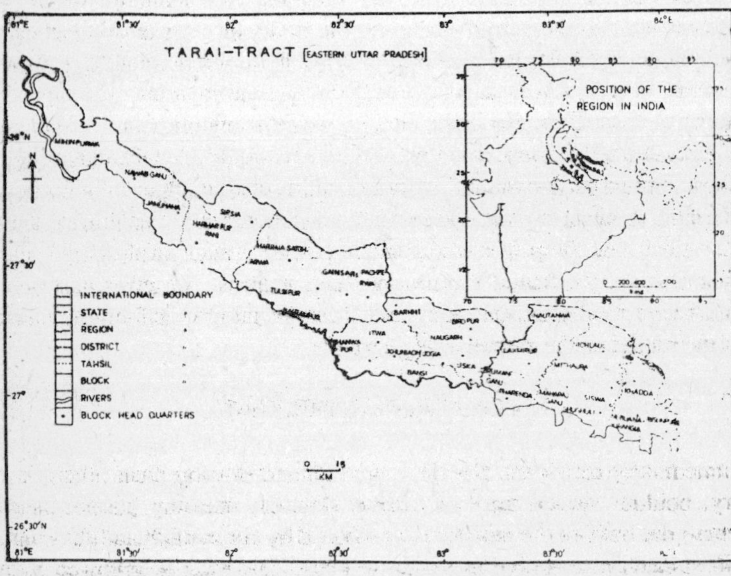

FIG. 1.1: Index Map

East-west extension of the Tarai tract together with its northern boundary, as mentioned, is quite explicit and poses no problem of delimitation. The northern boundary coincides with the Indo-Nepalese frontier. The eastern boundary is formed by the river Great Gandak which forms the eastern boundary of the district Deoria also for some distance and separates it from the Champaran district of Bihar. The western boundary is formed by the river Kauriala (Ghaghra) which is also a separating line between this and the district Lakhimpur Kheri.

It is the southern boundary which is quite erratic and which has throughout been changing, sometimes advancing and sometimes retreating. Originally, the Tarai covered a wider zone, perhaps, 80 to 96 km wide and extending in Oudh as far south as the middle Ghaghra, but much of this has been so altered by settlement that the true Tarai is now confined to a relatively strip parallel to the Bhabar.[9] Agriculture has pushed ahead the pioneer fringe by and by till today the dwellers of the northern most Basti Tarai would call the Tarai lying '*latthapar*' (beyond the pillar stones fixed on the Indo-Nepalese border).[10] The Tarai, thus, has been shrinking mainly due to the reclamation measures and the deforestation schemes, but sometimes the unfavourable natural conditions such as excessive rains, heavy floods attended by epidemics have caused retreat towards the south.

It may be mentioned here that there is no clear line of demarcation between the Tarai and the northern upland in the Gorakhpur and Deoria districts and the former merges imperceptibly with the latter. There is no fixed line but a transition zone separating them. Hence based on the main criteria of general configuration and drainage, presence of lowlying and moist land, high water-table, predominance of clay factor in the soil, vegetation which is Savana type par excellence with moist as well as dry deciduous tropical forest,[11] rice dominating crop combination, and making some adjustments in accordance with the boundaries of the development blocks which have been taken as the basic study units, tentative boundary has been drawn (Fig. 1.1) which passes through the northern portion of the Tahsil of Hata (Deoria) and the Tahsils of Maharajganj and Pharenda in the district Gorakhpur. Towards west, in the districts of Basti and Gonda, it follows the tortuous course of the river Rapti which continues in the Bhinga Tahsil of district Bahraich, further west, it passes through the northern portion of the Tahsil Nagpara in the district Bahraich. Thus it comprises of the development blocks of Khadda, Nebua Naurangia and Bishunpura of the Tahsil Hata of district Deoria; Nichlaul, Mithaura, Maharajganj, Ghughli and Siswa of the Tahsil Maharajganj and blocks Nautanwa, Laxmipur, Brijmanganj and Pharenda of the Tahsil Pharenda, district Gorakhpur; complete Tahsil of Naugarh with blocks Barhni, Birdpur, Naugarh, Jogia and Uska; blocks Itwa, Khuniaon and Bhanwapur of Tahsil Domariaganj and that portion of the block Bansi of the Bansi Tahsil, which lies north of the river Rapti, all belonging to Basti district; four complete blocks of the Tahsil Balrampur (district Gonda), i.e. Haraiya Satgharwa, Tulsipur, Gainsari and Pachperwa and the northern portion of the block Balrampur lying north of the river Rapti; complete Tahsil of Bhinga in district Bahraich with blocks Sirsia and Hariharpur Rani and the northern portion of the Tahsil Nanpara with the blocks Mihipurwa, Nawabganj and Jamunaha (Fig. 1.1).

GEOLOGY AND STRUCTURE

Structurally, the Tarai tract forms essentially a part of the vast Ganga Plains that lie interposed between two segments, i.e. the Peninsula of India in the south and the Extra-Peninsula in the north. It has been produced by the Pleistocene and subrecent alluvium deposition which commenced after the first upheaval of the Himalayas. Its geological history appears to be interwoven with that of the Bhabar and the foothills, although no final verdict could be given on the exact nature of their connections.

As regards the thickness of the alluvium, various estimates have been made and the authorities are not unanimous on the subject, although there is a general belief that the greater depth is in the north and it goes on decreasing towards south. A zone of 8,000-10,000 metres depth runs along the Himalayas. The Gorakhpur trough is over 8000 metres deep in the north.[12] On the basis of some geodetic data, Oldham[13] considers that the depth of the alluvium reaches a maximum, towards its northern edge, of 5000 to 6000 metres. Using the same data, Cowie[14] adopted even higher figures. But Glennie[15] considers it much less. Using the gravity results from Gonda and Gainsari stations, he calculates that the depth of the alluvium is only 2160 metres. This figure fits the geodetic data and it does not conform with the geological facts. Geologic survey of India estimates some higher figures.

As regards the origin of the Tarai, various views have been expressed. Hodgson[16] claims that the lower region owes its distinctive character as a whole to the vast masses of diluvial detritus which was shot from the mountains upon the plains, like gravel from a cart, at some great geological epoch, and which has been, since its deposit, variously and often abraded both in degree and in direction, by oceanic, and, in a far less degree, by ordinary floods. If at the time of descent of the debris, there existed a great dip in the Gangetic Plains from north-west to south-east, the lithologic character, as well as the distribution, of the debris would be materially affected thereby; for the subsiding oceanic currents would have a set from the former to the latter quarter, and would continue to lash the gravel into sand, and here to deposit both in a series of terraces, there perhaps utterly to displace both, in the latter quarter long after the former had emerged from the waves. According to Hodgson,[17] there did exist a longitudinal trough parallel to the mountains, and exclusively dominated the Tarai. This theory, however, is opposed to the results obtained by level operations and seems tantamount to a tempting speculation.[18]

Finch and others[19] hold that, broadly speaking, the origin of this sub-Himalayan region is like that of a piedmont alluvial plain. As already narrated, here the small streams of the immediate foothill zone, which are lost in the loose talus of the Bhabar, reappear by symphonic action in a line of copious springs which create swamps and marshes. The existence of an impervious stratum below the absorbent talus detritus also very well accounts for this special feature. The talus detritus gradually thins out and the finer and less permeable silt underlying it approaches nearer, and eventually reaches the surface, bringing with the water that has been absorbed by the shingle talus and has been retained by the impervious silt.[20]

Lawder[21] believes that the formation of the surface beds is solely due to fluvial action. Blanford and others[22] also hold the same view. Atkinson[23] explains in detail the procedure involved in the formation of the Kumaun Bhabar Tarai and this very well applies to the region under study. According to him, the mountain torrents along the foot of the Kumaun hills bring down every year a vast amount of debris which is spread out over the surface now on one side of their previous course and again on the other. This irregular deposition itself compels the torrents to change their beds from place to place until, as now obtains, the points where they debouch from the hills are marked by more or less irregular, great, fan-shaped boulder and gravel deposits. The clayey or semi-soluble particles are necessarily carried farthest and are readily deposited not only where there is a check in the slope, but where the current is impeded by the tortuous nature of the channel which itself naturally assumes that form under these conditions. Here, during the rains, the streams saturated with clayey silt, overflow their banks, form new channels, fill up old ones, and create the Tarai. Above the deposit of clay, we find one of clay combined with sand, in which, however, the latter predominates. From this bed issues the line of the

Tarai springs that flow uninterruptedly throughout the year, and its margin marks the northern boundary of the Tarai. Above we meet beds of sand and gravels or gravels and boulders as we approach the hills.

As regards the order of deposition of these materials, it has been seen that the boulders are generally deposited at a slope of 12.7 m to one kilometre, while the sand and gravel at a slope of 9.8 m to one kilometre, the sand and clay at a slope of 3.8 m to one kilometre and the clay at a slope of 1.6 m to one kilometre.[24]

SURFACE CONFIGURATION

The influence excerted on land utilisation by the form of the earth's surface is partly direct and partly indirect. The relief of the land exercises a direct influence particularly through slope, elevation and ruggedness, whilst on lower ground it introduces such questions as those of flooding. The indirect influence is that of elevation and land-forms on climate.[25] This necessitates a proper understanding of the relief and slope of the study area.

Singh[26] remarks that the geography of the Tarai is more of hydrography than topography. Water is the keynote in the time and space relationships: everywhere it reigns supreme. Surface water in marshes, swales, bogs, lakes, tals and sluggish streams, and underground water in copious springs and high water-tables give it a veritable character of a half 'solid' and half 'liquid' passive surface which raises an obstacle of pure inertia to human movements. These characteristics hold good in the case of the northern parts only, although banishing slowly and steadily with reclamation processes in its southern parts.

The Tarai tract, generally, is low with slight undulations here and there. The Dundwa range is the only outlier that approaches the northern Gonda-Bahraich frontier and affects the relief of the adjoining portion of the study region which rises to a height of 180 metres or even more (Fig. 1.2). The general slope is from north-west to south-east and it is very slight, i.e. hardly 0.3m to one kilometre, but north-south slope, as it is revealed by the transverse section, is generally higher with great local variations. The general slope is 1.5 to 3.8m to one kilometre. Starting from the west, we find that the gradient approximately is 1.5m to one kilometre in larger part of the Bahraich Tarai. This increases to 1.7m to one kilometre in the extreme east of this district. This trend continues in the Gonda district also to some distance, where it is 1.8m to one kilometre. But in its eastern part, it again decreases to 1.2m to one kilometre. The higher gradient in this central part is due to the existence of Dundwa ranges. Towards east in the Basti district, where the dead level is only broken by shallow valleys of the rivers, the gradient is, perhaps, the lowest, i.e. 0.2m to one kilometre. Further east, in the Gorakhpur district, it again rises to 0.59m to one kilometre. In the extreme east in the Deoria district, it drops down to 0.41m to one kilometre.

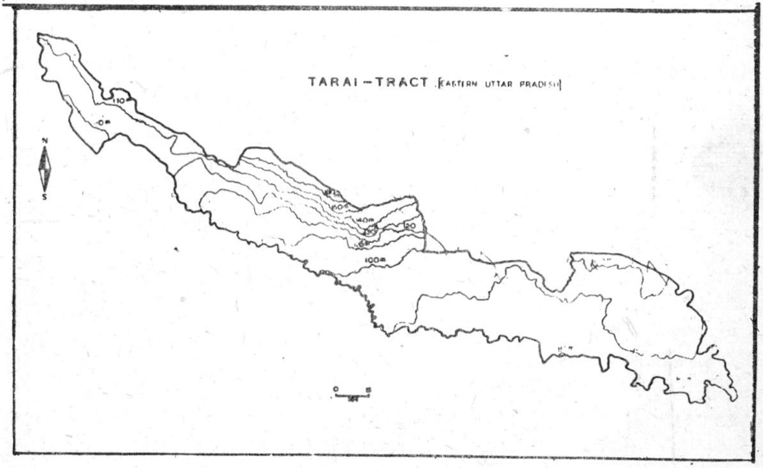

FIG. 1.2: Contours

Broadly speaking, the Tarai may properly be considered as a sub-division of the Bangar, which is entirely criss-crossed by small and shallow streams and *nalas*, infested, at places, with lakes and *tals* and the dead arms of the channels. They together with the river levees and alluvial terraces break the homogeneity of the level plain. In contradiction to this, there are depressed belts or low and broad valleys of the rivers known as *kachhar* or *khadar* as opposed to Bangar. The *Kachhar* of the river Kauriala in the west, the *Kachhars* following the rivers Rapti, Banganga and Burhi Rapti in the central part and that along the river Great Gandak in the east are such lands which are inundated in most of the years and the soils where are naturally moist.

DRAINAGE

Drainage lines hold particular significance in the region, not only do they provide redeeming topographic breaks in the general flatness of the plain and provide sub-regional or even local uniqueness and individuality to the different parcels of land, but they also govern to great extent, the human occupance of land, particularly the agricultural land and settlements.[27] This hold good in all its various aspects in the case of the study area.

The whole of the Tarai tract is a network of rivers and numerous streams which drain the area and follow its general slope. They belong to three drainage systems, viz., the Ghaghra in the west, the Rapti in the centre and the little and Great Gandak in the east, all of which form ultimately a part of the great Gangetic system (Fig. 1.3)

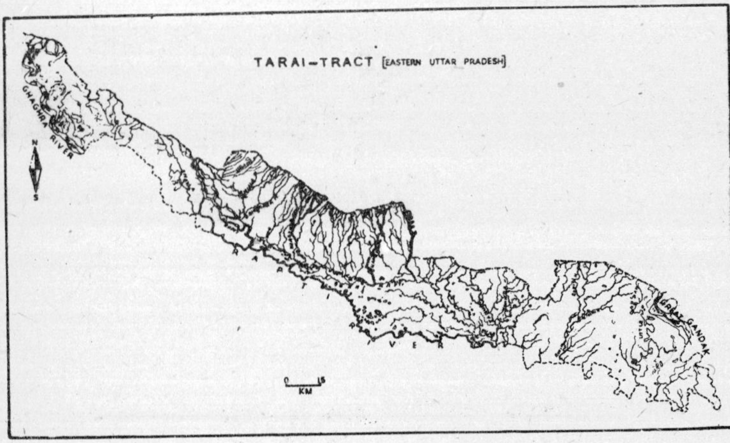

FIG. 1.3: **Drainage**

The Ghaghra and its tributaries form the western most drainage system. The Ghaghra assumes its name from the point where it is joined by the river Dahawar, above which it is known by the name of Kauriala river. The Kauriala issues from the mountains of Nepal and enters the area at its extreme north-western corner. It receives the waters of Girwa and some other hill torrents. The tortuous courses of the Sarju and a few others also join the Ghaghra. The Kauriala and the Girwa frequently change their courses and their channels are dotted with numerous islands.

The principal drainage line of the region is the Rapti which is the recipient of all the drainage of the area extending from the eastern part of the Nanpara Tahsil of the Bahraich district in the west to the Maharajganj Tahsil (excluding its extreme eastern part) of the Gorakhpur district in the east. Rising in the mountains of Nepal, the river enters the area in the Bahraich district and then flows in a very sinuous and tortuous course in the south and south-eastern direction, making well the southern boundary of the study area in Gonda and Basti districts. Its channel is a succession of loops and sharp bends, some of which after being cut off from the main stream are in the form of deserted channels. The Kain and the Bhakla are the main affluents of the river Rapti in the Bahraich district, which, in turn are fed by numerous streams. Another very important and, perhaps, making the largest sub-system of the Rapti river system is the Burhi Rapti. It makes its first appearance near the western border of the Gonda district and flows in an easterly direction roughly parallel to the main river up to a point of few kilometres ahead from its junction with Arrah in Basti district, whence it bends towards south-east to meet the river Rapti about 12 kilometres east of Bansi. The point of junction is at all times liable to change, depending especially on the action of the Banganga, and it is believed that the former course of the Burhi Rapti is now adopted by the main stream. All along its length is seen a maize of 'Nankhans' or old channels generally of a horse-shoe shape, similar to those along the Rapti.

Particularly in the Gonda district, the northern tributaries of the Burhi Rapti are exceedingly numerous, all of which bear a general resemblance to one another being hill torrents, with boulder-strewn beds, mostly disappearing in the dry season, but being transformed into rushing rivers and causing much damage to the land in their neighbourhood and frequently covering the fields with a deposit of barren sand during rains. In the Basti district, the northern streams joining Burhi Rapti are Arrah, Chhagriha, Ghurai, Awindar or Aundhi, Sarohi and several others whose channels are seldom well defined, especially, as they approach the Burhi Rapti and in many places their deserted beds have formed into marshes and lakes. The Parasi and Sikri join it on the right bank. The river Banganga is of considerable dimension, which rises in the Nepal Tarai and flowing in a south and south-east direction unites with Burhi Rapti. The point of junction is subject to change as the result of the heavy floods brought down by the streams.

The country to the east of the Banganga is drained by another series of hill streams which are no less perplexing in their ramifications and constant liability to change. Jamuwar is one of them. It is fed by a number of small streams and fall into the river Kumhra, a few kilometres south of the town Naugarh. The latter flows through the town of Uska to join the Rapti. Ghungi is another tributary of Kumhra, which forms boundary between the Basti and Gorakhpur districts. The united waters of the Ghunghi and the Kumhra generally go by the name of Dhamela which is merely an old channel of the Rapti and which after a course of about 16 kilometres in the district of Basti rejoin the original bed of the Rapti. The next affluent of the Rapti is the Rohin which also issues from Nepal and flows through the region.

The Little and Grèat Gandak drainage system drains the easternmost part of the area. The Great Gandak is the stream of the first magnitude. On its first entry into the Indian territory, the bed is stony and the water is cold and clear with a very rapid current. Its affluents are inconsiderable, but even in the hot weather, its volume is immense. the Little Gandak represents an old channel of the Great Gandak. It also originates in Nepal and takes a southerly course in the study area draining its extreme north-east corner. For the greater part of the year, it is an insignificant stream.

The drainage system of the Tarai is of dendritic pattern. It has been likened to the reticulations of a leaf, the streamlets on the edge of the moist country uniting to form larger channels which again feed the arterial lines of drainage and all eventually join the great mid-rib stream.[28]

Lakes and Jhils

As mentioned earlier, the Tarai abounds in lakes and *jhils* which are generally formed by the action of the rivers in changing their beds and thus represent the deserted channels horseshoe-shaped, while in other cases, they are deep natural depressions in which the surface drainage collects without finding an adequate outlet. There are temporary swamps and *jhils* which become extensive during rains and remain shallow marshes or even arable land in the dry season.

It is clear from the above discussion that natural drainage of the study area is far from perfect in most of the parts, especially in the basins of the Rapti, Burhi Rapti and their affluents. Extensive inundations occur during the rains and a large area is submerged, particularly along the courses of the rivers. These inundations are still more serious and obnoxious where, sometimes, the rivers change their courses and where by the rush of water, fertile clay soil is apt to be overlaid with a deposit of barren sand.

The main causes of the defective drainage of the Tarai are not far to seek. The Tarai streams are marked by their meandering courses, especially, in the upper reaches, duplicated or even triplicated channels and ox-bow lakes, obstructed cross-drainage, and excessive cross-section due to fluctuating regime outflows obtaining between years of normal and abnormal rainfall.[29]

As it is clear from Fig. 1.2, along the Dundwa ranges in the northern parts of the Bahraich and Gonda districts, the gradient of the hill torrents is abruptly reduced with the result that they deposit the debris transported by them at the foothills in the shape of alluvial fans which together present an appearance of a rolling plain. In the Tarai, the river channels are formed in depressions below the general level of the country. Thus with the excessive slope in the upper reaches of the streams and poor tenacity of the soils in the Tarai, the channels fail to accommodate the excessive volume of water. Consequently, with the carving-in of the banks and silting up of the river beds, the streams overflow their banks and carve out new courses. Such phenomena are very common here.

Tall grasses and forest patches also check the Tarai run-off in some areas, causing wide-spread surface pooling. Hutton writes that the Tarai streams are blocked in their courses with fallen trees and overhanging branches to an extent almost inconceivable to one not acquainted with them, and high flood the velocity is greatly diminished by their presence.[30]

SUB-SOIL DRAINAGE

Such as surface water, the underground water also affects the land-use of an area favourably and adversely both ways. It feeds springs, streams, wells, tube-wells etc, which control human activities in many ways.[31] The presence of underground water at low depth makes the sub-soil moist and enables one to grow crops even without irrigation, as it may be seen in the riverine tracts and in the ill-drained lowlying lands towards north. Sinking of tube-wells and digging of wells is also easier and cheaper here. If the sub-soil water is extremely low or some hard strata is found below the surface, as is in the case of the northernmost tract adjoining the Siwalik hills in the Gonda and Bahraich districts, it precludes the possibilities of irrigation from wells or tube-wells.

It is in the light of these facts that the study of the underground water seems very essential for the proper appraisal of land utilisation in the region under study.

Broadly speaking, the water-table in the study area is very high, particularly so during the rainy season and the real problem of the tract is not surface drainage, but sub-soil drainage,[32] but there are, as Fig. 1.4 exhibits, local and sub-regional variations in it.

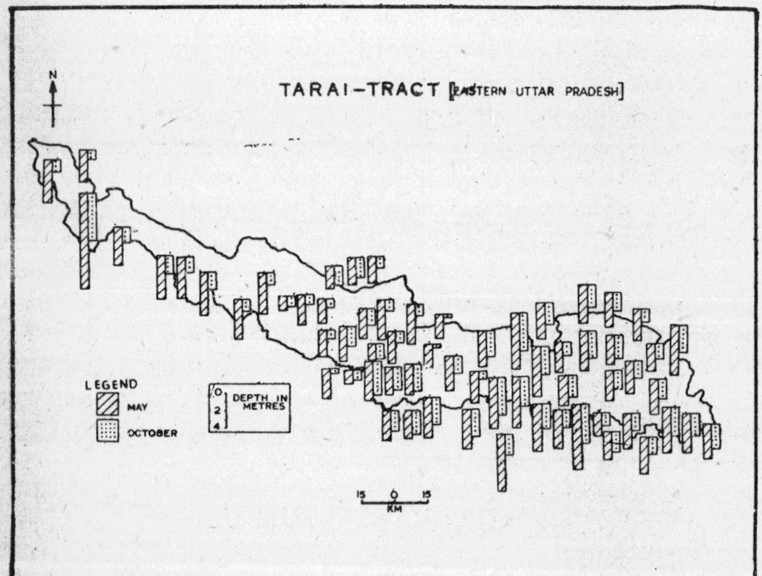

FIG. 1.4: Average Seasonal Depth of Water-table

The spring level is found at the lowest depth in the northern sloping and undulating forest-clad tract. It is generally between 6 and 10 metres and at some places even more than 20 metres, for example, in the Bankatwa, Suhelwa and Pipra forest ranges of Gonda district, it is between 20 and 30 metres. Sloppy elevated ground causing speedy run-off and less absorption of water, absence of irrigation channels etc. are some of the factors which may be held responsible for such a low water-table in this area.

In the upland region of the Bahraich district, the water-table is generally found at a greater depth than 5 metres reaching as low as 10 to 15 metres in the Mihipurwa forest range. In the upland tract of Gorakhpur and Deoria districts and in the alluvial flats of the Basti district, the depth of water-table generally ranges between 4 and 8 metres, being greater in the well-drained upland areas and lesser in the areas of defective drainage and in those with canal irrigation where the seepage from the canal water courses has been responsible for such a rise in the water-table.

In the riverine tracts and in the gently rolling ill-drained area lying south of the undulating forest belt in the Gonda and Bahraich districts and along the northern border in the Basti and Gorakhpur districts, the water-table is comparatively higher. It generally ranges between 2 and 4 metres. Defective drainage with impeded run-off, inundation or water-logging are some of the factors which are responsible for high water-table in the riverine tracts. The high water-table of the ill-drained lowlying areas may be ascribed to these factors and the symphonic influence that makes the water absorbed by the Bhabar soils to reappear here.

Changes in the Water-table

With the reclamation of the low marshy and swampy lands infested with reeds and tall grasses and with an improvement in the drainage conditions, undoubtedly, there has occurred a slight fall in the water-table in certain parts of the Tarai tract, but the areas where the canal irrigation has been introduced or where bunds and dams have been constructed across the streams or which are lowlying with defective drainage, the water-table has risen in the past. Seepage from canal water courses is playing a great havoc in the Gandak Command area of the region.

Seasonal Change

The water-table changes considerably with the change of the season also (Fig. 1.4). It rises in the rainy season coming very near to the surface, rather reaching it and lowers down in the summer season. This fluctuation in water-table is closely related with the rainfall which is an important source of sub-soil water. In the riverine tract, the Gandak command area and in the lowlying areas of the north where water-table is already high, the rise ranges between 0.4 and 2 metres. Although the rise is within 2 metres in these areas, it is very significant because water-table here is within 4 metres from the surface. It may be mentioned here that along the northern border in the Gonda and Bahraich districts and north-western border of the Basti district and in the upland areas, it ranges between 2 and 3 metres. Here the water-table is already at a greater depth and this rise in it is not so harmful as in the case of the former areas.

From the analysis of the local variations in the rise of the water-table with the seasons, it may be concluded, although with some exceptions, that higher the water-table, lesser is the rise in it and the lower the water-table, the greater is the rise in it, although the differences are not that much significant.

It is suggested that proper thought should be given to the improvement of sub-soil drainage conditions, prevention of extensive water-logging and maintenance of better hydrological balance in the areas where the sub-soil water is at shallow depths and is causing an alarming menace to the optimum utilisation of land. The following measures may be recommended for it:

1. Construction of a chain of tube-wells wherever possible, to pump out the undesirable ground water and thus lowering down the water-table and keeping a suitable hydrological balance.
2. Improvement of surface drainage system.
3. Elimination of excessive seepage from canal water courses by lining them.
4. Provision of drainage channels along with irrigation channels to individual holdings and then connecting the same with natural drains.
5. Judicious use of water for irrigation in accordance with the in take rate of the soil, effective root depth and evapotranspiration needs of the crops.

CLIMATE

Climate is one of the most important factors which determine the suitability of a particular type of land utilisation in a region.[33] It is the principal aspect of the physical environment affecting agriculture. The characteristics of the soil, the essential medium for plant growth, are largely the product of present and past climates and the vegetation that has flourished in them, and the effects of relief are to no small degree expressed through resulting climatic variations.[34] It (Agriculture) is still entirely at the mercy of the weather and climate.[35] The potential crop producing ability of a given area is dependent primarily upon the existing climatic and soil conditions under which the crop in question must be grown.[36] Climate can affect the choice of a farming system either indirectly through its influence on soil formation or directly through such factors as the length of the growing season, the occurrence of frost and the availability of water for crop growth.[37]

Climate limits the choice of the crops and affects their per acre yield and various agricultural practices. Agriculture in all its operations remain at the mercy of the weather, particularly so during certain periods. The vagaries of weather in India, for example, delay in the outbreak or early cessation of the summer monsoon, untimely and excessive rainfall or drought, spells of too high or too low temperatures, thunderstorms, hails, cold and hot winds etc. would seriously affect the growth and yield of crops. Analysis of existing data provides ample evidence to conclude that of all the controls, the climatic factor is one control which accounts for at least 50 per cent of the variability of crop yields over a series of years.[38]

It is, therefore, very essential to study the climatic conditions of the study region in detail, so that the correct appraisal of the present and past land utilisation and its future planning on sound lines be made feasible.

The climate of the tract, generally, is unhealthy, which together with some other local factors, viz., defective drainage, high water-table, inferior quality of water and uncleared land etc. has been responsible for chronic diseases such as malarial fever and cholera, particularly during and immediately after rains. It is on account of this unbracing, enervating and unhealthy climate that the region has been well known as 'Penal Settlement'.[39] The debilitating climatic factor has surely stood in the way of progress and has been responsible for its under-developed economy and cultural stagnation in no small measure,[40] but the matters have greatly improved with the reclamation of large areas of forests and the extension of cultivation in the recent past.

The region experiences a sub-tropical monsoon type of climate which is characterised by the seasonal rhythm, i.e. alteration of seasons resulting from the reversal of prevailing winds which have been termed as wet monsoon and dry monsoon, the former which are oceanic, blow from east to west and the latter which are of continental origin, blow from west to east.

Broadly speaking, the period of dry monsoon extends from November to June and on the basis of the changes in temperature and other phenomena, it can conveniently be sub-divided into cold weather and hot weather seasons, the former extending from November to February and the latter from March to Mid-June. The wet monsoon season, commonly known as the season of general rains, covers a period of about four months, i.e. from Mid-June to Mid-October or end of October. Accordingly, three distinct seasons may be recognised in the study area:

1. The cold weather season (November to February)
2. The hot weather season (March to Mid-June)
3. The season of general rains (Mid-June to October).

It may be noted that the division of the year into three distinct seasons is more useful in the study of the agricultural land utilisation, because these seasons more or less correspond to the crop seasons, for instance, the cold weather season corresponds to the period of Rabi crops, the seasons of general rains to that of Kharif crops, while the hot weather season to that of Zaid crops.[41]

The Cold Weather Season

The weather conditions of this season may be summarised under the following heads:

*Temperature**

The beginning of the cold weather season is marked by a fall in

* As no climatic data, except those of rainfall, are recorded at any station in the study area, these data of Bahraich, Gonda and Gorakhpur have been used and may be taken as close approximations of the climatic conditions of this area. These data have been compiled from the Climatological Tables of Observatories in India (1931–60), Government of India, Indian Meteorological Department.

temperature with the southern march of the sun. The mean monthly temperatures in November at Bahraich, Gonda and Gorakhpur are 21.0 °, 20.6°and 21.6°C respectively (Fig. 1.5). The mean maximum temperatures in this month at respective stations are 28.6°C, 28.6°C and 28.5°C, while the mean minimum temperatures in the same month are 13.4°C, 12.6°C and 14.6°C respectively (Fig.1.5). The mean diurnal ranges of temperature for this month are 15.2°C, 16.0°C, 13.9°C respectively. A further decrease by about 4.0°C is recorded in the mean monthly temperature in December. The mean maximum at all the stations is 24.3°C and the mean minimum ranging between 8.8°C and 10.8°C.

January is the coldest month as it records the lowest temperature of the year. The mean monthly temperatures at Bahraich, Gonda and Gorakhpur respectively are 15.7°C, 15.6°C and 16.5°C. The mean maximum and the mean minimum temperatures for this month for these stations are 22.6°C, 22.9°C and 23.0°C; and 8.8°C, 8.3°C and 8.9°C respectively. The diurnal ranges respectively are 13.8°C, 14.6°C and 13.1°C. The highest and the lowest recorded temperatures for this month respectively are 28.8°C 0.6°C for Bahraich, 28.9°C and 1.1°C for Gonda, and 30.0°C and 1.7°C for Gorakhpur (Fig.1.5). Although the frosts are not common in December and January, the heavy mists or fogs are very common.

In the month of February, the thermometre rises slightly. The mean monthly temperatures at Bahraich, Gonda and Gorakhpur are 18.3°C, 18.1°C and 19.1°C respectively, which are lower than those of the month of November. The mean maximum and minimum range from 25.6°C to 25.9°C and from 10.5°C to 12.3°C respectively.

Pressure and Winds

Pressure increases with the fall in temperature. The mean monthly pressures in the month of November at Bahraich, Gonda and Gorakhpur are 999.3mb, 1000.5mb and 1004mb respectively, which rise to 1001.2mb, 1003.4mb and 1006.6mb respectively (Fig. 1.5) in January which is the coldest month with the highest pressure. In February, there is, again, a slight fall in pressure, showing a correlation with temperature.

The Punjab high pressure controls the prevailing wind direction and the north-westerly winds which move down the Ganga valley, generally blow from west to east in this region, as their direction is determined partly by pressure distribution and partly by the trend of the Himalayan mountains. The pressure gradient remains moderately low. The mean wind speeds in the month of November at Bahraich, Gonda and Gorakhpur respectively are 2.7, 2.4 and 2.0 in November, 2.9, 2.8 and 2.3 in December, 4.0, 3.9 and 2.9 in January and 4.8, 5.1 and 3.9 km per hour in February. Occasionally, winds with higher velocity visit the area, but only for a day or so. The winds generally blow from west to east followed by north-west and south-west direction, although calm conditions prevail for a pretty long time.

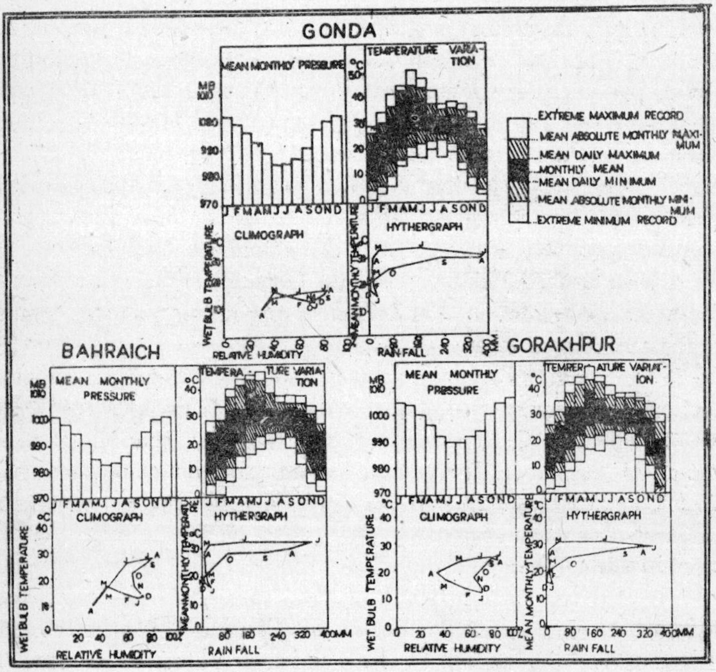

FIG. 1.5: Temperature, Hythergraphs etc.

Precipitation

The figures of monthly relative mean humidity in the month of November at Bahraich, Gonda and Gorakhpur are 61.5 per cent, 68.5

per cent and 62.5 per cent respectively. For December, January and February, these figures respectively are 67.5 per cent, 73.5 per cent and 69.0 per cent; 69.5 per cent, 72.5 per cent and 68.5 per cent; 60.5 per cent, 59.0 per cent and 57.0 per cent. This shows that the months of December and January have comparatively higher humidity and if the readings at 0830 hours are taken into account, the figures are still higher ranging between 80 per cent and 84 per cent in January. This is why fogs dominate the scene and often cause poor visibility in December and January. About three-fourths days in the months of November and December are marked with clear skies and in the rest of the days, the cloud amount ranges from 7.2 to 8.0 oktas. In January and February, the days with clear skies account only for about 50 per cent.

The serene atmosphere of this season is occasionally disturbed by some vigorous depressions from the west and also the nasty cold winds from the bills.

Figure 1.6 shows the average monthly and seasonal rainfall* and Table 1.1 shows the proportion of the winter rainfall to the total annual rainfall. It is clear from them that the total amount of rainfall during the season does not exceed 46.4 mm, and its contribution to the annual rainfall is in no case more than 3.9 per cent. It may be seen further from the Fig. 1.6 that January and February receive more rainfall than November and December.

Since the winter rainfall is mostly contributed by the cyclones coming from the west, the effectiveness of which decreases towards east, it generally decreases towards east. Nanpara receives 46.4mm while Pharenda 33.29mm, Maharajganj 36.06mm and Padrauna 36.83mm of rainfall. Although, the cold weather rainfall is small in quantity, it is highly beneficial to the winter crops, as it comes at a time when the plants are flowering. The effectiveness of the rainfall is further increased by the prevailing low temperatures.[42] These rains are of immense value, particularly in this tract where there is inadequate provision of irrigation facilities.

It may be noted here that the rainfall of the cold weather season even in this area is associated with shallow depressions whose nature and origin still provide something of a mystery.[43] Originating either in

* All the data regarding rainfall of the selected stations have been obtained from the Tahsil and District headquarters, Office of the Board of Revenue, Government of Uttar Pradesh, Lucknow and averages have been calculated for a period of 87 years (1901—1987), except a few stations which have these data only for a shorter period which have been taken into account for this purpose.

the Mediterranean sea or locally in the high lands in the west of India and being affected by the Asiatic high pressure and the Himalayas, these depressions take a course well to the south and travel eastward to reach the study region. Remarkable weather changes are experienced during their passage. The arrival of the storms is usually preceded by a rise in the temperature, which is followed by a considerable fall in temperature and cooler and stronger winds of the rear. The cloudy weather normally lasts a day or so and is followed by clear skies. The cold waves sometimes do a great damage to the standing crops. The damp cloudy weather whenever it persists causes various plant diseases. The cold weather depressions are sometimes accompanied by hot storms which again cause a considerable damage to the crops, particularly late in the season.

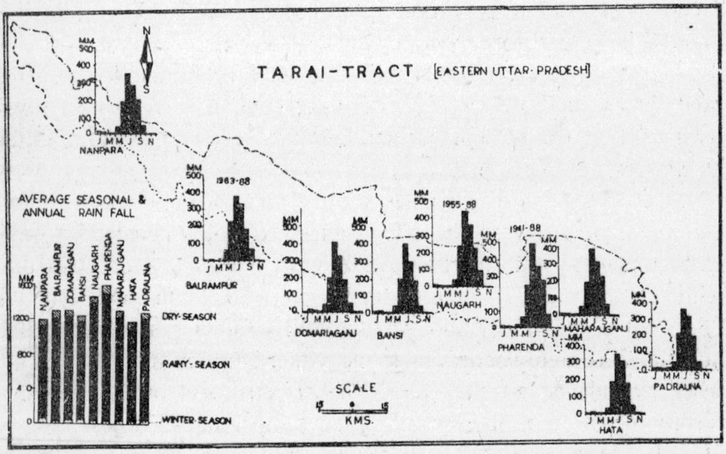

FIG. 1.6: Average Monthly Rainfall

The Hot Weather Season

The hot weather season extends from March to first half of June. It is the pre-monsoon season.

Temperature

After January as the sun marches northward, the temperature

constantly rises. The mean monthly temperatures for March at Bahraich, Gonda and Gorakhpur respectively are 23.7°C, 23.8°C, and 24.9°C, (Fig. 1.5), while the mean minimum and mean maximum temperatures for these stations respectively are 15.4°C, 15.4°C and 17.2°C; 31.9°C, 32.2°C and 32.6°C. The temperatures continue to rise during the coming months of April and May. The mean monthly temperatures in April at these stations respectively are 29.2°C, 29.3°C and 29.9°C, while these figures rise to 32.7°C, 32.8°C and 32.5°C in may which is the hottest month. The mean minimum and maximum temperatures in this month respectively are 25.6°C, 25.6°C and 25.9°C; 39.8°C, 39.9°C and 39.0°C. The diurnal ranges of temperature for these stations are 4.2°C, 4.3°C and 3.1°C respectively. The highest recorded temperatures for this month are 45.6°C at Bahraich, 49.9°C at Gonda and 48.3°C at Gorakhpur; and the lowest being 15.6°C, 16.9°C and 16.7°C respectively. There is slight decrease in temperature figures in June, the mean monthly temperatures ranging from 31.6°C to 32.3°C and the mean maximum from 36.4°C to 37.6°C and mean minimum from 26.7°C to 27.0°C.

TABLE 1.1: Rainfall in Cold Season

Stations	Total amount of rainfall in mm	Rainfall in cold weather season	% of total rainfall
1. Nanpara	1185.06	46.40	3.91
2. Balrampur	1301.99	32.52	2.50
3. Domariaganj	1310.10	41.82	3.27
4. Bansi	1243.75	43.78	3.52
5. Naugarh	1458.74	32.70	2.24
6. Pharenda	1585.84	33.29	2.10
7. Maharajganj	1295.77	36.06	2.78
8. Hata	1166.71	35.70	3.06
9. Padrauna	1268.31	36.83	2.90

Pressure and Winds

The increase of temperature during the hot weather season is accompanied by a fairly continuous decrease of pressure. The mean monthly pressure in March ranges between 995.2mb and 1002.2mb, in April between 991.4mb and 995.9mb, in May between 986.9mb, and 992.4mb and in June between 983.5mb and 989.0mb.

As regards the air movement, this period is undoubtedly transitional between the winter and the summer seasons, a time when local factors are able to make themselves felt above the planetary controls. The prevailing winds are from the west, the intensity of which increases with the end of March and which become stronger in April and May, lasting with frequent intermissions of easterly breezes till early June. The hot blasts of *loo* which are so marked a feature of the Ganga-Yamuna Doab, are not frequent here. Dust storms are also of rare occurrence, but in May and June, they are comparatively more frequent when some thunderstorms and rains also follow them. It may be remarked here that the heat in the summer months is less extreme and the west winds of the hot weather are but rarely experienced and lose much of their burning force.[44] In the early summer months, the air is often refreshed by storms from the hills.

The mean wind speeds in the months of March, April, May and June at Bahraich, Gonda and Gorakhpur respectively are 4.8, 6.5 and 5.4 km per hour; 6.0, 7.4 and 6.4 km per hour; 7.3, 7.7 and 7.1 km per hour; 8.4, 7.3 and 6.8 km per hour. This shows that the wind speed is the highest during the hot weather season.

Precipitation

These are the driest months of the year, the relative humidity and the cloud amount being the minimum. The mean relative humidity in March ranges between 40.5 and 43.5 per cent, in April between 30.5 and 34.5 per cent and in May between 38.5 and 46.0 per cent. In June, these percentages rise to 59.5 and 63.5 per cent. About 18 to 20 days in March, April and May have clear skies and the sky is overcast hardly for a day or so. In June, the cloud amount increases.

The rainfall in this season is sporadic, short-lived and subject to great local variations (Table 1.2). Generally, the rainfall increases from west to east with some exceptions. Nanpara, the westernmost station receives 66.56 mm (5.6% of the total) and Padrauna, the easternmost station 80.60mm (6.4% of the total). The rainfall decreases from north to south as well, as Hata which is situated a few kilometres south of the study area receives only 53.88mm (4.6 per cent of the total), while Pharenda receives 99.69mm (6.3 per cent of the total) and Maharajganj 78.47mm (6.1% of the total).

Although the amount of rainfall received in this season is very meagre, it is helpful in giving a temporary relief from the desiccating heat of the day, in the preparation of the fields for the sowing of some

Kharif crops as well as facilitating the growth of standing crops such as sugarcane.

TABLE 1.2: Rainfall in Hot Weather and Rainy Seasons

Stations	Rainfall in hot weather season in mm	% of the total	Rainfall in the rainy season in mm	% of the total
1. Nanpara	66.56	5.6	1072.10	90.5
2. Balrampur	84.59	6.5	1184.88	91.0
3. Domariaganj	77.61	5.9	1189.64	90.8
4. Bansi	64.84	5.2	1135.13	91.3
5. Naugarh	64.41	4.4	1361.63	93.3
6. Pharenda	99.69	6.3	1452.86	91.6
7. Maharajganj	78.47	6.1	1181.24	91.2
8. Hata	93.88	4.6	1077.13	92.3
9. Padrauna	80.60	6.4	1150.88	90.7

The Season of General Rains

This season normally sets in the region by mid-June with the arrival of summer monsoon and ends by the end of the month of October. It assumes rather an unpleasant outlook and represents the worst Tarai climatic conditions. The excessive humidity, often air-saturation 75-95 per cent, combined with high temperature become intolerable in the extreme and totally benumbs active human nerves.[45] The weather changes experienced during this season are enumerated as follows:

Temperature

The immediate effect of the advent of monsoon is a fall in temperature. The mean monthly temperatures in July at Bahraich Gonda and Gorakhpur are 29.7°C, 29.6°C and 29.6°C respectively. They remain almost the same in August. The mean maximum and mean minimum temperatures also range between 32.2°C and 33.0°C, and between 26.1°C and 26.4°C respectively. In September, the mean maximum temperature lowers down to 28.9°C at Bahraich, 28.7°C at Gonda and 29.1°C at Gorakhpur with further reduction in October. There is not much difference in the mean maximum figures, although mean minimum figures show greater reduction.

Pressure and Winds

With full intensification of the low pressure area over land the north-west of India by the end of May, the winds are attracted from the high pressure area created over the Indian Ocean. The mean pressure during this season also goes on increasing. It is 983.9mb in July at Bahraich and rises to 995.6mb in October. At Gonda and Gorakhpur, these figures rise from 987.7mb to 997.6mb; and from 989.2mb to 1000.7mb respectively.

The winds generally blow from the east. The hills in the north exercise great influence on their direction. The mean wind speed is 7.8 km per hour at Bahraich in July in it decreases to 3.7 km per hour in October. For Gonda these figures are 6.4 and 3.0 and for Gorakhpur 6.3 and 2.7 km per hour respectively. The wind velocity increases during the passage of cyclones and decreases after their passing away.

Precipitation

The cloud proportion in these months is fairly high. August shows the highest proportion when about half the days are over-cast with clouds or greater part of the sky is covered with clouds. The relative humidity is very high. The mean relative humidity in July is 77.0 per cent at Bahraich, 80.0 per cent at Gonda and 79.5 per cent at Gorakhpur; for August, these figures are 80.0 per cent, 82.5 per cent, and 81.5 per cent; for September, 76.0 per cent, 79.0 per cent and 77.5 per cent respectively. This shows that August is the most damp month followed by July and September. If 0830 hour readings are taken into account, these figures are still higher, i.e. 83 per cent and 85 per cent. In the month of October which is marked by the activities of the retreating monsoon, the relative humidity decreases to 65.0 per cent at Bahraich, 72.5 per cent at Gonda and 67.5 per cent at Gorakhpur.

The amount of rainfall is just in conformity with the cloud proportion and relative humidity. With the arrival of the moisture-laden winds, the cloud proportion and the relative humidity increase first and then diminish. Almost the same trend is observed fin the monthly distribution of rainfall with some minor differences. It may be seen that the average monthly rainfall for the month of June ranges between 197.17mm (Hata) and 251.09 mm (Pharenda). It rises abruptly in July which is the rainiest month in the season. In the west, Nanpara receives 347.81 mm of rainfall, while the stations of eastern part receive a bit higher rainfall, for example, Pharenda 479.46 mm, Maharajganj 386.83 mm, Padrauna 361.81 mm. August also receives a heavy rainfall, although it is a bit less than that of July. The rainfall slackens in September which receives almost the same amount of rainfall as is received in June. In October, the rainfall drops abruptly to a very meagre amount, viz., Nanpara 48.90mm, Pharenda 88.07 mm, Maharajganj 64.18mm, Padrauna 53.91 mm. As regards the spatial distribution of rainfall in this season, the Tarai follows the general pattern of the Doab, i.e., the rainfall decreases from north to south and from east to west (Fig. 1.6 and Table 1.2), although there are local irregularities within the region.

More than 80 per cent of the total rainfall is received during the rainy season, in which July and August together claim lion's share. It is noticeable here that the period of the wet monsoon is not of continuous rains. The bursts of general rains alternate with breaks of fine weather. These rainless intervals are shorter in July and August, but become greater in the months of September and October. This pulsatory activity of the monsoon is associated with the passage of depressions[46] which are formed in the Bay of Bengal and travel towards west. Their axis, when pushed towards the foot of the hills results in heavy rainfall in the region, when their passage is shifted towards south, the region receives little rainfall. The breaks with spells of fine weather after the outbursts of rain are usually very useful to the crops of the season, but if the rainless period exceeds 10 to 15 days, it affects the kharif corps adversely with catastrophic and disastrous consequences.

As already mentioned, the east to west decrease of rainfall is clearly seen, for example, Pharenda in the east receives 1452.86 mm of rainfall and Nanpara the westernmost station receives only 1072.10mm of rainfall. It may be mentioned here that there is some anomaly in this general trend, as Padrauna, the easternmost station receives only 1150.88 mm of rainfall which is less than that of Pharenda. This appears due to the presence of dense forest near the latter station. Maharajganj is the northernmost station which receives 1181.24 mm of rainfall, while Hata which is situated to the south of the region receives only 1073.13 mm of rainfall. Pharenda is again an exception, which, although, is situated a few kilometres west of Maharajganj, receives more rainfall, for which the dense forest belt near the former seems to be responsible.

RAINFALL

Of all the weather elements, rainfall is by far the most important single factor that affects the plant growth and crop production most vitally and directly.[47] The agricultural or pastoral utilisation of the land is the only real and lasting source of wealth, and both of these are in a large measure, dependent on rainfall. The yield of wheat, sugarcane, maize and numerous other crops varies with the rainfall of the year in such a way as to leave no doubt that the rainfall has been the real determinant.[48] In India, with her large population and predominantly agricultural economy, the farmer has ever prayed to the 'Rain Gods' for timely and adequate rains looking forward for a good distribution of rainfall during the year.[49]

This shows that rainfall plays a very important role in the land utilisation and the study area is no exception to it. This necessitates that this element of climate should be discussed in some more detail.

Distribution of Annual Rainfall

The Tarai region enjoys a moderately heavy rainfall because of the proximity to the mountain hills which virtually guide the westward course of the monsoon currents from the Bay of Bengal and also act as precipitation agent. The patches of forests join hands with the hills and augment a further increase in precipitation.[50] As such the rainfall generally decreases from east to west and from north to south (Fig. 1.7). The annual rainfall at Pharenda is 1585.84mm, at Naugarh 1458.74mm, at Balrampur 1301mm and at Nanpara 1185.06mm. Here it may be noted that although Padrauna is situated further east, it receives only 1268.31mm of rainfall which is much less than that of Pharenda and Naugarh. This appears due to its greater distance from the hills than that of the latter. If we examine the north-south trend, we see that generally the rainfall decreases from north to south, viz., Maharajganj receives 1295.77mm of rainfall, while Hata receives only 1166.71 mm of rainfall. Pharenda is the exception, which, though, is situated a few kilometres towards south-west, receives more rainfall. This anomaly may be ascribed to the patches of forests and dense vegetation cover in this part.

Annual Variability of Rainfall

Albeit rainfall is the most important climatic variable having volitional command over the use of agricultural land, the usual method of the cartographic representation of annual, monthly and weekly totals may give wrong impression and can be very misleading. It is the deviation of the rainfall from the mean that matters in sub-tropical and tropical agriculture.[51]

It may be seen from Figs. 1.7 and 1.8 that the rainfall in the area is not only unevenly distributed in space but also in time and these latter variations have far reaching consequences. The annual variability is sufficiently high. At the same station, the total incidence of rainfall in some years is far above the mean and in some years far below it. For example, in 1913, Domariaganj received 4605.23mm of rainfall against an average figure of 1310.10 mm, i.e. 251.5 per cent above the average rainfall, while in the year 1907, it received only 585.68mm of rainfall, i.e. 55.8 per cent below it. Such high vicissitudes in the amount of annual rainfall are noticeable with regard to other stations also.

Figures 1.7 and 1.8 show that there is great variation in the amount of annual rainfall from place to place within the study area. It may be seen that one station receives the annual rainfall far in excess of its average rainfall, but another station, a few kilometres away from it receives in the same year an amount well below its average. The dry conditions prevail at one station, while humid conditions are in excess at another. For example, in 1917, the rainfall at Domariaganj was 558.23 mm or 44.3 per cent of the average, while in the same year, the rainfall at Padrauna was 1752.83 mm; at Hata 1497.52 mm, i.e. 38.17 and 28.28 per cent respectively more than the average. In the year 1935, Domariaganj received 105.5 per cent more rainfall than the average; while at Nanpara and Padrauna, it was below the average by respectively 16.5 and 13.3 per cent. In the 1969, the rainfall at Pharenda was 37.8 per cent less than the average, while Padrauna received 41.2 per cent more rainfall than the average. In the year 1974, Pharenda and Naugarh received respectively 48.4 and 42.3 per cent more rainfall than the average, while Nanpara and Domariaganj received only 77.0 per cent of the average rainfall. Still numerous other examples may be seen from Figs. 1.7 and 1.8. Hardly a year is seen in which a uniform rainfall has been received at all stations.

The Co-efficient of Variability

The co-efficient of variability which is the standard deviation expressed as a percentage of the mean rainfall is given in Table 1.3 and

Fig. 1.9. The data for deriving co-efficient of variability of rainfall have been obtained from *Monthly and Annual Rainfall and Number of Rainy Days (Period 1901–1950)*, Pt. III-A, Indian Meteorological Department, and from the Tahsil headquarters. It may be seen from Fig. 1.9 that the co-efficient of variability of annual rainfall varies from 20.99 per cent (Maharajganj) to 44.18 per cent (Domariaganj). If Domariaganj is excluded, which is an extreme case, the variability at all the stations is found up to 26.73 per cent, which has far reaching consequences, as according to Blanford an annual variability of 12 per cent or more makes an area susceptible to famine.[52]

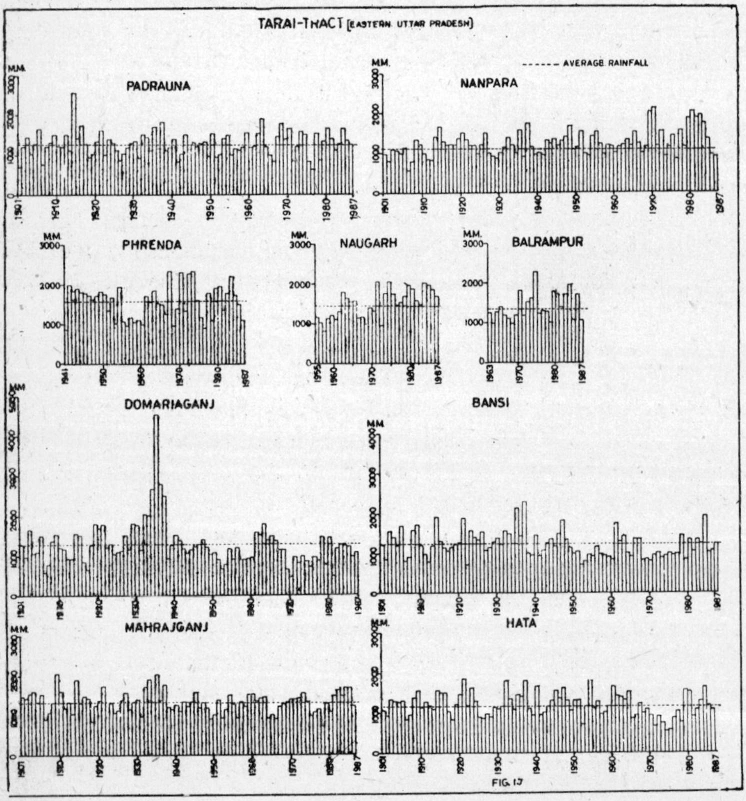

FIG. 1.7: **Annual Rainfall of Selected Stations**

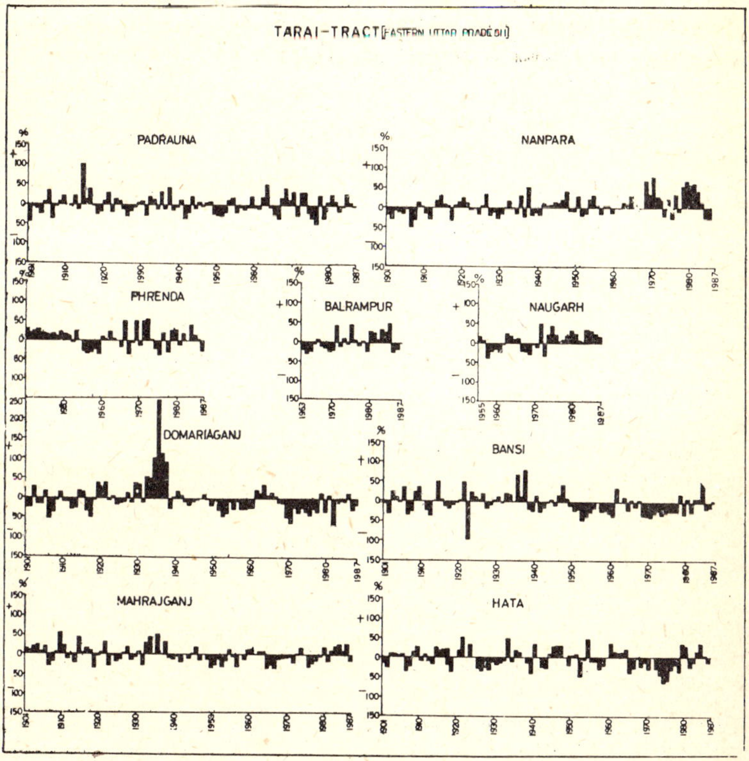

FIG. 1.8: Yearly Percentage: Departure of Rainfall

If we analyse the spatial distribution pattern of the co-efficient of variability in the area, we find that generally there exists an adverse relationship between the total amount of annual rainfall and the amount of variability, i.e. the higher the average annual rainfall, the lower is the co-efficient of variability, viz., at Hata which receives the lowest rainfall, thc variability is 26.26; at Nanpara with 1185.06 mm of rainfall, it is 25.19; at Padrauna with 1268.31mm of rainfall it is 25.27; while it is 23.72 per cent at Naugarh which receives an average annual rainfall of 1458.77mm; it is 20.99 per cent at Maharajganj which receives 1295.77 mm of rainfall. There are some discrepancies in this general relationship, as it is exceptionally high at Domariaganj which stands third as regards the total amount of average annual rainfall is concerned; it is 26.23 per cent at Pharenda which receives the heaviest rainfall in the region. It should have been the lowest here. Why this discrepancy does occur, is difficult to explain.

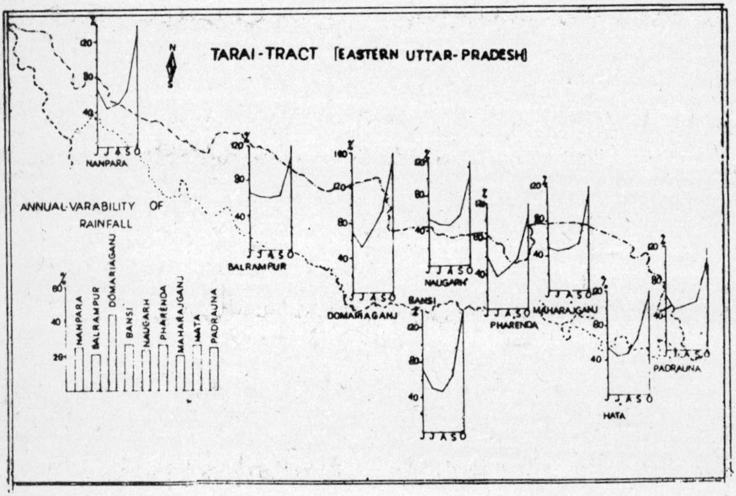

FIG. 1.9: Co-efficient of Monthly Rainfall

Monthly and Seasonal Variability of Rainfall

From the standpoint of agricultural operations, the seasonal and monthly variability of rainfall, particularly in the wet monsoon months, is more significant than the annual variability, for even a sub-normal rainfall, if well distributed produces a good yield. As a matter of fact, timely distribution is more important than the annual total.[53] For example, if the amount of rainfall is inadequate or excessive in the month of June, it delays the sowing of kharif crops and affects their outturns adversely. The heavy downpours, if they occur, cause floods damaging the standing kharif crops and the prolonged breaks in the months of July and August cause the failure of the various kharif crops, particularly rice. Early or late retreat of the summer monsoon does harm the sowing of the rabi crops and the quality and yield of the late kharif crops. Heavy falls in these months may lead to water-logging, while premature cessation of the rains may cause postponement of or restriction on the sowing of the crops.[54]

Table 1.3 and Fig. 1.9 show the co-efficient of the monthly variability of some selected stations for the months of the wet monsoon season.

TABLE 1.3: Co-efficient of Monthly and Annual Variability

Stations	Months					
	June%	July%	August %	Sept. %	Oct. %	Annual
	±	±	±	±	±	±
1. Nanpara (87 years)	65.30	44.89	50.24	63.99	126.32	25.19
2. Balrampur (25 years)	56.16	52.50	50.81	52.18	94.27	21.45
3. Domariaganj (87 years)	68.09	52.45	62.18	92.44	147.67	44.18
4. Bansi (87 years)	72.85	49.47	46.12	64.62	134.19	26.73
5. Naugarh (33 years)	53.12	47.45	46.94	57.13	100.90	23.72
6. Pharenda (47 years)	59.90	37.02	48.00	58.04	107.63	26.23
7. Maharajganj (87 years)	49.77	46.38	49.33	54.44	115.93	20.99
8. Padrauna (87 years)	65.75	46.45	51.99	57.33	113.26	25.27
9. Hata (87 years)	55.63	45.81	46.93	64.25	113.81	26.28

It is clear that the co-efficient of variability is the least in the months of July and August which are the rainiest months of the year. In July, it ranges between 37.02 and 52.50 and in August between 46.72 and 62.18 per cent. Next come the months of June and September with almost equal amount of monthly rainfalls. In these months, the figures of variability vary form 49.77 to 72.85 and 52.18 to 92.44 per cent. It is the maximum in the month of October, i.e. between 94.27 and 147.67 per cent. This is the month which receives the lowest rainfall in the wet monsoon season. The high variabilities in the months of June, September and October are of immense importance. The highest variability of the month of October denotes that the rainfall in this month is the most erratic and capricious and has the far reaching consequences as regards the agricultural operations are concerned. So is the case with the month of June in which the sowing of the kharif crops is done.

Variability from the Median

It has been experienced that the medians, and their percentile values derived from the diagrams are of greater importance than mean or average values of their co-efficients in arriving at a more rational estimation of rainfall variability.[55] According to Cowie, the ordinary average rainfall of a place is sometimes insufficiently sensitive and at other times over sensitive to extreme variations.[56] He is of the opinion that variability should be judged from the median or the middle value which is an amount of rainfall actually experienced at a station at least in any particular year, and it is an indication of the probability that in given series of years half the numbers of years will receive rainfall above the median and half the numbers of years below it.

Therefore, an attempt has been made here to analyse the variability of rainfall of the wet monsoon months by the median value method, which is shown in Fig. 1.10 A and B.

It may be seen that dispersion of rainfall at all stations is the greatest in the month of July with the lowest inter-quartile distance. This is the rainiest month of the season as well as with the highest median value. August follows it as regards the median value of rainfall, inter-quartile range and dispersions are concerned. This shows that the rainfall in these months is highly variable, but this high value is least significant for agricultural operations in view of the high totals in these months.

The inter-quartile range and the dispersion of rainfall are, perhaps, the lowest in the month of October, but they are very significant, because the number of October with extremely low rainfall is greater than the number with excessive rainfall and this affects the cropland use adversely.

The variability in the month of June is quite high, as is shown by the inter-quartile range and the wide dispersion of rainfall. It is more likely to be critical for agricultural activities, because the monthly mean is comparatively low and any reduction in this amount delays or hampers the agricultural operations.

The inter-quartile range in the month of September at most of the places is wider than that of August and even June. This, when considered in view of its median value, becomes all the more significant with regard to agricultural operation.

It may be concluded on the basis of the above analysis that rainfall in the study area is quite variable in those months of the year when its regularity is most needed. As such, the months of June, September and October are very critical having significant impacts.

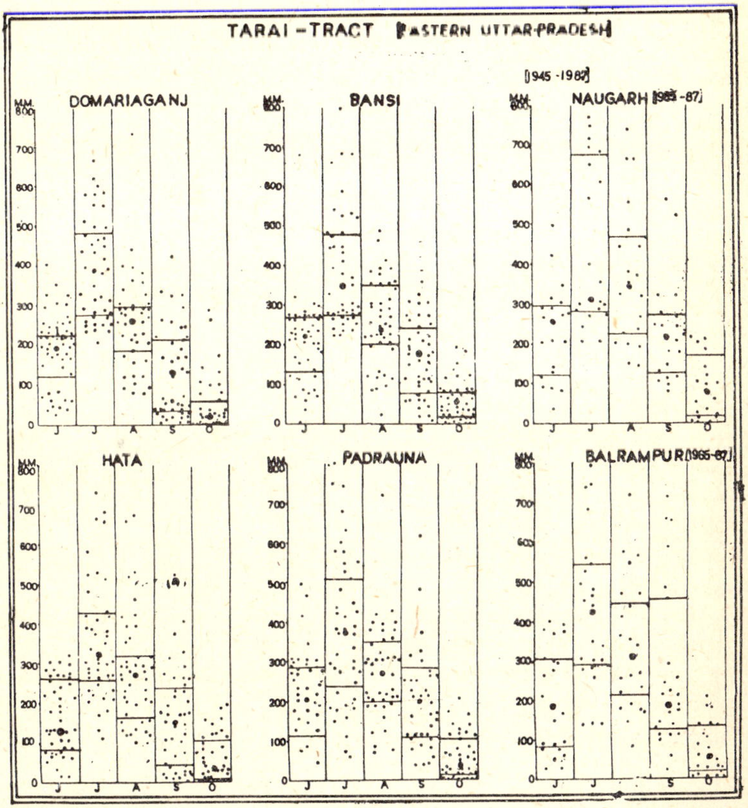

FIG. 1.10A: Dispersion of Rainfall

SOIL

Soil is the basic factor in the study of the land utilisation. Any successful land-use planning for a region cannot be suggested without a proper and systematic study of its soils.[57] The formation of land-use plans demands that the soil pattern should be mapped and the physical and chemical characteristics of different soils described.[58] This should precede fertilizer use, intensive agricultural development and high yielding varieties programmes, the agricultural colonisation of the cultivable wastelands, the creation of permanent intensive agriculture and the rational planning of the land-use and cropping pattern.[59]

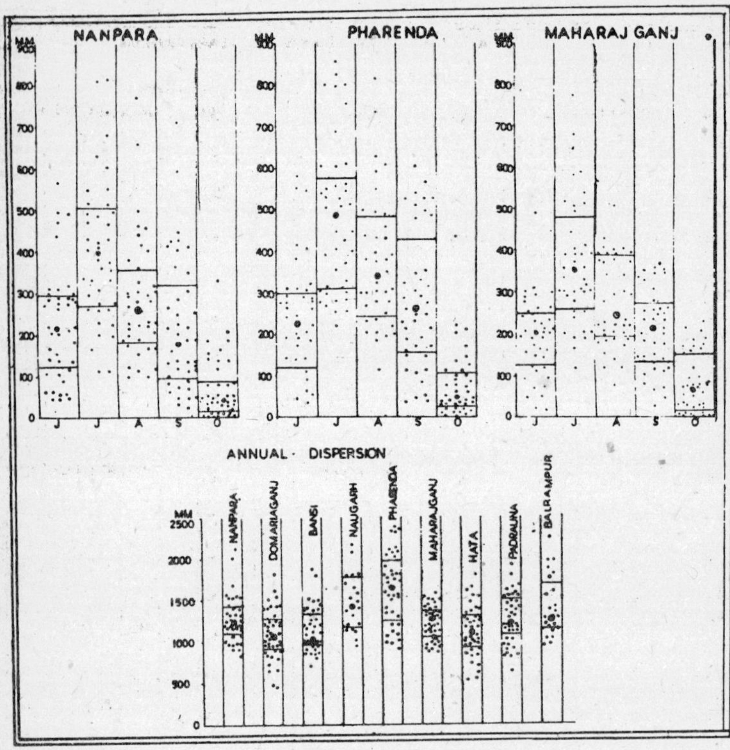

FIG. 1.10B: Dispersion of Rainfall

Since the soil is the essential material on which agriculture is based, any comprehensive survey of the geography of agriculture should include a fairly, thorough treatment of soils.[60] Therefore, an attempt has been made in the forthcoming pages to describe and discuss the soils of the study region.

The soils of the Tarai region differ strikingly from those found in other parts of India. Viewed geologically, they are of sub-recent origin with alluvial formation having substantial proportion of undecomposed plant residue. This feature together with the climatic and biotic factors, give these soils a homogeneous character and make it a difficult task to attempt a proper classification of these soils. However, topography which affects soil morphology may easily serve as a criterion for a proper classification of the Tarai soils on a sample survey basis.[61]

Thus on the basis of the general configuration of the region, the soils of the Tarai may be classified into the following groups (Fig. 1.11):

1. Soils of the undulating topography,
2. Soils of the gently sloping to nearly level and lowlying land,
3. Soils of uplands,
4. Soils of the flat but slightly elevated topography, and
5. Soils of the lowlying riverine topography.

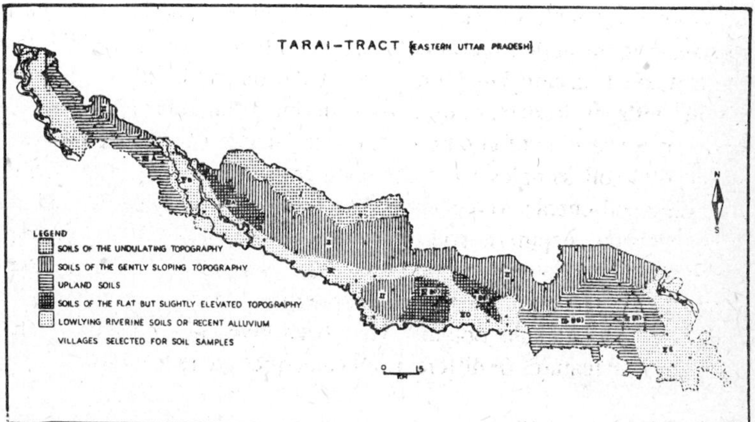

FIG. 1.11: Soils

It may be mentioned here that adequate scientific data on the soils of the study area are not available, as scientific soil surveys which aim at providing comprehensive information about soils and making an inventory of soil resources of this area, studying and determining the properties, characteristics and qualities of each soil in order to understand their genesis, development and classification etc., grouping them for various uses, planning and making applicable soil and water management programmes, have not been undertaken, except in the eastern part in Gorakhpur and Deoria districts which come under the Gandak Command Area Development Project, where detailed soil survey following the procedure laid down in the *Soil Survey Manual*

(1970) issued by All India Soil and Land-use Survey Organisation, I.A.R.I., New Delhi and *Technical Guide for Soil and Land-use Survey* (1974) issued by the Department of Agriculture (Soil Conservation Section), Krishi Bhawan, Lucknow, has been undertaken. Since these soil survey reports relating to the study area have yet not been published, complete information could not be obtained, although some results of soil analysis have been obtained and utilised as far as possible.

Undoubtedly, the Settlement Reports and the District Gazetteers contain a fairly good deal of information with regard to the textural classification of soils arrived at by empirical methods and taking into consideration texture, colour, availability of water and the level of land etc. Soil characteristics have been discussed in these reports according to the physiographic divisions. On the basis of this and other existing information, the author has attempted a classification of the soils of the Tarai tract of eastern Uttar Pradesh. At the outset, he selected some sample villages from various parts differing from each other on the basis of surface configuration, drainage, depth of water-table etc. About 400 soil samples were collected from different depths. While recording soil characteristics in the field, information regarding relief, slope, erosion, depth of soil, drainage, colour, texture and present land-use etc. was also collected. The physical, chemical and mechanical analysis of these soil samples was done in the laboratory and some of the results obtained have been given in Appendix I. The characteristic features of different soil categories are as follows:

Soils of Undulating Topography

Extending from the Nepal boundary at the foot of the hills to a varying distance to the south, average breadth being little more than 6 kilometres, there is a torrents tract in the northern part of the district of Gonda and in the north-eastern part of Bahraich district. The ground here is undulating being slightly higher towards west. It is intersected by numerous hill forests which are confined by cliffs varying in heights. The innumerable torrents bring down boulders and debris from the hills and their broad beds are covered with shingle and sand.[62] The soil analysis shows that the surface texture of the soils is loamy sand with high proportion of coarse sand in the areas between the river beds underlain with the soils of the same texture. They are ash-grey or brownish-grey in colour. They are well or at places excessively drained soils, as the water-table is lower and the surface drainage is good to excessive. The water-holding capacity is low and the permeability is rapid. They are mildly alkaline in reaction. The natural fertility in respect of organic carbon is very low, in the case of phosphorous low and medium in the case of potash.

Soils of the Gently Sloping to Nearly Level and Lowlying Lands

This type comprises the soils of a belt some 16 to 24 km in width, running along the northern border of the area which is largely lowlying and imperfectly to moderately well drained, liable to inundation during the rains by the reason of the constant overflows from the numerous hill torrents by which it is intersected, where water is quite near to the surface and the soils predominantly are clayey loam or silty loam to silty clay loam. They are underlain with similar type of soils. The soils are light to dark grey in colour with high content of clay at the surface. The water-holding capacity is generally high and the permeability slow. The soils are mildly alkaline in reaction. Phosphorous and organic carbon status are low and potash status is medium.

There are some local variations within this belt and accordingly, it may be sub-divided into the following parts:

(a) *Western Region*

These soils comprise of the areas lying in the northern fringe of the block Mihipurwa in Nanpara tahsil, northern half of the block Sirsia in Bhinga tahsil in Bahraich district; Haraiya Satgharwa, Tulsipur, Gainsari and Pachperwa blocks of Gonda district. The peculiar characteristics of this sub-division is that its northern portion adjoining the undulating forest belt in the north is different from the southern portion. In the north where the gradient is a bit more due to the vicinity of the hills, the innumerable torrents bring down quantities of boulders and debris from the hills and their beds are covered with shingle and sand. These deposits of sand and gravel cover the neighbouring fields also, with the result that the areas of coarse type soil are relatively more here, which are moderately well to well drained and the water-holding capacity is low and the permeability is rapid. They are more suited to the crops like arhar. Fertility status is low.

Towards south, the soil is generally a stiff heavy clay or clayey to silty loam with certain proportion of loam, as we find towards the eastern part of Bahraich and western and eastern parts of Gonda district. The whole tract is subject to inundation during the rains, whereby it is being gradually raised due to deposits of white sand and the lowlands which formerly used to produce fine rice, have been converted into wheat fields. The eastern part is less subject to such an inundation and comprises of some very rich soils, although south-eastern part towards Burhi Rapti has some very poor soils.

The soil analysis data also reveal that the surface texture of the soils of this region is mainly loam to clay loam underlain with clay loam or silty loam soils. Water-table is found at a depth of 1 to 2 metres during the rainy season and 3 to 5 metres during summer season. The soils are imperfectly to moderately well drained. The water-holding capacity is generally high and the permeability is slow. The soils are mildly to moderately alkaline in reaction. As regards the fertility nutrients, potash is adequate, but organic carbon and phosphorous are low.

(b) *Eastern Region*

These soils are found in the northern portions of the Basti and Gorakhpur districts comprising the blocks of Barhni, Naugarh and Birdpur (Tahsil Naugarh, Basti district); Nautanwa and the larger part of the blocks Laxmipur and Brijmanganj (Tahsil Pharenda) and block Nichlaul (Tahsil Maharajganj, district Gorakhpur). This also includes a rich stretch of predominantly clayey or silty clay loam soils which belong to the lowlying areas of Itwa and Bhanwapur blocks of Domariaganj Tahsil of Basti district.

Throughout this region also, clay predominates and surface texture, as revealed by the soil analysis, is mostly clayey loam, silty loam or silty clay loam or clayey with certain area of loamy soil. The water-holding capacity is moderately high to high and permeability is moderately slow to slow. The soils are mildy alkaline in reaction. The fertility is low in respect of organic carbon and phosphorous and medium in that of potash.

Soils of the Upland

These soils have developed in areas of older alluvial deposits and may be sub-divided into a number of topographical and textural sub-divisions with varying textures. The texture of the soils varies from sandy loam to sandy in high lying areas and loam to clayey or silty loam in low lands. In depressions, the surface soils get enriched with fine particles transported from the adjoining uplands. Under conditions of extreme marshiness, dampness and water-logging, the soils occasionally undergo humification which imparts a grey colour to them. Such soils often crack on drying. In uplands, they are brown to dark brown or greyish brown in colour, lime contents are low in uplands, but higher in lowlands. Clay contents are from low to moderately high. The soils are well to poorly drained depending upon the surface and underground drainage and the texture of the soil. The water-holding capacity is moderate to high and in certain cases even low. The permeability is also from slow to rapid accordingly. The fertility status is from low to medium. The spatial distribution of these soils is as follows:–

(a) *Bahraich Upland Soils*

The central tableland of Bahraich, which is of almost uniform character being a level plain, dotted here and there with patches of scrub jungle and broken only by slight undulations and unimportant drains of channels, extends to the study area and occupies the greater parts of the blocks of Mihipurwa, Nawabganj and Jamunaha. The prevailing soil is a light but fertile loam with clayey soils in the depressions. On the edge of the forests and on the slopes between the upland and lowlying portions, the soil is light and sandy.

The average results of the soil samples show that the surface texture of the soils is generally loam or silty loam. They generally are moderately well drained to well drained and at some places even excessively drained. The soil reaction is neutral to mildly alkaline. The water-holding capacity is moderate to moderately high and the permeability moderate to moderately slow. As regards the natural fertility, it is medium in respect of potash, low in that of phosphorous and very low to medium in that of nitrogen.

(b) *Gorakhpur Upland Soils*

According to the presence and /or absence of calcium carbonate in the soil profile, these soils have been sub-divided into two sub-divisions, viz., (i) Partially leached calcium soils and (ii) Leached calcium soils.[63]

(i) *Partially Degraded or Leached Soils*: This type represents soils in the tract lying on the western side of the river Little Gandak and on a strip of land bordering the zone of the Great Gandak and Little Gandak basin, occupying parts of the Maharajganj and Mithaura blocks and parts of the Siswa and Ghughli blocks of Maharajganj Tahsil of Gorakhpur district. In this area water-table generally varies from 1 to 1.5 metres during the monsoon and 2.5 to 3 metres during the summer.

These soils are called 'Partially degraded or leached calcium soils' because the calcium carbonate originally present in the soil has yet not been completely removed from the profile and shows various stages of leaching. However, one feature is common. There is an accumulation zone of lime in the sub-soil at a depth of about 0.5 to 1.5 metres from the surface where the calcium carbonate has been deposited in the form of *kankar* nodules of various sizes varying from groundnut to big potato size and even bigger. But the *kankar* has not formed the pan. However, there is hidden salinity hazard if drainage is not improved in the area.[64]

The surface texture of these soils generally is loamy or silty loam at higher elevations and silty clay loam with greyish brown colour in lowlying flats or in nearly level terraces. The soils are mildly alkaline in reaction. The drainage condition is not good. The permeability and intake rate is very slow. The water-holding capacity is moderately high to high. The fertility status is low to medium. The soil is well adapted to medium and late paddy, sugarcane and mustard etc.

(ii) *Degraded or Leached Calcium Soils*: They occur further west of the former tract and embrace portions of the blocks Maharajganj, Nichlaul, Siswa in Maharajganj Tahsil and Pharenda, Brijmanganj and Laxmipur blocks of Tahsil Pharenda in Gorakhpur district.

These soils have been described as 'the leached calcium soils', since they show considerable leaching and the exchange complex (the portion of the soil which takes part in base exchange) has also been affected as a result of this leaching.[65]

The water-table is generally lower in this type, being generally, on an average, 1.5 to 2 or 3 metres during the monsoon and 3 to 6 metres during the summers.

The texture of these soils is mostly sandy loam to silty loam or silty clay loam in lowlying areas. The silty loam soil is found on level to nearly level land with sandy loam well drained soils on very gently sloping or gently sloping land. The silty clay loam poorly drained soils are found on flat or lowlying terraces. They have a yellowish grey colour tending to be redder in the sub-soils with a laminor structure and are sufficiently porous. Although the soil is characterised by two distinct layers varying in colour and the sub-soil appears heavier than the surface soil, there is at present no evidence of any hindrance to ordinary root development. They lack greatly in bases specially lime and consequently at places develop an acidic character.[66] The water-holding capacity is moderate to moderately high or high and permeability is slow to rapid. The soil reaction is mildly alkaline and the natural fertility of the soil is generally low. Nitrogen is the crying need of these soils. Some form of manuring with lime is essential. Green manuring with Sanai and use of cakes may be advantageous.[67]

Soils of Flat but Slightly Elevated Topography

This soil type occurs at slightly elevated levels which are generally outside the fluvial action of the rivers. The texture of the soils varies from sandy loam or loam in high lying areas and loam to clayey loam or silty or silty clay loam in lowlying areas. Clay contents are from average to high. In uplands, they are brown to dark brown and in low lands, they are grey to dark grey in colour. They are moderately well drained to imperfectly drained soils, as the water-table during the summer remains during 4 and 6 metres and during the rains comes nearer to the surface (2 to 3 metres). The water-holding capacity is moderate to moderately high or even high and the permeability from moderate to moderately slow or slow. Potash status is generally medium, and that of nitrogen and phosphorous very low or low.

The important areas of this type of soil are as follows:

(a) *Trans-Rapti Flatland of Bahraich District*

North-east of the river Rapti and lying in the block Hariharpur Rani and Southern portion of Sirsia (Bhinga Tahsil, Bahraich district), there is a stretch of upland standing fairly high, the western part of which is occupied by 'Reserved Forest', while the eastern one is said to possess stiff soils which soften into rich loam. It is clear from the soil analysis results that the surface texture of the soils is loam underlain with loam in sub-soil. The water-table lies at a depth of 2 metres during the rains, while it recedes to 3 to 4 metres during the summer. The soils are moderately well drained. The water-holding capacity is moderately high to high and permeability moderately slow to slow. Potash status is high and that of nitrogen and phosphorous low.

(b) *Flatland of the Lower Rapti – Burhi Rapti Doab (Basti District)*

In this part which lies east of the widespread lowlying areas of Domariaganj Tahsil, the ground rises gradually towards east in block Khuniaon and portions of block Itwa and a flat and slightly elevated topography appears. Here the soils become lighter, though clay still predominates. In the north along the streams, the ground is poor, but towards south, the soil is said to be a mainly good loam stiffening into clay in the depressions. It is revealed from the soil analysis data also that the surface texture is loam in general. The sub-soil is loam or clay loam. The soils are well to moderately well drained. The water-table remains within 2 to 3 metres during during the rains and 4 to 5 metres during the summers. The water-holding capacity is moderate to moderately high and the permeability moderate to slow. The soils are mildly alkaline in reaction. Potash status is medium, organic carbon and phosphorous status are low to very low.

(c) *Banganga-Jamuwar Flatland of Basti District*

They are found mainly in Naugarh and portions of Jogia block in Tahsil Naugarh. The surface texture of the soils towards north is generally loam to silty clay loam. The sub-soils are also of the same texture. The water-table is found at a depth of 2 to 3 metres during the rains and 3 to 6 metres during the summers. The water-holding capacity is moderately high and permeability slow. The fertility status is low.

Soils of the Lowlying Riverine Topography

These soils are developed on recent alluvial deposits subject at times to inundation during high floods or heavy rains. They are coarse sandy, sandy to silty loam in texture with light or ash grey to greyish brown colour at the surface. The sub-soils are also sandy to silty loam. Clay content is low or average at the top in the case of the sandy loam soils and medium in the case of the silty loam soils. The soils are mildly to moderately alkaline in reaction. Some soils are very rich in lime, while others have low to average proportion of this component. The drainage of the area is restricted due to high water-table and they are thus poorly to moderately well drained soils. The water-holding capacity is low to moderate in some cases and generally moderately high to high and permeability rapid or moderately rapid or moderately slow to slow. Nitrogen and phosphorous status are generally low, while potash status is medium. The soil profile shows the presence of a number of stratified layers with practically no profile development, i.e. the profile development is immature. The region may be sub-divided into the following river basins:–

(a) *Kauriala or Ghaghra Basin*

This is the westernmost river basin of the region, occupying the western and north-western part of the block Mihipurwa and comprising the lowlying portions along the river Kauriala or Ghaghra, which are subject to constant danger of flooding and possess poor soils containing undue proportion of sand which makes it the soil of comparatively inferior quality with fluctuating and precarious cultivation and occasional patches of poor forests. The soil is thus of recent origin having been deposited from the annual flooding of the land along the river. It is, at places, pure sand, the water-holding capacity of which is low. The water-table is high. During the rainy season, it remains virtually on the surface. Drainage is restricted due to this high water-table. The sub-soil is also sandy. It is away from the banks that the soil becomes sandy loam or proportion of silt increases. The soils are mildly calcareous with slight calcareous reaction. The colour of the soil is grey to ash grey. The fertility status is low.

(b) *Bhakla-Rapti Basin*

The lowlying lands between the rivers Bhakla and Rapti in the blocks Jamunaha, Hariharpur Rani and southern portion of block Sirsia belong to this basin which is said to be one of the most fertile portions. The soil is generally loamy of a distinctly alluvial character. The deposits brought down by the Rapti in times of flood has a fertilising effect, so that the damage done by the river is generally little.[68] The western end of the tract, however, is composed of clay soil with a few small patches of loam. The soil of the Rapti valley is naturally moist and thus irrigation is not that much required.

The analytical data of the soil samples collected from this region reveal almost the same characteristics. The surface texture of the soils is generally loam and sandy loam or silty loam underlain with loam or silty loam soils. They are poorly to imperfectly drained, as they remain flooded during rainy season and even during summers. The water-table is very high. The water-holding capacity is moderately high to high. The permeability is moderately slow to moderate. The soils are mildly alkaline in reaction. As regards the fertility status of the soils, it is low to medium in the case of organic carbon and potash and low in the case of phosphorous.

(c) Rapti-Burhi Rapti Basin

It occupies predominantly the block Balrampur and the southern fringe of the block Haraiya Satgharwa, Tulsipur, Gainsari and Pachperwa in Tahsil Balrampur of Gonda district; and extends in a long strip across the whole breadth of the district and widening towards the Basti frontier. This tract suffers from the annual floods of both rivers which join in rains. The soil generally is a fair loam or rather alluvial loam underlain with loam to silty loam sub-soil. The water-table is generally high and the soils are imperfectly drained. The water-holding capacity is moderately high to high and the permeability is moderate to slow. The soils are mildly alkaline in reaction. The organic carbon and phosphorous status are low and the potash status is medium.

(d) *Burhi Rapti-Banganga and Rapti Basins*

The Burhi Rapti-Banganga Kachhar comprises the soils lying in the blocks Jogia, Uska and portions of the blocks Barhni and Naugarh in tahsil Naugarh, and portions of the blocks of Itwa and Khuniaon of tahsil Domariaganj. The Rapti basin occupies the portions of the block Bhanwapur and Khuniaon, and also the northern parts of the block Bansi. In the *Kachhar* of Burhi Rapti and Banganga, the soil is mainly '*Bhat*' but of inferior character, along the river Rapti also, particularly, towards east, there is a strip of '*Bhat*' or silt, but it is characterised by great fertility. The surface texture of the soil is generally silty loam underlain with silty loam or silty clay loam soils. The water-table is high and the soils possess high water-holding capacity with slow permeability. They are mildly alkaline in reactions. The fertility status is low as regards the organic carbon and phosphorous and medium in the case of potash.

(e) *Great Gandak and Little Gandak Basin*

This type comprises the soils lying between the Great Gandak and the Little Gandak in the blocks of Bishunpura, Khadda and Nebua Naurangia (tahsil Hata, district Deoria) and the portions of the blocks Nichlaul, Ghughli and Siswa (tahsil Maharajganj, district Gorakhpur).

The water-table is generally very much near the surface, being at places 1 to 2 metres in summer and remains on surface in the rainy season. They are imperfectly to moderately well-drained soils. The surface texture is silty loam to silty clay loam and near the riverside from sandy loam to fine sandy loam. The sub-soil is almost of the same texture. The soils are greyish brown to light brownish grey in colour. The clay content is generally low to medium and rarely high.

They are highly calcareous being very rich in calcium carbonate which is present in a finely precipitated form throughout the profile.[69] Sometimes, the lime forms as much as 50 per cent of the bulk soil material. This has imparted some peculiar properties to the soils, that are reflected in cropping pattern, crop rotation and agronomic practices prevalent in the area. The moisture requirement of the soil is less than that of other soils. The pH of the soil is mildly to moderately alkalines.[70] The water-holding capacity generally is high, except that of sandy loam soils where it is comparatively low. The permeability is slow, although in the case of the latter soils it is rapid.

The organic carbon content and available phosphate is low, while potash content is satisfactory. Soluble phosphate applied to the soil becomes unavailable by leaching with lime of the soil and poses a problem of phosphate deficiency. Deep cultivation with soil turning plough in the soil is not desirable since the sub-soil which is entirely deficient in organic carbon is exposed.[71]

The soil management should aim at building up a higher humus status, which can be attained only by repeated application of heavy doses of farmyard manure or well-rooted compost year after year or through the ploughing in of the green manures like *sanai* and wherever possible the use of oil-cakes as manure. Green manuring with *sanai* or any other suitable crop seems to offer immense possibilities for this type of soil. The high lime content of the soil coupled with its favourable moisture status will greatly stimulate the growth of *sanai* and a good yield of green matter on well drained soils can be expected. The subsequent decomposition of this green matter will also be facilitated by an adequate presence of lime and moisture. In places where water-logging takes place during Kharif, it may be difficult to get good crop of *sanai* and alternating green manuring crops such as *dhaincha* may be tried. Ammonia sulphate can be used to make up the deficiency of the nitrogen.[72]

In the water-logged areas, drainage conditions should be improved and in the area where soluble salts of calcium and magnesium are present on the surface, levelling and bunding should be taken up, so that accumulated rain water may leach the soluble salts to lower horizons. It is not desirable to adopt cultivation practices which might further impede drainage and lead to the injurious salt accumulation and thus damage the crop and the soil.[73]

NOTES

1. Singh, R.P., (1981), 'Concept of Land-use' in Mohammad, N. (ed.), *Perspective in Agricultural Geography*, Vol. III, Concept Publishing Company, New Delhi, p. 167.
2. Bhattacharya, K.C., (1965), 'Methodology for Preparing Land-use Maps' *All India Congress Committee Review*, December 1, Vol. 17, Nov. 12, Whole No. 372, New Delhi, p. 13.
3. Singh, R.L. and Singh, K.N., (1971), 'Middle Ganga Plan' in Singh, R.L., et al. (eds), *India, Regional Study*, Varanasi, p. 247.
4. Atkinson, E.T., (1882), *The Himalayan Districts of N.W Provinces of India*, Vol. I, p. 85 (Quoted in Singh, L.R., (1965), *The Tarai region of Uttar Pradesh - A Study in Human Geography*, Allahabad, p. 1.
5. Batten, J.H., 'A Few Notes on the Subject of the Kumaon and Rohilkhand Turaee', (Quoted in Singh, L.R., *Ibid*).
6. Madden, E., 'The Turaee and Outer Mountains of Kumaon', (Quoted in Singh, L.R., *Ibid*).
7. Singh, L.R., *Ibid.*
8. *Ibid.*
9. Spate, O.H.K., (1960), *India and Pakistan*, p. 498.
10. Singh, L.R., *op cit.*
11. *Ibid.*, p. 20.
12. Singh, R.L. and Singh, K.N., *op cit.*, p.190.
13. Oldham, R.D., (1917), 'The Structure of the Himalayas and the Gangetic Plains', *Memoires of the Geological Survey of India*, Vol. XLII, Pt. II, Calcutta, p. 82.
14. Cowie, H.M., (1921), 'A Criticism of R.D. Oldham's Report on the structure of the Himalayas and of the Gangetic Plains as Elucidated by Geodetic Observations

in India', *Memoires of the Geological Survey of India*, Professional Paper No. 18, (Dehradun) p. 6.

15. Glennie, E.A., (1932), 'Gravity Anomalies in the Structure of Earthic Const.' *Memoires of the Geological Survey of India*, Professional Paper No. 27 (Dehradun),p. 18.
16. Hodgson, B.H., (1849), 'On the physical Geography of the Himalayas', *J.A.S. Ben*, Vol. XVIII, pp. 782-773 (Quoted in Singh, L.R., *op.cit.*, p. 4.
17. Hodgson, B.H. *Ibid*, p. 783.
18. Singh, L.R., *op cit.*, p. 4.
19. Finch, *et al*, (1957),*Elements of Geography*, McGraw-Hill Co., N.Y., pp. 287-290 (Quoted in Singh, L.R., p. 5).
20. Batten, J.H., (1844), 'A Few Notes on the Subject of the Kumaon and Rohilkhand Turaee', *J.A.S. Ben* Vol. XIII, pt. 2, pp. 912-913, (Quoted in Singh, L.R., *op cit.*).
21. Vide Proceedings, N.W.P., P.W., Deptt., May, 1864, July, 1869 and Feb. 1872, (Quoted in Singh, L.R., *Ibid*).
22. Blanford, W., *Manual of Geology of India* IX, p. 393 (Quoted in Singh, L.R., *op cit.*).
23. Atkinson,E.T., (1982), *op cit.*, pp. 92-93, (Quoted in Singh, L.R., *op cit.*).
24. Singh, L.R., *op cit.*.
25. Stamp, L.D., (1950), *The Land of Britain, Its Use and Misuse* London, p. 216.
26. Singh, L.R., *op cit.*, p. 6.
27. Singh, R.L. and Singh, K.N., (1971), *op cit.*, pp. 192-93.
28. Nevil, H.R., (1904), District Gazetteer (Nainital), p. 11.
29. Singh, L.R., *op cit.*, p. 8.
30. Hutton, C.H., (1901), 'Report on the Tarai Canals', p. 5, (Quoted in Singh, L.R., *op cit.*., p. 9).
31. Sharma, S.C., (1968), 'Land Utilisation in Etawah District of Uttar Pradesh,' An Unpublished Thesis, p. 22.
32. Singh, L.R.,*op cit.*.
33. Sharma, S.C., (1968), *op cit.*, p. 27.
34. Symons, L., (1970), *Agricultural Geography*, London, p. 21.
35. Miller, A.A., (1956), *Climatology*, p. 1.
36. Singh, J., (1976), *An Agricultural Geography of Haryana*, p. 44.
37. Shirlaw, D.W. Gilchrist, (1971), *Agricultural Geography of Great Britain*, Oxford, Pergamon Press, p. 20, (Quoted in Singh, J., 1974-75), *An Agricultural Atlas of India: A Geographical Analysis*, p. 11.
38. Government of India, Indian Meteorological Department (1962) 'Weather and the Indian Farmer', Poona, p. 4.
39. Nevil, H.R., (1903), *Bahraich—A Gazetteer*, Vol. XLV, pp. 29-30.
40. Singh, L.R., *op cit.*, p. 10.
41. Sharma, S.C., *op cit.*, p. 29.
42. Shafi, M., (1960), *Land Utilisation in Eastern Uttar Pradesh*, Aligarh, p. 19.
43. Miller, A.A., *op cit.*, p. 139.
44. Nevil, H.R., (1926), *Basti—A Gazetteer*, Vol. XXXII, p. 18.
45. Singh, L.R., *op cit.*, p. 13.
46. Spate, O.H.K., *op cit.*, p. 44.
47. Sharma, S.C., *op cit.*, p. 46.
48. Miller, A.A., *op cit.*, p. 16.

49. Government of India, Meteorological Department, *op cit.*, p. 11, (Quoted in Sharma, S.C., *op cit.*, p. 47.).
50. Singh, L.R., *op cit.*, p. 10.
51. Singh, J., (1976), *op cit.*, p. 58.
52. Blandford, H.F., (1986-88),'Rainfall of India', *Memoires of the Indian Meteorological Department*, Vol. III, p. 130, (Quoted in Sharma, S.C., *op cit.*, p. 51).
53. Sharma, S.C., *Ibid.*, p. 52.
54. Clark, K.G.T., (1932), 'The Vicissitudes of the Summer Rainfall of the Indo-Gangetic Plain and the Assam Valley', *Geography*, Vol. XVII, pt. IV, p. 288
55. Sharma, S.C., *op cit.*, p. 54.
56. Crowie, P.R., (1933), 'The Analysis of Rainfall Probability, A Geographical Method and Its Application to European Data', *Scotish geographical Magazine*, Vol. XLIX, p. 73, (Quoted in Sharma, S.C., *Ibid*).
57. Sharma, S.C., *Ibid.*, p. 57.
58. Pullan, (1969), 'The Soil Resource of West Africa', in Thomas, M.F. and Whittington, G.W. (eds.), *Environment and Land-use in Africa*, London, Methuen, p. 148.
59. Singh, J., (1976)., *op cit.*, p. 82.
60. Symons, L., (1970), *op cit.*, p.32.
61. Singh, L.R., *op cit.*, p. 15.
62. Nevil, H.R., *Gonda District Gazetteer*, Vol. XLIV, pp. 258-63.
63. Mukerjee, B.K., Agarwal, R.R. and Mukerjee, S.K., '*Sugarcane Cultivation in Relation to the Soils of the District Gorakhpur*, United Provinces, pp. 4-6.
64. Soil and Water Management Pilot Project, Gandak Command, Gorakhpur, U.P., 'Soils of Gandak Command and Their Management' *Technical Bulletin* No. P.I., p. 9.
65. Mukerjee, B.K., *et al.*, *op cit.*, p. 6.
66. *Ibid.*
67. *Ibid.*, p. 7.
68. Nevil, H.R., (1903), *Bahraich-A Gazetteer*, Vol. XLV., p. 4.
69. Mukerjee, B.K., *et al.*, *op cit.*, p. 3.
70. Soil and Water Management Pilot Project, Gandak Command, *op cit.*, pp. 11-12.
71. *Ibid.*
72. Mukerjee, B.K., *et al.*, *op cit.*, pp. 3-4.
73. *Ibid.*

TWO

CULTURAL FACTORS AFFECTING LAND-USE

CULTURAL FACTORS are of no less importance in the study of land utilisation. The areal distribution of different types of land-use, writes Hartshorne is significantly related to other geographical features, both cultural and natural. . . .[1] Man and his cultural attainments play a very important role in the extent and degree of land utilisation and hence it appears very essential to describe and discuss the effects of some important factors of cultural environment for proper land-use analysis of an area. An attempt has been made in this direction in the ensuing pages.

POPULATION

There exists a close relationship between land and people, for they affect each other. Few relationships in life are more significant or more fundamental than that exists between man and land. Man is the chief recipient of the benefits and fruits of land. Land provides him with living space, with organic elements so necessary for the production of most of the foods and fibres, and with the raw materials he uses in his home and industries. Not only does man benefit from the use of land in the production process, but also he enters into the process. Man supplies the ingenuity, integration and knowhow that make land valuable in the production process.[2] Nearly all of man's food comes from the soil. Thereby the land-use is modified by the various aspects of population, the density of cultivators in particular. The relation between population and land-use is reciprocal, for instance, the changes in farm population pattern influence not only the utilisation of the land, but the latter, with changes in the agricultural controls to a great degree, also determines the pattern of farm population.[3]

Besides the above factors that warrant a detailed study of

population in the land utilisation of a region, the rapidly increasing pressure of population on land resources also makes it essential to undertake on one hand a factual objective study of population in its many aspects and on the other hand for a similar objective study of the land, its use and misuse, its actual and potential resources, their exploitation and conservation.[4] Parallel with the need for mapping use and non-use of land is that for the mapping of the present distribution of population.[5] Therefore, it seems logical to make an attempt to discuss the population of the study area.

Growth of Population

The population* of the study area is 33,99,709 (1981 Census). It was 21,96,125 in 1951, 24,46,081 in 1961 and 27,51,865 in 1971. Thus during the period of three decades, the population in the study area has increased by 54.8 per cent, as against 80.4 per cent in the state of Uttar Pradesh (11.4 during 1951-61, 12.5 during 1961-71 and 23.5 per cent during 1971-81, as against the State's respective figures of 16.7, 19.8 and 25.5 per cent). Figure 2.1, Table 2.1a and Appendix II show that the rate of growth has not been uniform throughout the region. There are sub-regional variations in it. About 58.1 per cent of the blocks are below the regional average. The blocks with 40-50 per cent or with 30-40 per cent growth rate number 16 (Jamunaha—46.5, Sirsia—32.8, Haraiya Satgharwa—44.4, Tulsipur—34.4, Pachperwa—42.0, Balrampur—45.3, Itwa—40.4, Bhanwapur—40.4, Khuniaon—42.4, Barhni—45.1, Naugarh—32.8, Jogia—45.7, Uska—45.7, Brijmanganj—49.6, Nantanwa—46.7 and Siswa—44.2). Some of these blocks are border blocks impregnated with unfavourable conditions such as inadequate means of transport and communication, low agricultural development, mass poverty and backwardness, unhealthy climatic conditions and male-selective out-migration. In some of them, the population pressure was already sufficiently high and there has been a little increase in the net sown area. There are 5 blocks in the category of the growth rate of 50-60 per cent. They are Nawabganj (57.2), Gainsari (53.0), Birdpur (52.1), Mithaura (57.2) and Nebua Naurangia (58.4).

* The population data have been compiled from the District Census Handbooks of the districts Bahraich, Gonda, Basti, Gorakhpur and Deoria.

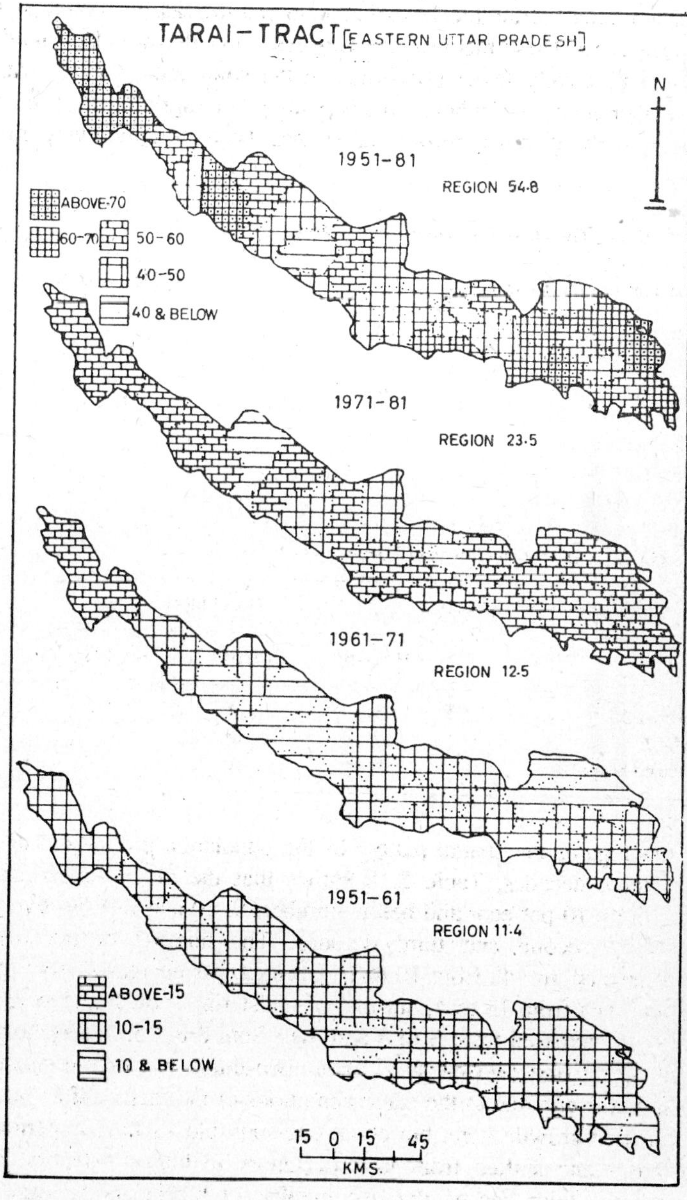

2.1: Growth of Population

The number of the blocks experiencing an increase of more than 60 per cent is 10. These blocks either possessed low density of population and comparatively lower proportion of net sown area, both of which have increased, or where intensification of agriculture and the development of other infrastuctures has been comparatively more rapid.

TABLE 2.1a: Growth of Population (1951—1981)

Percentage growth rate categories	No. of blocks
Above 70	4
60-70	6
50-60	5
40-50	13
30-40	3

TABLE 2.1b: Decennial Growth of Population

Percentage growth rate categories	No. of blocks		
	1951—61	1961—71	1971—81
Above 15	3	8	25
10-15	19	13	4
10 and below	9	10	2

As regards the spatial pattern of the population growth during the past three decades, Table 2.1b shows that the blocks experiencing growth by 10 per cent and below numbers 9, 10 and 2 respectively in the first, second and third decades. The number of the blocks experiencing growth from 10 to 15 per cent, has decreased from 19 to 13 and then from 13 to 4, but that of registering growth above 15 per cent has increased from 3 to 8 and then from 8 to 25. It may be seen that during first two decades, about two-thirds blocks did maintain their *status quo* due to the static conditions of the factors affecting the population growth there, but about one-sixth blocks have experienced increases and shifted from lower category to higher category. It is during the third decade that the number of blocks in the past two categories has decreased immensely and that in the first category, it has increased abruptly, which may be attributed to the decrease in the magnitude of unfavourable conditions causing arrested growth of population and an increase in the facilities and amenities attracting population in-migration.

It may be remarked that the notion that the Tarai has throughout been a scene of accelerated growth of population in the post-Independence period is not completely true. Undoubtedly, the reclamation measures and consequent agricultural colonisation, extensification and intensification of agriculture, amelioration in the living conditions and various developmental activities have attracted in-migration and increased growth rate, but not so fast as claimed, particularly in the former two decades. As mentioned earlier, the regional growth rate is lower than that of the state and the country as a whole. But if the district growth rates of the rural population for the decade 1971-81 are compared (which are: Bahraich – 26.9, Gonda – 20.8, Basti – 17.2, Gorakhpur – 21.3 and Deoria – 19.6 per cent), it may be observed that about 58.8 per cent of the blocks, i.e. Mihipurwa and Hariharpur Rani in Bahraich; Haraiya-Satgharwa and Gainsari in Gonda; Itwa, Bhanwapur, Khuniaon, Uska and Birdpur in Basti, Nautanwa, Laxmipur, Pharenda, Nichlaul, Maharajganj and Ghughli in Gorakhpur and all the blocks of the Deoria district show somewhat higher rates than that of the districts in which they fall. In rest of the blocks, the growth rates have been slower. This may partly be due to the slow process of agricultural colonisation and slow rates of developmental activities and existence of unfavourable factors like mass poverty, unhealthy climatic conditions etc. discouraging in-migration and encouraging out-migration in the latter blocks and the speedy process of agricultural colonisation and faster developmental activities together with certain other factors in the former blocks.

Figure 2.2 and Appendix III exhibit the distribution of population in the study area. It may be seen that the population is unevenly distributed. Towards the northern border, the distribution comparatively is sparse, as the percentages of the total area occupied by the blocks and the percentages of the total population possessed by them are not identical. In most of the blocks, the percentages of the total area exceed the percentages of the total population. For example, in the blocks Mihipurwa, Nawabganj, Jamunaha, Hariharpur Rani, Sirsia, Haraiya-Satgharwa, Tulsipur, Gainsari, Pachperwa, Barhni, Laxmipur and Nautanwa, there lives only 42.9 per cent of the total population of the region on 52.8 per cent of the total area. The factors affecting the sparse distribution are very clear, viz., high concentration of forested land and corresponding lower proportion of cultivated land, unfavourable and unhealthy climatic conditions, configuration intersected and criss-crossed by narrow streams, inadequate means of transport and communication and other infrastructures and poor and backward economy. Some of the blocks, viz., Hariharpur Rani in Bhinga tahsil, Birdpur in Naugarh tahsil, Nichlaul in Maharajganj tahsil and Khadda in Hata tahsil are the exception to this general trend, where the population percentages exceed the area percentages, i.e. greater population is living on comparatively lesser area. This appears either due to the reduction of the total area of these blocks or to some extent due to the higher growth rate which, in turn, is the result of a decrease in the magnitude of unfavourable factors. In rest of the blocks, almost everywhere there is higher concentration of population on the land, as 57.1 per cent of the total population of the region lives on only 47.2 per cent of its total area. These are blocks which occupy generally the central and southern parts of the region with fertile soils, better means of transport and communication and other infrastructural facilities attracting in-migration. Even in these blocks, there are local variations in population distribution, for example, in the flood affected and forested areas or where some of the other unfavourable conditions prevail, the distribution is sparse.

It may be remarked here that the distribution pattern has remained almost static in the past, i.e., the areas of high concentrations or sparse population have hardly changed their categories and this shows that the efforts to develop some of the unfavourable and backward pockets of this region and to make them more habitable have not been fruitful everywhere and they still suffer from stagnation and dogmatism.

Density of Population

The population distribution map shows simply whether the population is sparse or clustered in an area and reveals nothing about the relationship between man and land area, while land-use policy and practice and through it the contribution of land to the overall well-being of the economy, is determined to a considerable extent by the ratio of population to land resources.[6] Hence an attempt has been made here to discuss the man-land ratio in different parts of the region.

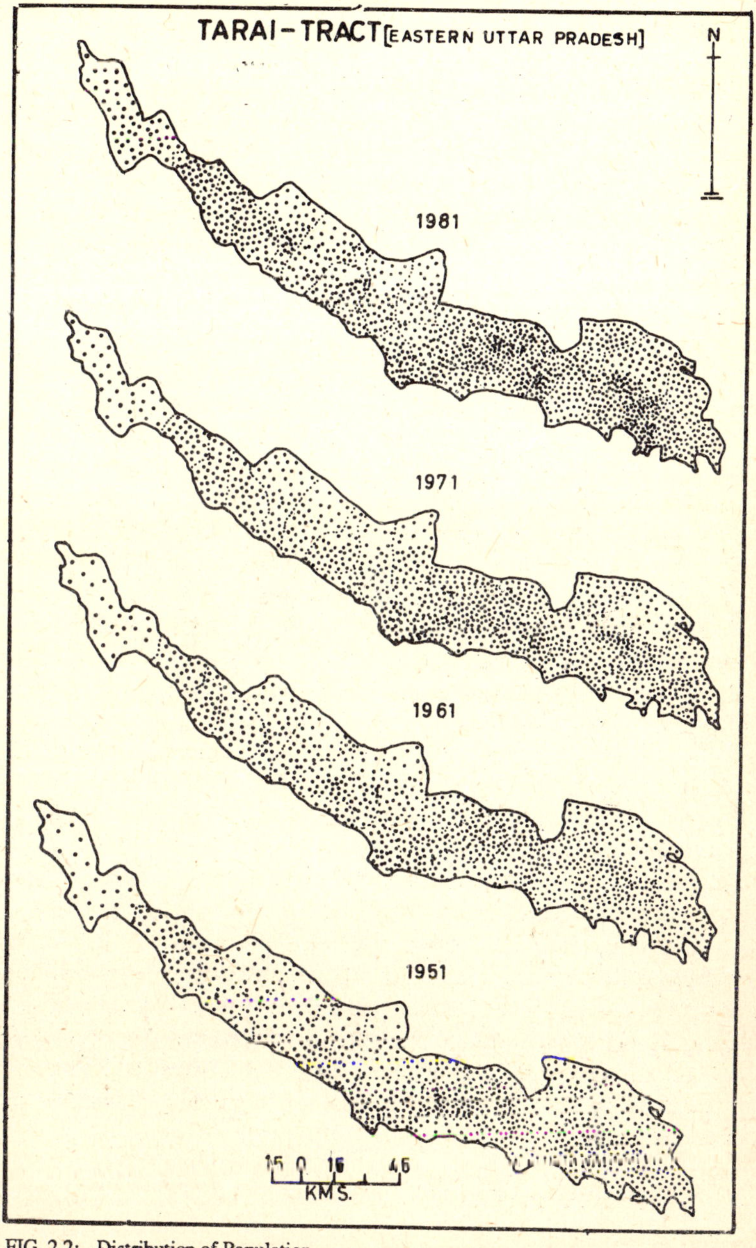

FIG. 2.2: Distribution of Population

Arithmetic Density of Population

It may be seen from Fig. 2.3, Table 2.2a and Appendix IV that the average density of population in the study area is 376 persons per sq. km (1981), as against 377 and 216 in the State of Uttar Pradesh and India respectively. The density pattern in the study area, as is apparent from the map, is characterised by remarkable sub-regional variation. About 55 per cent of the blocks record density higher than that of the regional average. This magnitude of diversity in the density pattern seems due to the physical as well as socio-economic attributes of different parts.

The density varies from 260 (Sirsia) to 621 (Ghughli). Generally speaking, the density is higher towards east than that of the west. Almost all of the 21 blocks having densities varying from 350 to 450 and above hail from eastern and middle parts and they are: Itwa, Bhanwapur, Khuniaon, Bansi, Naugarh, Birdpur, Jogia, Uska, Brijmanganj, Pharenda, Nichlaul, Mithaura, Maharajganj, Siswa, Ghughli, Khadda, Nebua Naurangia and Bishunpura. Almost in all these blocks, extensification and intensification of cultivated land with better irrigation facilities, particularly in the Gandak Command Area, and better developmental activities have contributed to the higher concentration of population. There is high positive correlation between population and net sown area.

In the density range 250–350, there are 10 blocks, only one of which belongs to the eastern part and one to the middle part, while 7 (Mihipurwa, Jamunaha, Sirsia, Haraiya, Satgharwa, Tulsipur, Gainsari and Pachperwa) belong to the western part. These are the blocks where the forests occupy sufficiently large areas, adversely affecting the extent of net sown area which ranges between 65.8 and 79.0 per cent.

It emerges from the above discussion of the spatial pattern of population density that it is higher in those areas where the proportion of net sown area, irrigated and double-cropped area is higher, means of transport and communication are better developed and the schemes of agricultural and rural development have been launched with fair amount of success. It decreases with the increasing magnitude of adverse factors.

Changes in Arithmetic Density

Table 2.2a and Fig. 2.3 show that there was no block having arithmetic density above 450 persons per sq. km at the beginning of the first and second decade, while there appeared one block at the beginning of the third decade, which rose to 9 by the end of the third decade. The number of the blocks in the range 350–450 went on increasing from 2 to 13 and that in the range 250–350 remained static up to 1971 and then fell to 10. In the range 150–250, there were 15 blocks at the beginning of the study period, which went on decreasing to become zero in the end. In the last category, therc remained one block throughout, except in the end.

Appendix IV and Fig. 2.5 show that in the region as a whole, the arithmetic density increased factually by 136 persons per sq. km during the span of three decades, as against the State's average of 163 persons. Within the region, it varies from 64 (Sirsia) to 242 (Ghughli). The number of the blocks (Table 2.3b) experiencing very low changes, i.e., by 80 persons and below is 2 and that by 80 to 130 is 13. The number of the blocks registering an increase by 130 to 180 and 180 to 230 is 9 and 5 respectively and that by more than 230 is 2. A comparison of the Figs. 2.3 and 2.5 shows that generally the blocks of the eastern part have witnessed higher increases than that of the blocks in the west. As regards the decennial changes, they have been more rapid in the third decade, as the number of blocks in the higher categories has increased and that in the lower categories decreased (Table 2.3a).

Physiological Density of Population

The simple man-land ratio, since it expresses the pressure of population on the unit of the total area, is not a satisfactory measure of the real density or the pressure of population on the resource base and thus unrealistic and misleading. Therefore, physiological density which is expressed as the number of persons per unit of cultivated area has also been considered here.

Figure 2.4 and Table 2.2b and Appendix IV show that the average physiological density in the study area is 485 persons per sq. km. It varies from 334 (Sirsia) to 726 (Ghughli). The spatial pattern is largely similar to that of the arithmetic density. For example, the lowest density is again in the block Sirsia. Of the 11 blocks in the range of 350–450, 7 belong to the western part, 2 to the central and 2 to the eastern part. Of the 8 blocks in the range 450–550, 2 belong to the western, 3 to the central and 3 to the eastern part, while of the 11 blocks in the category with more than 550, only 4 belong to the central part and remaining all 7 to the eastern part.

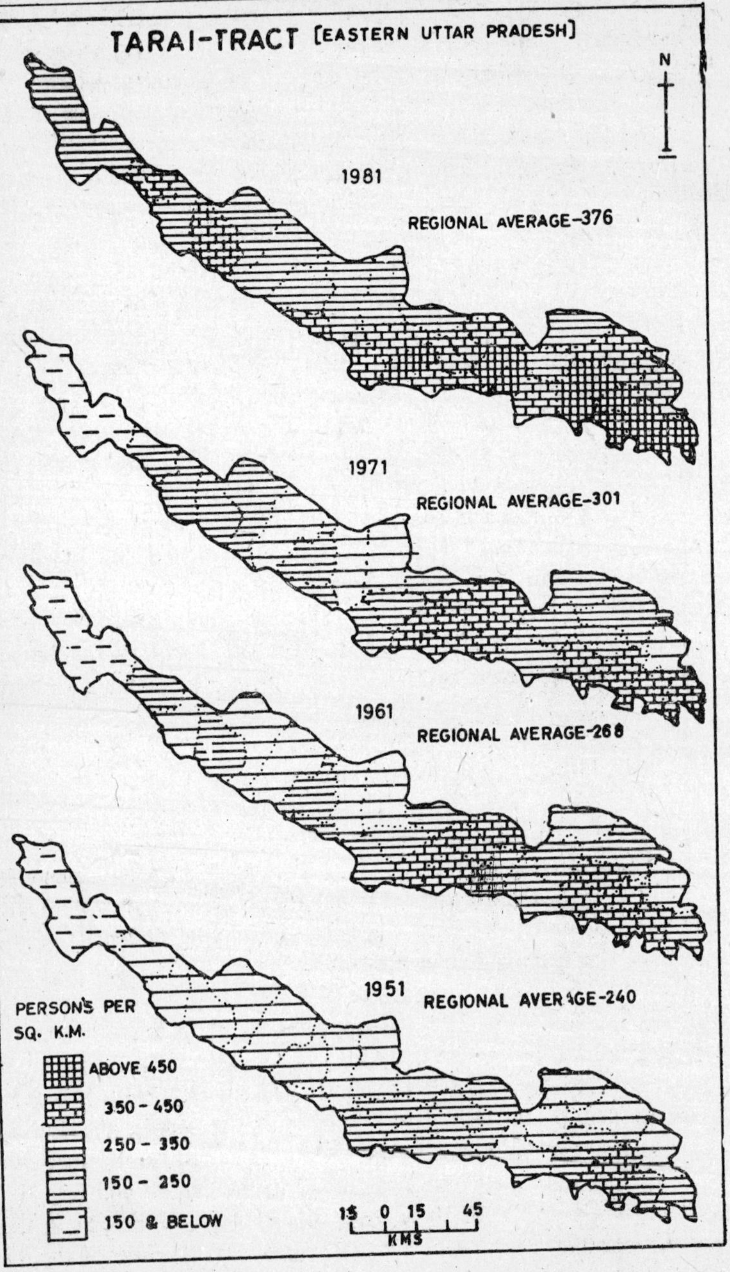

FIG. 2.3: Arithmetic Density

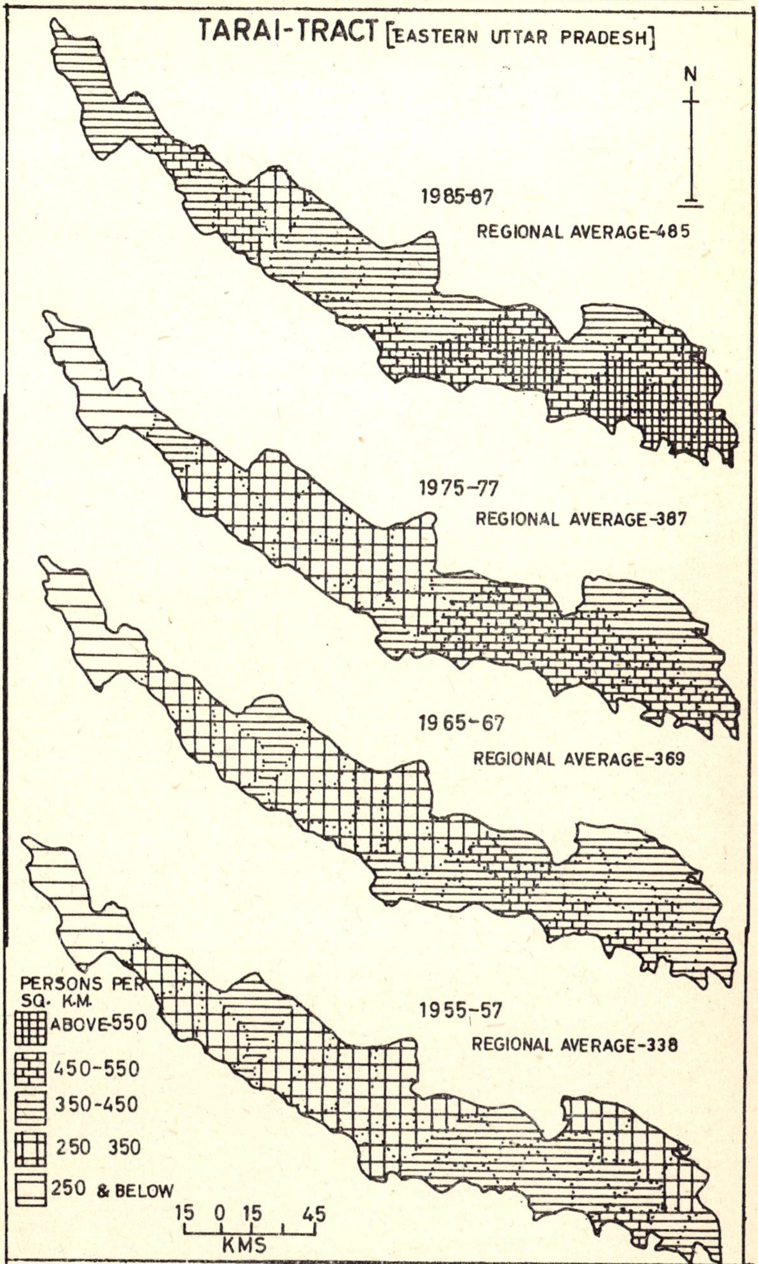

FIG. 2.4: Physiological Density

Changes in Physiological Density

A glance at Fig. 2.4 and Table 2.2b shows that physiological density has also increased in almost all the blocks except Sirsia where it has decreased slightly. The number of the blocks in the range 450–550 and above has increased from 1 to 19. In the range 350–450, it has decreased from 14 to 11 only, but in the 250–350, there is abrupt decline, i.e. from 15 to 1 only. When factual increases and decreases in the physiological densities are considered, we find that the region on an average registered an increase of 147 persons per sq. km (Appendix IV), although there are sub-regional variations (Fig. 2.5), increases varying from 20 to 314. About two-thirds of the blocks have experienced increases above 130 and about one-third below it (Table 2.3b). This exhibits that in sufficiently large number of the blocks, the increases in the net area sown have lagged behind the increases in the population. The decennial changes display that in the latter decades, the increases have been more rapid, as the number of the blocks in the higher categories has increase greatly and correspondingly it has decreased in the lower categories.

Pressure on Land

Since agriculture is the main occupation in the study area, total land supporting one person is of principal concern. The per head share of the total, net sown, double cropped and total cropped area in the region respectively is 0.27, 0.21, 0.10 and 0.31 hectare.

There are spatial variations in per head share of different categories of land (Appendix V). The pattern is almost the same as is seen in respect of population distribution and density (Figs. 2.2 - 2.4). The per head shares of different categories are higher than the regional averages almost in all the northern blocks with the exception of three blocks (Haraiya Satgharwa, Birdpur and Nichlaul) where they are below the regional averages. In these blocks, the per head total area varies from 0.27 to 0.38, net sown area from 0.21 to 0.31, double cropped area from 0.7 to 0.16 and the total cropped area which is actually supporting one person, from 0.38 to 0.45 hectare. In the remaining blocks which generally occupy the central and southern parts, per head shares are lower than the regional averages. They range between 0.18 and 0.40 (Total area), 0.14 and 0.23 (Net sown area), 0.03 and 0.12 (Double cropped area) and 0.22 and 0.33 (Total cropped area) hectare.

TABLE 2.2a: Arithmetic Density of Population

(*Per sq km*)

Density category	No. of blocks			
	1981	1971	1961	1951
Above 450	9	1	—	—
350-450	12	12	10	2
250-350	10	13	13	13
150-250	—	4	7	15
150 & below	—	1	1	1

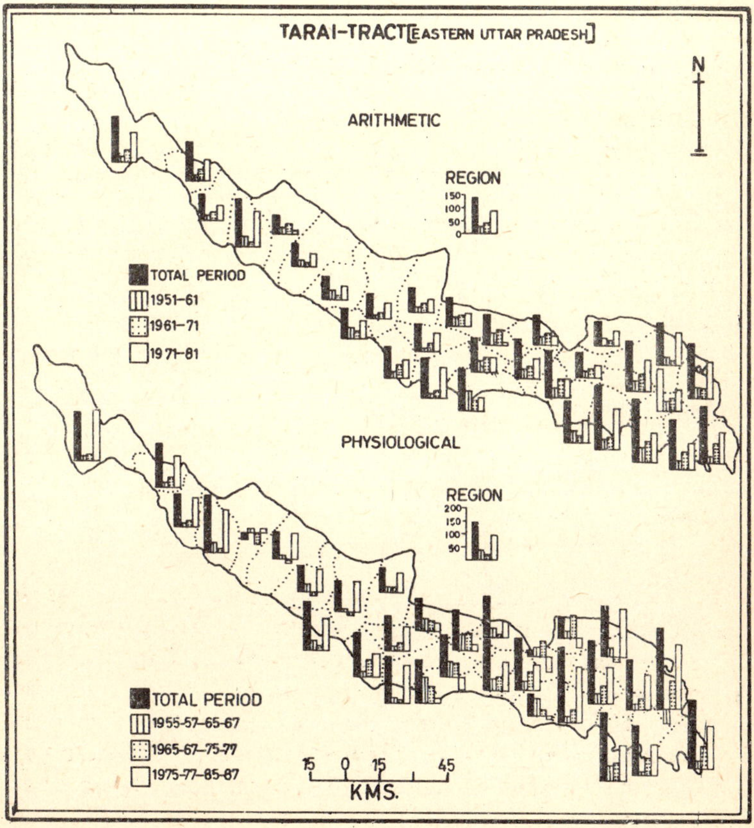

FIG. 2.5: Variation in Density of Population

TABLE 2.2b: Physiological Density of Population*

(Per sq km)

Density category	No. of blocks			
	1985—87	1975—77	1965—67	1955—57
Above 550	11	2	—	—
450—550	8	13	5	1
350—450	11	7	13	14
250—350	1	8	11	15
250 & below	—	1	2	1

* Population data are based on 1951, 1961, 1971 and 1981 censuses, but the data pertaining to net cultivated area are based on the averages of the village records for respective three consecutive years, i.e. 1955—57, 1965—67, 1975—77 and 1985—87. These latter years have been taken for showing the physiological densities.

TABLE 2.3a: Decennial Variation in Density of Population

Increase/Decrease Persons per sq km	No. of blocks					
	Arithmetic density (Decades)			Physiological density		
	1st	2nd	3rd	1st	2nd	3rd
Increases						
Above 45	4	8	25	7	15	23
35—45	4	9	2	4	2	2
25—35	11	6	2	6	2	—
15—25	7	4	—	6	2	1
15 & below	5	4	2	8	4	1
Decreases						
Above 25	—	—	—	—	1	3
15—25	—	—	—	—	2	—
15 & below	—	—	—	—	3	1

It may be concluded that the pressure of population on almost all types of land is higher in the blocks generally belonging to the central and southern parts and is lower in the northern blocks. This shows that opportunities for accommodating greater number are brighter in the latter than that of the former parts, provided effective efforts are made for the extensification and intensification of the cultivated land, for the development of means of transport and communication and other infrastructures anu introducing new innovations for agricultural development.

TABLE 2.3b: Overall Variation in Density of Population (1955-57 to 1985-87)

Increase/Decrease categories (persons per sq.km)	No. of blocks	
	Arithmetic	Physiological
Increases		
Above 230	2	6
180—230	5	6
130—180	9	10
80—130	13	7
80 & below	2	2
Decreases		
80 & below	—	1

It may be remarked here that per capita land in different categories is gradually decreasing in most of the blocks with increasing population. The per capita share in the total and net sown area respectively has decreased from 0.41 and 0.29 (1955–57) to 0.27 and 0.21 hectare (1985–87). This indicates that although there have been increases in the net sown and total cropped areas, but they could not keep pace with the rapidly increasing population. Within the region, there are enough variations, for example, in some blocks *status quo* has been maintained exhibiting simultaneous increases in the above categories of land and the population. In the majority of the blocks, the former has lagged behind the population growth and hence their proportion per head has decreased.

Irrigation

Irrigation exercises enormous influence on land-use in a country such as India where rainfall is erratic and uncertain and agriculture is said to be a gamble in monsoon, although the magnitude of this influence varies in time and space. The region under study enjoys moderately high (average annual rainfall from 1185mm to 1585mm) rainfall and possesses widespread humid soils with comparatively less thirst for additional water supply, but irrigation is necessary here also as an insurance against droughts for long dry spells or for reducing the dependence upon the vagaries and vicissitudes of the monsoon (The co-efficient of variability of annual rainfall generally varies from 20.99 to 26.73 per cent, the monthly variability is very high, particularly in the months of June, September and October). It is needed for extending the net area sown, the area that can be sown more than once, for the successful growth of water loving crops, for encouraging the tillers of the soil to use the inputs like HYV of seeds, manures and chemical fertilizers and pesticides etc., all of which require adequate timely supply of water.

Figure 2.6 and Appendix VI show that net irrigated area in the study region accounts for 2,28,758 hectares, i.e., 32.6 per cent of the net area sown, as compared to 57.1 per cent in the state of Uttar Pradesh and 29.7 per cent (1984-85) in the country.[7] The spatial distribution pattern exhibits that there are considerable micro-level variations (i.e. from 1.5 per cent in Pachperwa to 82.4 per cent in Maharajganj). The overall pattern shows that higher concentrations of irrigated area are found towards east and low to very low towards west, which is mainly due to the abundance of irrigation facilities in the Gandak Command Area and also from other sources in other areas.

There are 6 blocks, viz., Sirsia, Maharajganj, Mithaura, Ghughli, Brijmanganj and Khuniaon in high intensity group with the proportion of net irrigated area exceeding 60 per cent. In 6, i.e., Itwa, Bhanwapur, Naugarh, Uska, Laxmipur and Pharenda, this varies between 45 and 60 and in 8, i.e., Balrampur, Birdpur, Jogia, Bansi, Nautanwa, Nebua Naurangia, Bishunpura and Nichlaul between 30 and 45. These blocks abound either in canal irrigation or irrigation from other sources. There are only 4 blocks, i.e., Nawabganj, Jamunaha, Barhni and Khadda, in the 15-30 per cent intensity group. Rest of the blocks record very low proportion of irrigated area, which may be attributed mainly to the presence of more humid soils in their sub-montanne tract and to the lack of irrigation facilities which are being developed now. The decreasing number of blocks in the lower categories (Table 2.4) and that increaseing in the higher categories show that the proportion of net irrigated area is increasing.

Sources of Irrigation

There are two main sources of irrigation in the study area, which are as follows:

(i) Surface water

(a) Canals

(b) Ponds and depressions
(c) Other sources like rivers etc.

(ii) Underground water

(a) Tube-wells
(b) Wells

It may be mentioned here that irrigation has since long been practised in this region, although not so significantly and widely as at present. Ponds, lakes and depressions etc. have been utilised for this purpose. Some canals were also constructed, but the area irrigated by them and other sources was very meagre as compared to the present net irrigated area which has increased enormously in the recent past with the expansion of the network of canals and distributaries in the Gandak Command Area and in the number of tube-wells in other parts. Due to the non-availability of the data at the village level for the period 1955–57 and 1965–67, it could not be possible to give the changing pattern from the base year, but decadal changes for the period 1975–77 to 1985–87 have been studied.

Table 2.4 and Fig. 2.6 depict the relative importance of various sources of irrigation in the region for the period 1975–77 and 1985–87, from which it is clear that the surface water is the chief source of irrigation, as it accounts for 56.6 per cent of the net irrigated area as against 38.5 per cent in the State of Uttar Pradesh and 52.1 per cent (1984-85) in India[8] and the rest is accounted for by the underground water sources.

TABLE 2.4: Intensity of Irrigation

Intensity class (Net irrigated area as per cent of net area sown)	No. of blocks	
	1975–77	1985–87
Above 60	3	6
45–60	7	6
30–45	6	8
15–30	3	4
15 & below	12	7

The canals occupy the most important position among the surface water sources, as they claim 32.8 per cent of the net irrigated area of the region, although this proportion in the State of Uttar Pradesh is

33.8 and in India 38.0 per cent. There are micro-regional variations in their relative importance in the study area.

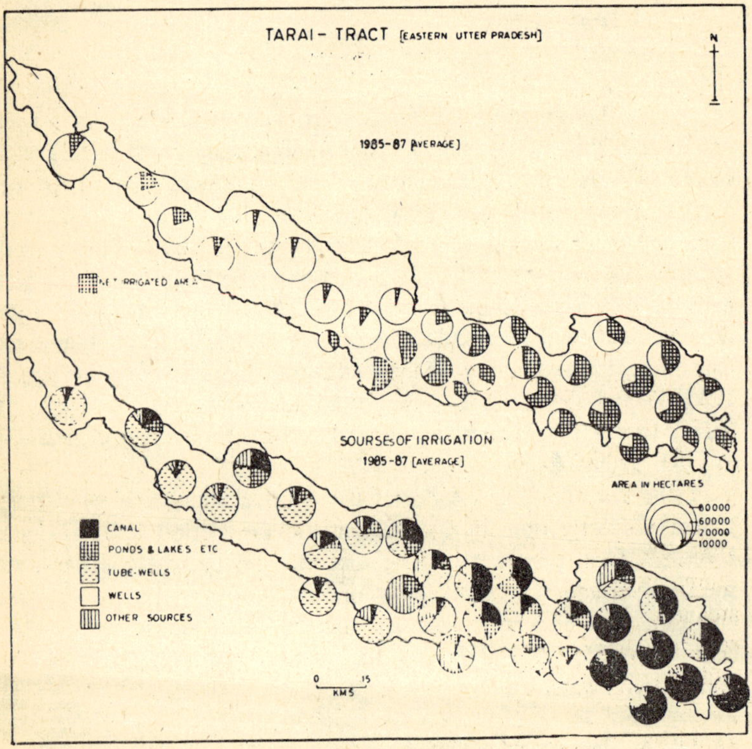

FIG. 2.6: Net Area Irrigated and Sources of Irrigation

The high concentration of the canal irrigated area is in the eastern part, particularly in the Gandak Command Area. It goes on decreasing towards west. All the 6 blocks with high proportion of net irrigated area (Above 60 per cent) belong to the eastern part. Of the 4 in the range 40–60, 2 belong to the eastern and 2 to the middle part. There is higher concentration of the blocks with lower proportion in the west, so much so that of the 6 blocks with no canal irrigation, 4 belong to the western part. It is clear from Table 2.5 that the canal irrigation is expanding slowly. There were 13 blocks in 1975–77 with no canal irrigation and now this number is 6 only. Similarly, there were 6 blocks in the range 20 per cent and below and now this number is 11. With the completion of the Saryu canal system, the proportion of canal irrigation will increase immensely in the west also.

TABLE 2.5: Relative Importance of Various Sources of Irrigation

Category (percentage of net irrigated area)	No. of blocks									
	Canals		Ponds etc.		Other sources		Tube wells		Wells	
	1975-77	1985-87	1975-77	1985-87	1975-77	1985-87	1975-77	1985-87	1975-77	1985-87
1	2	3	4	5	6	7	8	9	10	11
Above 80	2	2	—	—	—	—	—	3	—	—
60–80	4	4	—	—	6	1	6	5	—	—
40–60	3	4	—	—	4	2	6	6	1	—
20–40	3	4	6	3	2	6	4	11	2	2
20 & below	6	11	25	28	16	22	10	6	28	29
Nil	13	6	—	—	3	—	5	—	—	—

Next in importance among the surface water sources come the other sources claiming 14.9 per cent of the net irrigated area, as against 3.2 per cent in the state of Uttar Pradesh and 6.2 per cent in India. This percentage was 27.6 in 1975–77. In 3 blocks (Itwa, Barhni and Bansi), their proportion is above 40 per cent, in 6 (Sirsia, Pachperwa, Khuniaon, Uska, Nautanwa, Khadda) blocks, it is between 20 to 40 and in the rest of the blocks it is 20 and below. In 1975–77, there were 10 blocks with proportion higher than 40 per cent. This indicates that the importance of other sources is decreasing in many of the blocks.

Ponds and lakes contribute to only 8.9 per cent of the net irrigated area, as against 1.5 in the state of Uttar Pradesh and 7.9 (1984-85) per cent in India. This percentage was 11.4 in 1975–77. This indicates that the importance of this source is also declining. As regards the spatial distribution, it may be seen from Appendix VI and Table 2.5 that in 28 blocks the percentage is 20 and below and only in 3, it is between 20 and 40. In 1975–77, these numbers were 25 and 6 respectively.

Among the underground water sources, tube-wells are by far more important claiming 37.4 per cent of the net irrigated area, as against 55.3 per cent in the state of Uttar Pradesh and 27.0 per cent (1984-85)

in India. This percentage was 21.3 in 1975–77, which shows that the tube-well irrigated area has increased in the past. Figure 2.6, Table 2.5 and Appendix VI show that in 8 blocks (Mihipurwa, Jamunaha, Hariharpur Rani, Nawabganj, Balrampur, Bhanwapur, Brijmanganj and Pharenda), the proportion is above 60 per cent in 6, (Haraiya Satgharwa, Tulsipur, Gainsari, Khuniaon, Tulsipur, Laxmipur) it is between 40 and 60. All these blocks possess the extensive areas with high water-table and favourable conditions for sinking tube-wells. In 17 blocks, the concentration is either low or very low, which is either due to the predominance of canal irrigation or presence of unfavourable strata below the surface making the sinking of tube-wells an uphill task. Table 2.5 shows that the number of tube-wells in the range 20–40 and above has increased from 16 to 25 indicating an expansion in the tube-well irrigated area.

Wells account for 6.0 per cent of the net irrigated area, as against 6.2 per cent in the state of Uttar Pradesh and 20.9 per cent (1984-85) in India. This proportion was 8.7 per cent in 1975–77. This shows that their importance is decreasing slowly. In 29 blocks, their proportion is 20 per cent and below and only in 2, between 20 and 40. The position was almost the same in 1975–77. This appears due to some unfavourable local conditions for the excavation and maintenance of wells, particularly non-masonary wells.

It emerges from the foregoing discussion that the optimum utilisation of agricultural land resources warrants sufficiently high increase in the net irrigated area, for which there is enough scope. The canal irrigation has already occupied eastern part and is being spread over the western part, although it is not very much suited to the hydrological conditions of some areas and has very severe hazards in its womb. It is, therefore, suggested that the possibilities for extending the tube-well irrigation, for harnessing the waters of ponds and depressions and other surface sources be explored and genuine efforts be made to develop them.

NOTES

1. Hartshorne, R., (1948), *The Nature of Geography*, Landaster, p. 363.
2. Johnson, V.W. and Releigh, B., (1954), *Land Problems and Policies*, p. 166 (Quoted in Sharma, S.C., *op. cit*., p. 92).
3. Singh, J., (1974-75), *op. cit*., p. 75.
4. Stamp, L.D., (1960), *Applied Geography*, Penguin Book Limited, London, p. 26, (Quoted in Sharma, S.C., *op. cit*., p. 93).

5. Stamp, L.D., (1960), *Our Developing World*, London, p. 157, (Quoted in Sharma, S.C., *op. cit.*, *Ibid.*).
6. Johnson, V.W. and Releigh, B., (1954), *op. cit.*, p. 101.
7. Directorate of Economics and Statistics, Deptt. of Agriculture and Cooperation, Ministry of Agriculture, Govt. of India, (April, 1988), Indian Agriculture in Brief, (22nd edition), p. 221.
8. *Ibid.*, pp. 272-274.

PART II

PATTERNS AND DYNAMICS OF GENERAL AND CROPLAND USE

THREE

PATTERNS AND DYNAMICS OF GENERAL LAND-USE

THE EXISTING and past land-use pattern in the study area is exhibited in Fig. 3.1 and Table 3.1. The land-use statistics have been abstracted from the revenue records maintained by the revenue department and in order to allow for the annual erratic fluctuations three years averages have been used, viz., 1955-57, 1965-67, 1975-77 and 1985-87. Land-use classification in the present study is based mainly on the standard classification as recommended by a Technical Committee of Coordination of Agricultural Statistics (TCCAS) in 1950 and further revised by the Committee on Improvement of Agricultural Statistics.

The major types of land-use of the study area may be discussed under the following heads:

FORESTS

This category includes all areas actually under forests and scrubland whether reserved, unreserved or protected or administered as a forest under legal enactment.

Forests occupy 42,264 hectares of land, i.e., 4.8 per cent of the total area of the region, as compared to 17.3 per cent in the state of Uttar Pradesh and 22.1 per cent in the country as a whole.[1]

Figure 3.2 and Appendix VII show that there are marked variations in the spatial distribution of forested area in the region. Table 3.2a shows that in 11 blocks, the proportions of the area under forests are higher than that of the regional average. These blocks are: Mihipurwa (15.5%), Sirsia (8.8%), Haraiya Satgharwa (12.3%), Gainsari (8.7%), Pachperwa (7.6%), Nawabganj (6.0%), Nautanwa (6.0%), Nichlaul (10.6%), Maharajganj (6.1%), Laxmipur (5.8%) and Pharenda (6.3%). In all these blocks, there are widespread patches of reserved forests, for example, Dharampur range in Mihipurwa, Bhinga and Suhelwa ranges

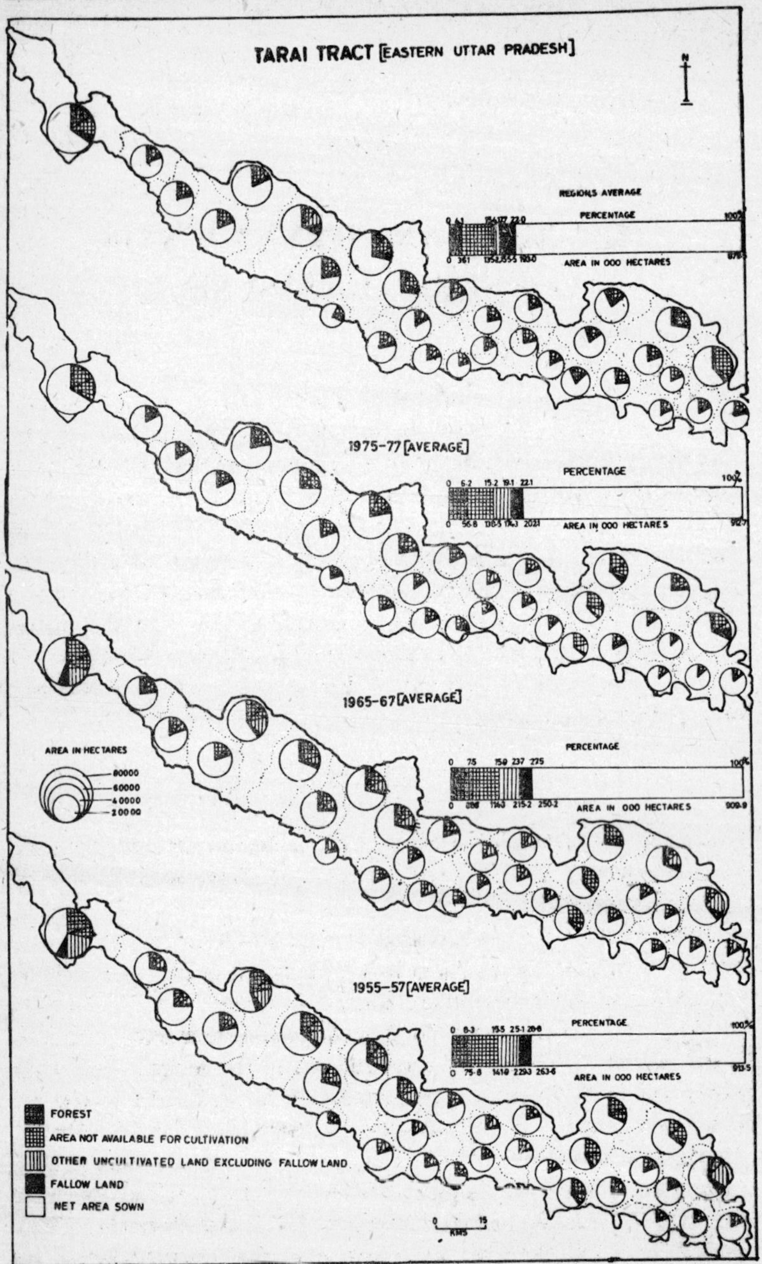

FIG. 3.1: General Land-use

TABLE 3.1: General Land-use

Major land-use categories	1985—87		1975—77		1965—66		1955—57	
	Area in hectares	% of the total area	Area in hectares	% of the total area	Area in hectares	% of the total area	Area in hectares	% of the total area
1	2	3	4	5	6	7	8	9
A. *Forest*	42264.0	4.8	56840.0	6.2	68860.0	7.5	78318.0	8.6
B. *Area not available for cultivation*	99890.0	11.1	82084.0	9.0	71242.0	7.8	66098.0	7.2
(i) Area under water	44561.0	4.9	39655.0	4.3	N.A.	—	N.A.	—
(ii) Area under settlement etc.	42674.0	4.7	31645.0	3.5	N.A.	—	N.A.	—
(iii) Barren & other land	12655.0	1.5	10784.0	1.2	N.A.	—	N.A.	—
C. *Other uncultivated land excluding fallowland*	20674.0	2.3	35543.0	3.9	70134.0	7.7	86918.0	9.5
(i) Orchard & groves	5758.0	0.6	11523.0	1.3	9956.0	1.1	10308.0	1.1
(ii) Pastures & grasslands	1667.0	0.2	2626.0	0.3	(Included in forested area)			
(iii) Culturable wasteland	13249.0	1.5	21394.0	2.3	60178.0	6.6	76610.0	8.4
D. *Arable land*	738738.0	81.8	738440.0	80.9	702718.0	77.0	682169.0	74.7
(i) Fallowland	37824.0	4.2	27781.0	3.0	35072.0	3.8	34159.0	3.8
(ii) Net sown area	700914.0	77.6	710659.0	77.9	667646.0	73.2	648010.0	70.9
Grand total	903560.0	100.0	912907.0	100.0	912954.0	100.0	913503.0	100.0
Double cropped area (as % of net sown area)	343918.0	49.1	286070.0	40.2	199563.0	29.9	207864.0	32.1
Net irrigated area (as % of net sown area)	228758.0	32.6	186577.0	26.6	N.A.	—	N.A.	—

N.A. = Not available.

in Sirsia-Charda range in Nawabganj, Bankatwa range in Haraiya Satgharwa and Tulsipur, Bhambhar range in Gainsari and Pachperwa, and Domakhand range in Nichlaul. Such ranges occur in Laxmipur, Pharenda and Nautanwa blocks also.

Table 3.2a shows that in 20 blocks (two-thirds of the total), the proportions of the forested area are below 5 per cent, in 15, even less than 1 per cent. In 8 blocks, the proportions vary between 5 and 10 per cent, and in 2, between 10 and 15 per cent. There is only one block in the range 15–20 per cent. It may be seen from Table 3.2a and Fig. 3.2 that the number of the blocks in the low concentration groups has increased and in the higher concentration groups, it has decreased. This indicates a declining trend in the forested area.

TABLE 3.2a: Area Under Forest as Percentage of Total Area

Percentage category	No. of blocks			
	1985–87	1975–77	1965–67	1955–57
1	2	3	4	5
Above 20	—	2	3	3
15–20	1	2	3	3
10–15	2	1	—	3
5–10	8	6	6	4
5 & below	20	20	19	18

TABLE 3.2b: Volume of Change in the Area Under Forest

Change category (% of the total area involved in change)	No. of blocks			
	1955–57 to 1965–67	1965–67 to 1975–77	1975–77 to 1985–87	1955–57 to 1985–87
1	2	3	4	5
Increase				
2–4	—	—	1	—
2 & below	—	—	15	8
Decrease				
Above 4	—	2	3	7
2–4	5	3	1	5
2 & below	26	26	8	8
No change	—	—	3	3

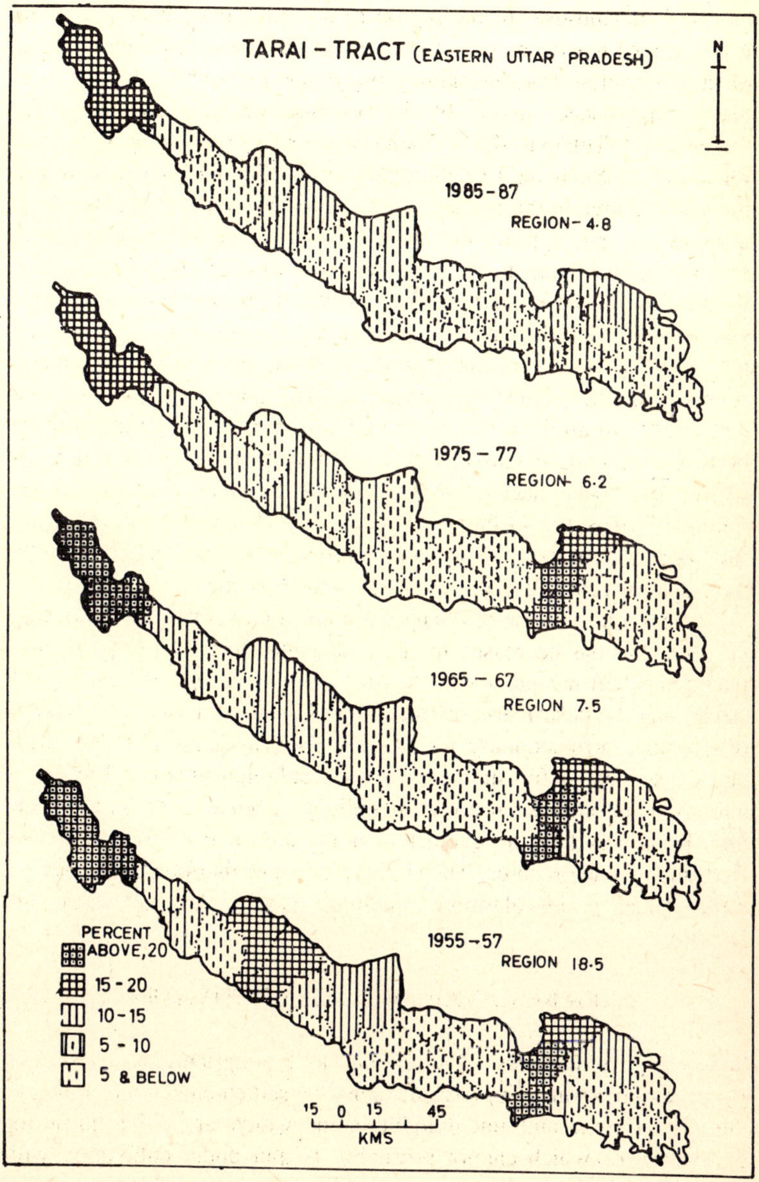

FIG. 3.2: Area Under Forest

Table 3.2b and Fig. 3.3 discern the volume of change in this category of land-use. In the region as a whole, the percentage of the forested area has decreased to the tune of 3.8 per cent during the span of three decades. The outstanding decreases, i.e., over 4 per cent have been experienced in 7 blocks (Mihipurwa–6.5, Jamunaha–5.5, Sirsia–7.7, Tulsipur–4.3, Nautanwa–13.4, Laxmipur–19.4 and Pharenda–22.8). It may be mentioned here that the high reduction in the forested area in the blocks – Laxmipur, Pharenda and Nautanwa, to a greater extent, appears due to the exclusion of the area of certain reserved ranges from the revenue records, which is being maintained by the forest department. In 5 blocks, the decreases range between 2 and 4 (Nawabganj, Haraiya Satgharwa, Gainsari, Pachperwa, and Khadda) and in 8 (Hariharpur Rani, Itwa, Bhanwapur, Khuniaon, Jogia, Ghughli, Nebua Naurangia and Bishunpura) 2 per cent and below. Almost all the blocks with high decreases lie along the northern border, possessing extensive patches of forests, the peripheral areas of which have been sheared due to the agricultural colonisation and felling of trees for meeting the increasing demands for timber and firewood. In 8 blocks, the area under this category has increased to the tune of 2 per cent and below and in 3, it has remained static.

If decennial volume of changes are taken into consideration, it may be seen that the decreases in the area under forest have been more phenomenal in the latter decades than that of the former one, which shows that the destruction of forests has been more rapid and severe in this period, corresponding well to the faster agricultural colonisation and consequent deforestation processes, although in a few blocks, as already mentioned, such a high magnitude of change, to some extent, may be ascribed to the exclusion of the area under reserved forests from the revenue records. The slashing down of the forests with such a rapid speed is an alarming situation which calls for immediate attention.

AREA NOT AVAILABLE FOR CULTIVATION

This category includes (i) Area put to non-agricultural uses, viz., (a) Area under water, (b) Area occupied by settlements, roads, railways etc., (ii) Barren and unculturable lands which are either unfit for cultivation or which cannot profitably be put under cultivation with immediate effect.

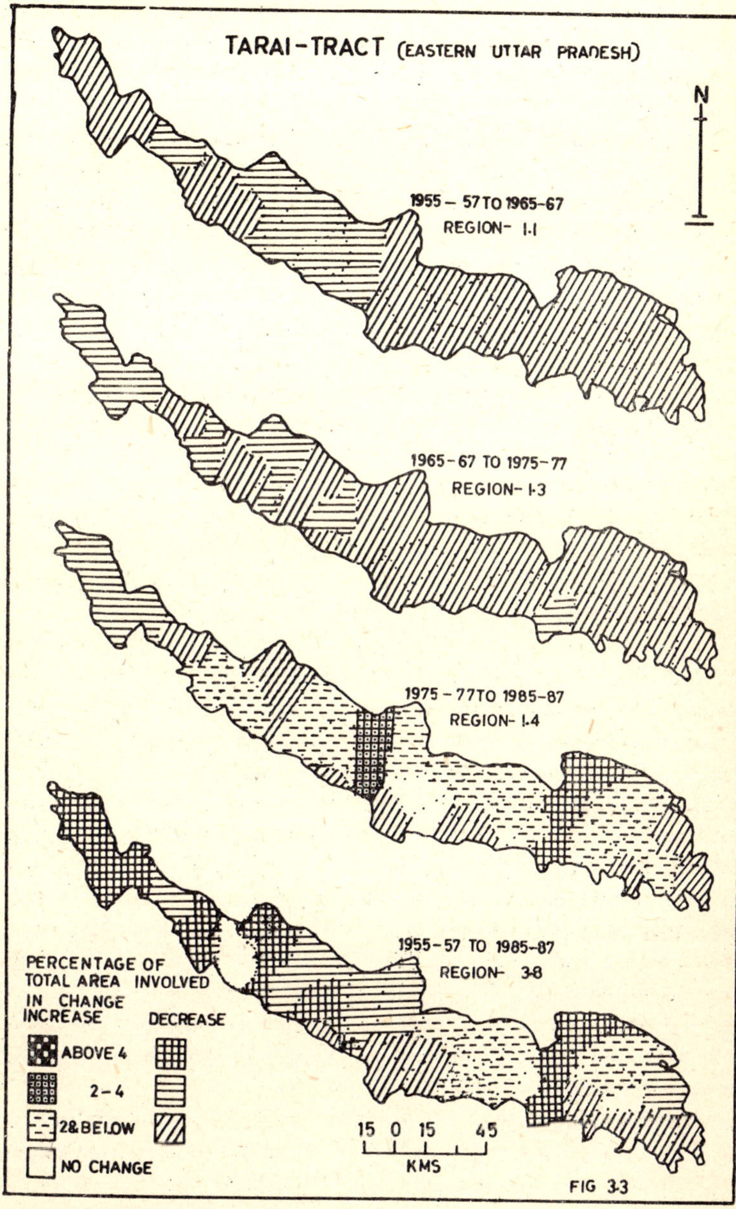

FIG. 3.3: **Volume of Change in Area Under Forest**

Table 3.1 and Fig. 3.4 shows that 11.1 per cent of the total area of the region is not available for cultivation, as compared to 11.6 per cent in the State of Uttar Pradesh and 13.3 per cent (1984-85) in the country as a whole.[2]

There are marked variations in the spatial distribution of the area under this category. Table 3.3a and Appendix VII show that 18 blocks fall below the regional average and only 13 blocks are above it. The blocks with higher proportions are: Mihipurwa, Jamunaha, Hariharpur Rani, Tulsipur, Balrampur, Bhanwapur, Uska, Naugarh, Birdpur, Maharajganj, Siswa and Khadda. In most of these blocks, the area under river beds and that put to non-agricultural uses is comparatively higher.

Under this broad category, special mention may be made of two sub-categories, viz., area under water and that under settlement and paths etc. The percentage of the area under water is higher than the regional average in 9 blocks (Mihipurwa–6.1, Jamunaha–6.3, Balrampur–6.4, Bhanwapur–7.8, Barhni–5.3, Birdpur–5.7, Uska–6.6, Bansi–5.7, Khadda–17.6) most of which have higher proportion of areas under river beds. It is lower in the remaining blocks, but it is very low in Nawabganj (2.1) and Pharenda (1.5) where there are fewer ponds and river beds. Table 3.3c shows that there are 10 blocks in the range 2–4 per cent. There is only one block in the range 2 per cent and below. In 14 blocks, the proportion varies between 4 and 6 per cent and in the remaining 6, it is higher than 6 per cent. This shows that a good number of blocks have fairly large area under water.

As regards the land put to non-agricultural uses such as settlements, roads etc., significant concentration of this category of land is in those blocks which have higher densities of population. Thus there is high positive correlation between the density of population and the area under this category. In 16 blocks, the proportion under this category of land is higher than the regional average of 4.7 per cent. There are only 4 blocks in the range of 6 per cent and above, while in the range 4–6 per cent, there are 19 blocks. The number of the blocks in the range 2–4 per cent, is 8. The population density in these blocks ranges between 239 and 490 persons per sq. km.

The area under barren land and unculturable land is 1.2 per cent of the total area of the region. It is above the regional average in 12 blocks, while it is below even 1 per cent in 17 blocks. In the block Khadda, the percentage is exceptionally high (10.2%), which appears due to the predominance of rough topography along the river Gandak in this area.

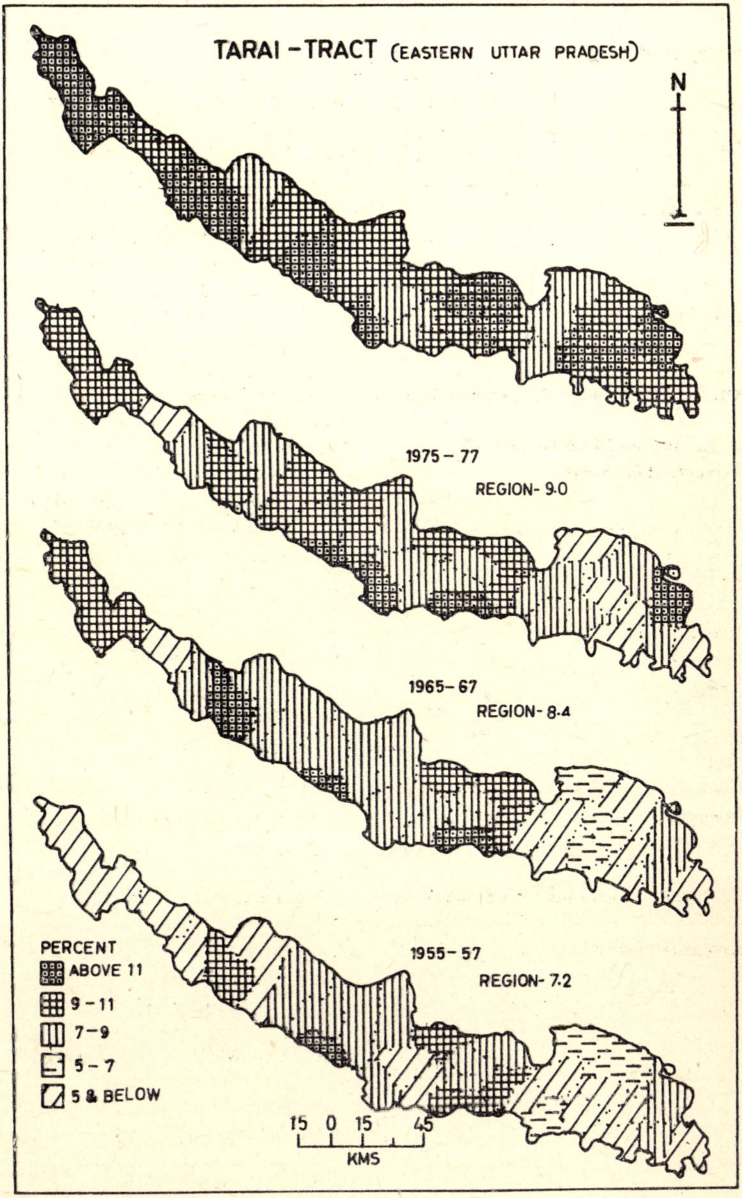

FIG. 3.4: Area Not Available for Cultivation

TABLE 3.3a: Area not Available for Cultivation as Percentage of Total Area

Percentage category	No. of blocks			
	1985–87	1975–77	1965–67	1955–57
1	2	3	4	5
Above 11	13	5	3	1
9–11	12	7	5	4
7–9	6	13	12	10
5–7	—	6	8	13
5 & below	—	—	3	3

TABLE 3.3b: Volume of Change in Area not Available for Cultivation

Change category (% of the total area involved in change)	No. of blocks			
	1955–57 to 1965–67	1965–67 to 1975–77	1975–77 to 1985–87	1955–57 to 1985–87
1	2	3	4	5
Increase				
Above 4	—	1	5	8
2–4	1	5	7	15
2 & below	23	17	10	6
Decrease				
2–4	—	1	—	—
2 & below	7	7	6	2
No Change	—	—	3	—

TABLE 3.3c: Area Under Water and Settlements etc. as Percentage of Total Area

Percentage category	No. of Blocks			
	1975–77		1985–87	
	Water	Settlements etc.	Water	Settlements etc.
1	2	3	4	5
Above 6	3	1	6	4
4–6	12	6	14	19
2–4	14	24	10	8
2 & below	2	—	1	—

TABLE 3.3d: Volume of Change in the Area under Water and Settlement etc. (1975–77 to 1985–87)

Change category (% of the total area involved in change)	No. of blocks	
	Area under water	Area under settlements etc.
1	2	3
Increase		
Above 3	3	1
2–3	—	4
1–2	7	15
1 & below	12	8
Decrease		
1–2	3	—
1 & below	6	3

There have occurred significant changes in the area not available for cultivation and in its sub-categories. It may be noted here that due to the non-availability of the data of different sub-categories for the early periods, land not available for cultivation has been considered as a whole for analysing the changes during the span of three decades and in the case of sub-categories, the changes occurred during one decade, i.e., 1975–77 to 1985–87 (Table 3.3c and 3.3d) have been analysed. The volume of changes in the area not available for cultivation have been summarised in Table 3.3b and represented in Fig. 3.5.

There has been an overall increase of 3.9 per cent in the area of this category during the span of three decades (1.2 in the first, 0.6 in the second and 1.1 per cent in the third decade). There are some spatial variations in the distribution pattern of the changes. In 8 blocks (Mihipurwa, Jamunaha, Mithaura, Maharajganj, Siswa, Khadda, Nebua, Naurangia and Bishunpura), the increases are to the tune of above 4 per cent, in 15, the increases range between 2 and 4 per cent, and in 6, they are 2 per cent and below. Two blocks have experienced decreases also (Balrampur and Bansi). It may be remarked that generally high plus changes are in those blocks where due to some development schemes, the area under settlements, roads and irrigational channels etc. has increased or due to floods, water-logging and erosion hazards, some area has gone out of cultivation.

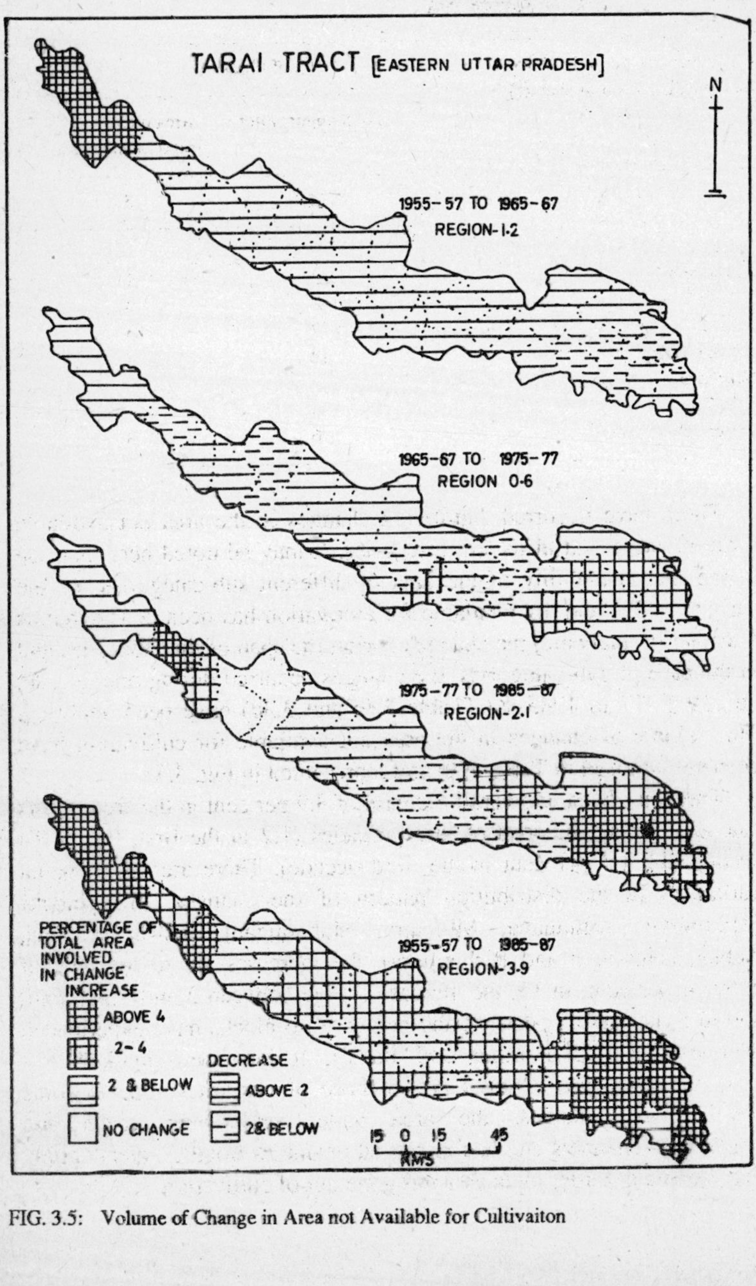

FIG. 3.5: Volume of Change in Area not Available for Cultivaiton

As regards the decennial changes in the area under water and settlements, it may be seen from Table 3.3c and Table 3.3d that the area under these categories has also increased during 1975–77 and 1985–87, which is well clear from the increasing number of blocks in the higher categories. The percentage of the area involved in change in the case of the area under water and settlements respectively is 0.6 and 1.2. The number of the blocks experiencing plus changes to the tune of above 3 per cent is 3 and 1, in the range 2-3, it is nil and 4, in the range 1-2, it is 7 and 15 respectively in these categories. This number is respectively 12 and 8 in the range 1 per cent and below. This also indicates that the increases in the case of settlements and paths etc. are more phenomenal. Some blocks have experienced decreases also. This number is higher in the case of the area under water (9), which may be attributed, to some extents, to the reclamation of waterlogged areas in these blocks.

OTHER UNCULTIVATED LAND EXCLUDING FALLOWLAND

This category consists of: (i) Permanent pastures and grasslands, (ii) Miscellaneous trees and groves, and (iii) Culturable wasteland.

Culturable Wasteland

Culturable wasteland denotes land which is cultivable but actually it has not been cultivated for the past five or more years. It is left uncultivated on account of various physical and socio-economic limitations. This is the land which is of great significance for the extension of the cultivated area.

The culturable wasteland accounts for 13249 hectares or 1.5 per cent of the total area of the region, as compared to the State's average of 3.7 and country's average of 5.2 per cent. This apparently shows that the extent of the culturable wasteland is small, which is indicative of the fact that future extension of cultivated land is limited, and for raising agricultural production, primary reliance ought to be placed on more intensive use of the net area sown.

Appendix VII, Table 3.4a and Fig. 3.6 show the distribution of the culturable wasteland in the region, from which it is clear that there are sub-regional variations in the extent of the land under this category.

TABLE 3.4a: Area Under Cultural Wasteland as Percentage of Total Area

Percentage categories	No. of blocks			
	1985–87	1975–77	1965–67	1955–57
Above 8	—	—	4	8
6–8	—	2	7	7
4–6	—	1	5	7
2–4	8	12	11	7
2 & below	23	16	4	2

Out of 31 blocks, 13 are below the regional average of 1.5 per cent and the remaining above it. At present, no block is above the range 2–4, 8 are in this range and the remaining 23 below it. Broadly speaking, the proportions of the area under this category are lower in those blocks which have higher proportion of cultivated land or where the area under forest and other similar uses is more.

It may be seen from Table 3.4a and Fig. 3.6 that the number of blocks in the lower categories has increased and that in the higher categories has decreased, so much so that there is no block now in the range higher than 2–4, while this number was higher in earlier years.

The distributional pattern of volume of change (Table 3.4b and Fig. 3.7) discerns that only one block (Bansi) has experienced a slight increase, while all the remaining blocks have experienced decreases. Very high (above 8%) decreases have been experienced in six (Mihipurwa, Sirsia, Haraiya Satgharwa, Pachperwa, Nichlaul and Khadda) blocks. The changes in 2, 6 and 11 blocks respectively are in the range 6–8, 4–6 and 2–4 per cent and in 4 in the range 2 per cent and below. The position has remained static in one block.

It may be remarked that outstanding and high changes have been experienced in those blocks where due to more unfavourable malarial conditions, deep-rooted grasses, extensive water-logging and sparse population, the culturable wasteland was very extensive. The magnitude of decrease appears a spontaneous reaction to the influx of population, rapid reclamation measures and resultant stepping up of cultivation. Most of these blocks have registered high increases in the net sown area.

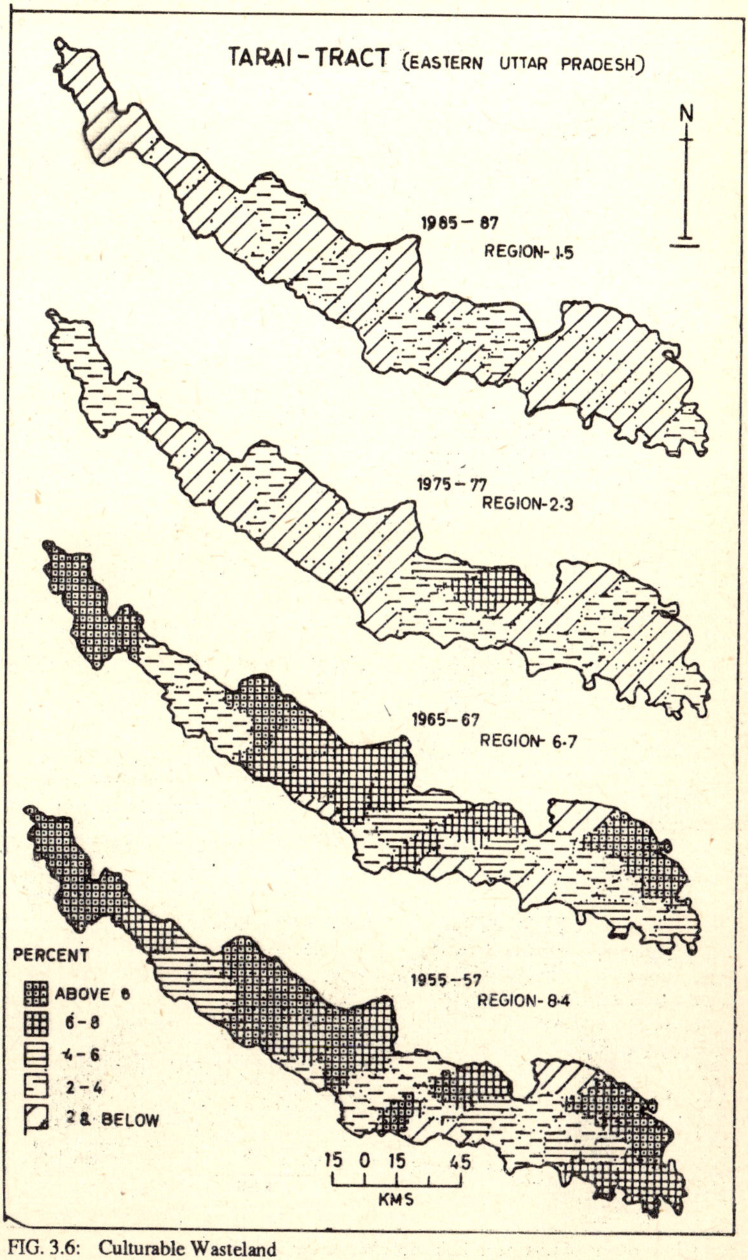

FIG. 3.6: Culturable Wasteland

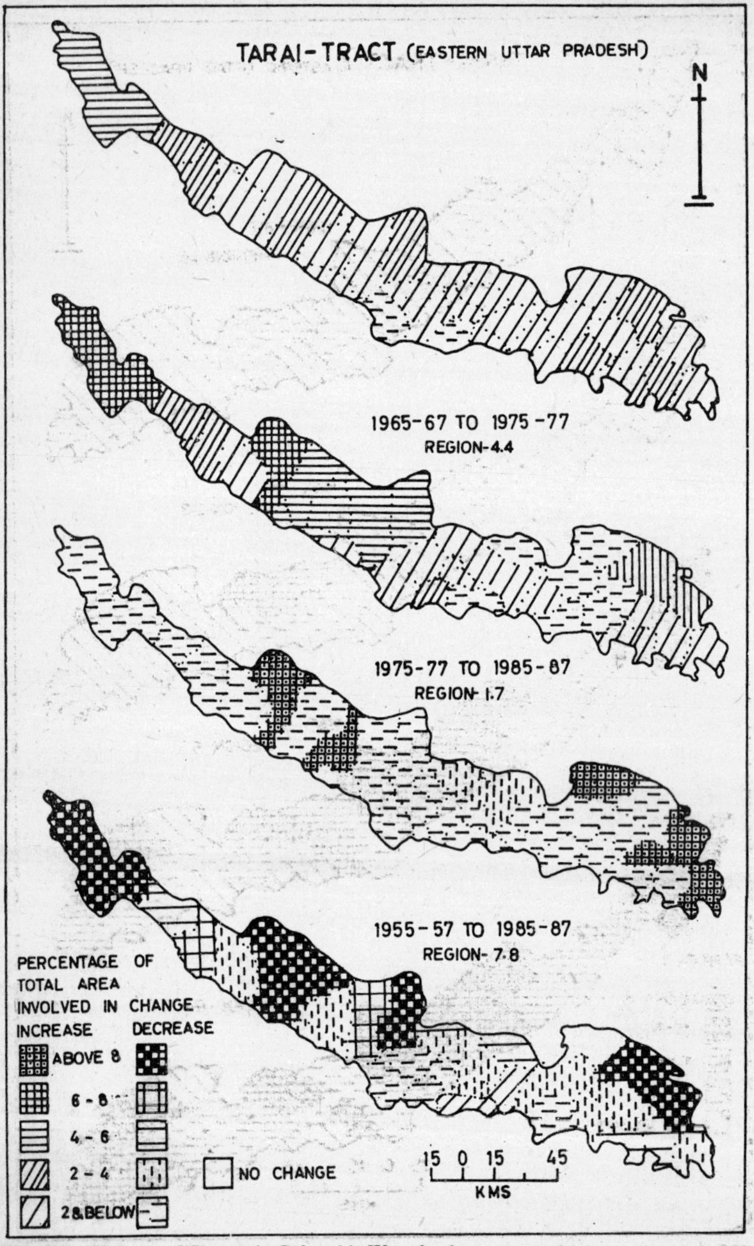

FIG. 3.7: Volume of Change in Culturable Wasteland

TABLE 3.4b: Volume of Change in the Area under Culturable Wasteland

Change category (% of the total area involved in change)	No. of blocks			
	1955–57 to 1965–67	1965–67 to 1975–77	1975–77 to 1985–87	1955–57 to 1985–87
Increase				
2 & below	2	7	6	1
Decrease				
Above 8	—	3	—	6
6–8	—	—	—	2
4–6	1	5	—	6
2–4	7	4	3	11
2 & below	21	12	21	4
No change	—	—	—	1

The decennial changes show that the second decade has experienced more rapid changes. In the first decade, 30 blocks are in the low intensity group and only one block slightly above it, while in the second decade, this number decreased to 23 and that in the higher group rose to 8. The increasing magnitude of decreases may be ascribed mainly to the faster growth of population pressure, speedier reclamation of land to cultivation thereby increasing the net area sown. In the third decade, the number of the blocks in the low intensity group has again risen to 30. This shows a declining trend in the changes, which appears due to the reclamation of the greater part of this category of land in the previous years and due to the measure to check the growth of this category of land in future.

Land under Miscellaneous Trees and Groves

The study area possesses only a negligible area (0.6 per cent of the total area) under this head, as against 1.9 per cent in the State of Uttar Pradesh and 1.1 per cent in the country as a whole. The percentages at the block level range from 0.4 to 1.8. The proportions are above the regional average in 14, equal to it in 5, below it in 7 and almost zero in five blocks.

Since the proportion of the area under this category has been very small, the volume of change is almost negligible, which shows that the percentage share of this category of land has remained almost static. Out of 10 blocks which have registered an increase, in 8 the increases have been 0.5 per cent and below. The decreases have been experienced in the remaining 21 blocks, in 8, above 1 per cent, in 3, between 0.5 and 1 and in 9, 0.5 per cent and below. In one block, the position remained static. As regards the decennial changes, in the first decade, 19 blocks experienced an increase ranging between 0.1 and 1.0 per cent, while in the second decade, 26 blocks experienced an increase between 0.1 and 1.2 per cent and decreases were experienced by only 5 blocks in the range between 0.3 and 1.3 per cent. In the third decade, only 7 blocks experienced increases and that too in the range 0.1-0.4 per cent only, while 21 experienced decreases in the range 0.1 and 1.9 per cent and in 3, the position remained static.

No specific reasons can be ascribed for these minor changes. In some areas, the groves have been destroyed and the area occupied by them reclaimed to agriculture, in other areas, they have been maintained and some big landholders have started the work of trees plantation.

Permanent Pastures and Grasslands

This category claims only 1667 hectares or 0.2 per cent of the total area of the region, as against 1.2 and 3.9 per cent in the State of Uttar Pradesh and India respectively. This shows the insignificance of this category of land which hardly needs any separate treatment and hence has been included in the forested area and dealt with there.

ARABLE LAND

The word 'arable' which has been derived from Latin word '*arabilis*', means land fit for ploughing, that is, ploughable. As such, it includes both tillage (net cropped) and fallowlands. This has been termed as cultivated land in the national scheme of land classification. The fallowlands are such lands as have been sown previously, but are not being sown in the year of reporting. They are, therefore, considered as good as the land actually tilled and can very well be included in the arable land. The cultivated land is recorded fallow, when it is put out of cultivation for a certain period. For the first three years, it is recorded as 'New' fallow and for the next two years as 'Old' fallow. The term used here includes both of them.

The arable land accounts for an area of 738738 hectares, i.e. 81.8 per cent of the total area of the region, as compared to 64.3 per cent in the State of Uttar Pradesh and 54.3 per cent in the country as a whole. This includes both fallowland and the net area sown.

Fallowland

The fallowland accounts for an area of 37824 hectares or 4.2 per cent of the total area of the region, as compared to 6.5 per cent in Uttar Pradesh and 8.1 per cent in India. The sub-regional disparities in the spatial distribution of fallowlands is exhibited in Table 3.5a, Appendix VII and Fig. 3.8. It may be seen that only 8 blocks are above the regional average. In 4 (Haraiya Satgharwa, Tulsipur, Gaiṅsari and Balrampur), this category of land claims more than 6 per cent of the total area. This high percentage may be ascribed either to the river actions causing floods and water-logging or to the insufficient irrigation facilities. In 8 blocks, the proportions vary between 4 and 6 per cent, in 17, between 2 and 4 per cent and in 2, it is 2 per cent and below. A comparative study of the spatial distribution of net area sown and fallowland (Figs. 3.6 and 3.8) very well exhibits that there is a reverse relationship between the proportion of the net area sown and the fallowland. The higher the proportion of the net area sown, the lesser is the proportion of the area under fallowland and *vice versa.*

It may be seen from Fig. 3.8 and Table 3.5a that the number of the blocks in the low concentration groups has increased and that in the higher concentration groups decreased.

The percentage of the total area involved in change in the case of fallowland in the study area is 0.4 during the span of three years. Table 3.5b and Fig. 3.9 show the spatial distribution of the volume of change. In 18 blocks, increases have been experienced and in 13 decreases. Of the blocks experiencing increases, about 40 per cent are in the range 2–4 per cent and above and 0.5–2 per cent each. Of those experiencing decreases, about 64 per cent are in the range 0.5–2 per cent. This shows that increases in some of the blocks are of greater magnitude than the decreases.

As regards the decennial variations in the volume of change, it may be seen that the position reversed in the second decade, as the number of blocks was reduced to about half in the increases, and increased to double in the decreases, but in the third decade, the position of the first decade was restored, showing that in this decade, the increases in the fallowland were again phenomenal, which may be ascribed to the rainfall variabilities causing widespread inundations and drought conditions in some parts of the region.

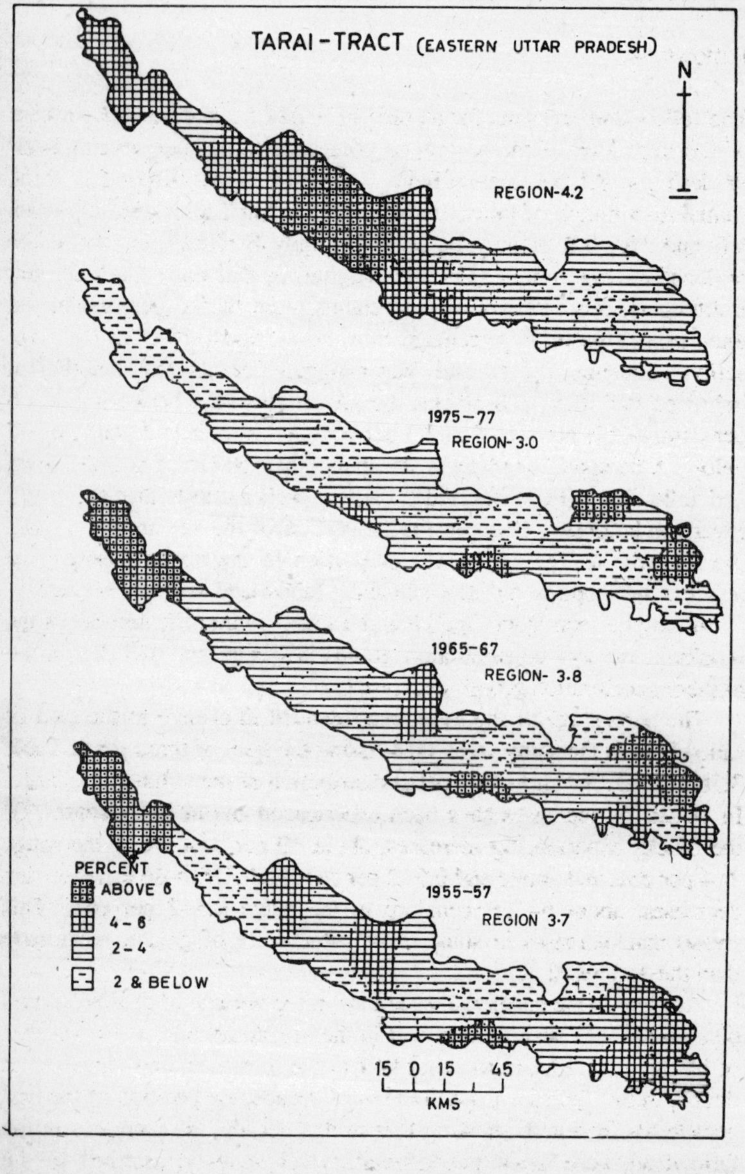

FIG. 3.8: Fallow Land

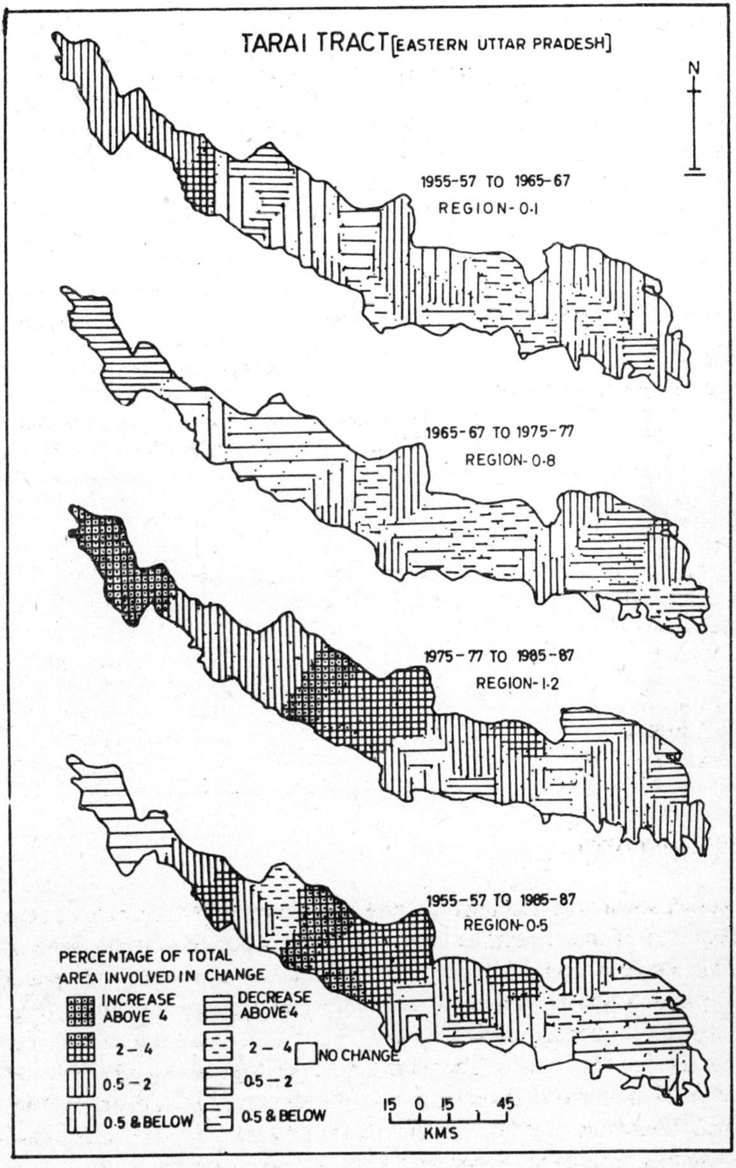

FIG. 3.9: Volume of Change in Fallow Land

TABLE 3.5a: Fallow Land as Percentage of Total Area

Percentage categories	No. of blocks			
	1985-87	1975-77	1965-67	1955-57
Above 6	4	3	4	3
4–6	8	1	3	6
2–4	17	18	19	15
2 & below	2	9	5	7

TABLE 3.5b: Volume of Change in Fallow Land

Change category (% of the total area involved in change)	No. of blocks			
	1955-57 to 1965-67	1965-67 to 1975-77	1975-77 to 1985-87	1955-57 to 1985-87
Increases				
Above 4	—	1	2	2
2–4	1	1	5	6
0.5–2	10	5	16	7
0.5 & below	9	4	2	3
Decreases				
Above 4	—	1	3	2
2–4	2	5	—	1
0.5–2	2	9	3	8
0.5 & below	7	5	—	1
No change	—	—	—	1

Net Sown Area

The net sown area accounts for 700914 hectares or 77.6 per cent of the total area of the region, as compared to 57.8 per cent in the State of Uttar Pradesh and 46.2 per cent in India as a whole. The spatial distribution shows that there are wide sub-regional variations (Fig. 3.10). Its proportion varies from 62.3 per cent (Khadda) to 86.1 per cent (Nebua Naurangia). Therc are 3 blocks (Brijmanganj, Ghughli and Nebua Naurangia) in the range above 85 per cent, 13 (Hariharpur Rani, Itwa, Khuniaon, Barhni, Birdpur, Jogia, Bansi, Mithaura, Laxmipur, Nautanwa, Pharenda, Siswa and Bishunpura) in the range 80-85 per cent, 9 (Nawabganj, Jamunaha, Tulsipur, Bhanwapur, Naugarh, Uska, Maharajganj, Sirsia and Balrampur) in the range 75-80 per cent, 3

(Gainsari, Pachperwa and Nichlaul) in the range 70-75 per cent and only 3 (Mihipurwa, Haraiya Satgharwa and Khadda) in the range 70 per cent and below. The blocks with high proportion of net sown area generally possess very small proportion of land under fallow land and culturable wasteland or forest, and there are better irrigation facilities and high pressures of population on land. It may be mentioned here that there is high positive (+0.71%) correlation between population pressure and net sown area. There is negative moderate correlation between the net sown area and culturable wasteland. In almost, all the blocks with low proportion of net sown area, there is higher proportion of land under forest and there is very high negative correlation between net sown area and the area under forest. In certain blocks of low concentration group, the proportion of land under water or barren land is also higher. For example, in Khadda 32.0 per cent of the total area is not available for cultivation, greater part of this is under water. It may be pointed out here that in the blocks Pharenda and Laxmipur, the proportion of net sown area would be still lower, if all the forested land is included in their total area.

TABLE 3.6a: Net Sown Area as Percentage of Total Area

Percentage categories	No. of blocks			
	1985-87	1975-77	1965-67	1955-57
Above 85	3	6	2	1
80-85	13	14	13	11
75-80	9	4	5	6
70-75	3	2	4	4
70 & below	3	5	7	9

It may be seen from Table 3.6a and Fig. 3.10 that the number of blocks in the higher concentration groups (80-85% and above) has increased from 12 in 1955-57 to 20 in 1975-77, although there is slight decrease to 16 in 1985-87. In the range 75-80 per cent also, there is an increase from 6 (1955-57) to 9 (1985-87). The number of the blocks in the lower concentration groups (70-75% and below) has decreased from 13 (1955-57) to 6 (1985-87). This shows that there is an increasing trend in the net sown area.

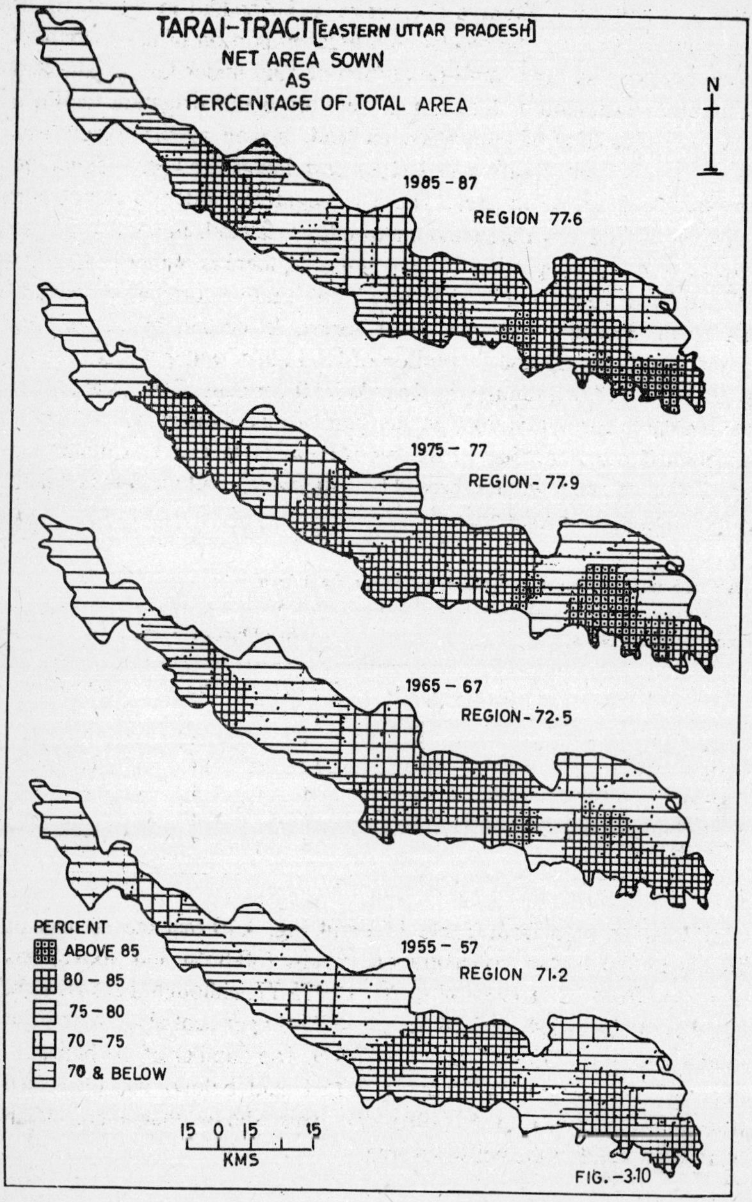

FIG. 3.10: New Sown Area

TABLE 3.6b: Volume of Change in Net Sown Area

Change category (% of the total area involved in change)	No. of blocks			
	1955-57 to 1965-67	1965-67 to 1975-77	1975-77 to 1985-87	1955-57 to 1985-87
Increase				
Above 15	—	—	3	5
10-15	—	2	—	—
5-10	3	7	1	4
5 & below	24	16	2	14
Decrease				
5-10	—	1	4	—
5 & below	4	5	21	8

As regards the volume of change, there has been an overall increase 6.7 per cent in the area under this category (2.3% in the first, 4.7 per cent in the second, but a decrease of 0.3 per cent in the third decade). There are sub-regional variations in the volume of change (Table 3.6b and Fig. 3.11). In 8 blocks, there have been decreases ranging from 0.7 to 4.3 per cent. These blocks are: Itwa, Bhanwapur, Birdpur, Naugarh, Jogia, Uska, Mithaura and Bishunpura. In most of these blocks, the area under fallow land has increased. The rest of the blocks have experienced increases ranging from 0.5 to 24.5 per cent. Five blocks (Mihipurwa, Sirsia, Laxmipur, Pharenda and Nautanwa) have recorded increases above 15 per cent. In 3 of them (Mihipurwa, Laxmipur and Pharenda), the population has also increased greatly, the increases in population ranging from 61.2 to 162.5 per cent. There is great hike in the population densities also in these blocks, for example, in Mihipurwa from 93 to 270, in Laxmipur from 249 to 347 and in Pharenda from 233 to 395 persons per sq km.

There have been corresponding decreases in the area under forest in all the five blocks (Mihipurwa–6.5, Sirsia–7.7, Nautanwa–13.4, Laxmipur–19.4 and Pharenda–22.8 per cent), either due to the reclamation of the forested area to cultivation or in some cases due to the exclusion of this area from the revenue records. The decreases of the area under culturable wastelands have been experienced in 4 blocks (Mihipurwa - 20.7, Sirsia - 12.8, Laxmipur - 2.3 and Pharenda - 1.9 per cent) and that under fallowland in 3 (Mihipurwa - 1.2, Sirsia - 2.2 and Laxmipur - 0.4 per cent). In 4 blocks, the increases in the net sown area have occurred in the range 5 - 10 per cent and in the remaining 14,

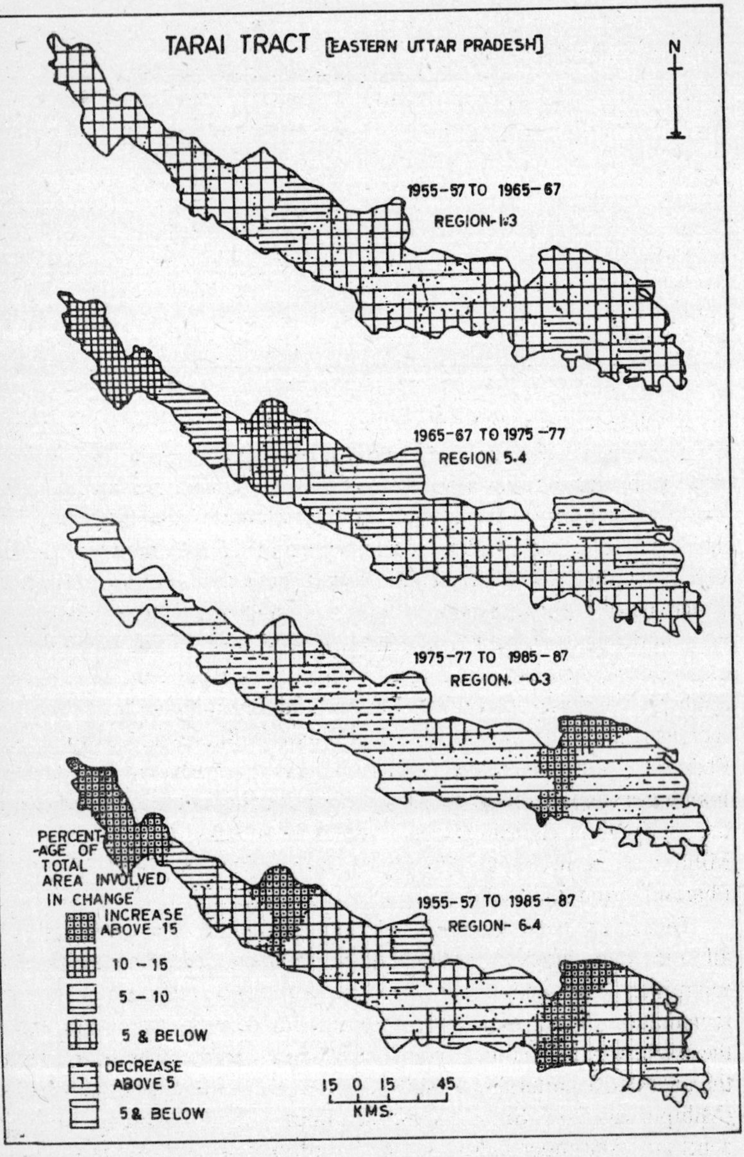

FIG. 3.11: Volume of Change in Net Sown Area

in the range 5 per cent and below. In most of these blocks, the decreases in the area under culturable wasteland, fallowland and forested land have accordingly been moderate to low. It may thus be remarked that the percentage increases in the net sown area largely correspond to the respective decreases in the area under culturable wasteland, fallowland and forest, which have been donors of the land to this category. The blocks with high increases have registered high decreases and with low and medium increases with the corresponding decreases in these relevant categories. There, thus, exists, as already observed, a negative correlation between the net sown area and culturable wasteland, fallow land and forest covered area.

As in the case of other categories, there are decennial variations in the volume of change in this category also. The changes in the first decade were slower than that of the second decade. In the former, of the 27 blocks experiencing increases, 24 were in the range 5 per cent and below; and all the blocks experiencing decreases were in this category. In the latter decade, of the 25 blocks experiencing increases, only 16 were in the range 5 per cent and below and the remaining were in the ranges above it, and of the 6 blocks experiencing decreases, 5 were in the category 5 per cent and below and one in the range 5-10 per cent. During the decade 1975-77 to 1985-87, of the 6 blocks registering an increase, 3 were in the range above 15 per cent, 1 in the range 5-10 per cent and 2 in the low range of 5 per cent and below, but of the 26 blocks registering decreases, 21 were in the low range of 5 per cent and below and the remaining 4 in the range 5-10 per cent, showing that the decreases were minor and that, too, due to the minor increases in the extent of fallowland.

OVERALL VOLUME OF CHANGE*

Table 3.7 and Fig. 3.12 show the overall volume of change in the study area. Although, the region has experienced an overall change of 11.0 per cent during the span of three decades, there are sub-regional

---

* For the quantitative measurement of the overall volume of change in land-use, Weaver's Index (1954) which he applied in his study of ascertaining the volume of change in cropland use in Middle West (U.S.A.) has been employed.

Index for determining the 'Volume of Change' in Land-use = A/B

Where *A* is the difference of percentages of land-use categories of increases and *B* the difference of the percentages of land-use categories of decreases for different periods.

variations. For example, 6 blocks (Hariharpur Rani, Itwa, Barhni, Jogia, Uska and Brijmanganj) represent stable areas where the changes have been very slow, i.e. to the tune of 5 per cent and below. These are the blocks where there were already high concentrations of net sown area and the area under culturable wasteland, fallowland and forest was not very extensive and consequently, there were minor transfers of areas among these categories. The volume of change in these blocks in the case of net sown area varies from 0.5 to 2.7 per cent, in the case of culturable wasteland from 1.8 to 4.0 per cent, in the case of fallowland from 0.4 to 2.9 per cent, in the case of forested area from 0.1 to 0.2 per cent, and in the case of area not available for cultivation from 0.8 to 2.9 per cent. Land-use has been moderately dynamic (5-10%) in 13 blocks (Nawabganj, Balrampur, Bhanwapur, Khuniaon, Birdpur, Naugarh, Bansi, Maharajganj, Ghughli, Siswa, Nebua Naurangia, Mithaura and Bishunpura). In these blocks, the changes in the net sown area vary between 0.7 and 8.6 per cent. The corresponding changes in the area under fallowland, forest and culturable wasteland vary between 0.3 and

---

The summation of the numerator and denominator should be the same. The technique of calculation is as follows:

Land-use in Block Jamunaha (Bahraich)

Periods	Percentages of the total area					
	Area not available for cultivation	Groves	Culturable wasteland	Forest land	Fallow land	Net sown area
1	2	3	4	5	6	7
1955-57	6.32	1.03	5.97	9.05	1.83	75.80
1985-87	8.20	1.20	0.90	2.90	2.20	84.60
Difference	+1.88	+0.17	-5.07	-6.15	+0.37	+8.80

$$\text{Index of volume of change} = \frac{NA\ 1.80 + G\ 0.17 + F\ 0.37 + NSA\ 8.80}{CW\ 5.07 + FO\ 6.15}$$

$$= 11.22/11.22$$

Where *NA* is area not available for cultivation, *G* is the area under groves, *F* is the area under fallowland, *NSA* is the net sown area, *CW* - culturable wasteland and *FO* - forest. In the case of this block, 11.22 is an index figure. It is actually the percentage of land which is involved in the transfer of area from one category to another.

7.3, 0.1 and 3.7, and 0.1 and 5.7 per cent respectively. The variations in the area not available for cultivation vary generally between 0.3 and 5.9 per cent.

There are 6 blocks (Jamunaha, Haraiya Satgharwa, Gainsari, Pachperwa, Tulsipur and Nichlaul) which have been grouped in the category of dynamic land-use (10-15%), and 6 (Mihipurwa, Sirsia, Nautanwa, Laxmipur and Pharenda) which have registered outstanding changes (above 15 per cent) have been put in the high dynamic category. These are the blocks which touch the northern border of India, and which formerly possessed extensive lowlying and swampy areas infested with reeds, tall grasses, widespread forests, culturable wastelands and fallowlands, unhealthy climatic conditions and sparse population. Agriculturable colonisation inviting great influx of population which has increased by 34.4 to 68.2 per cent in the blocks of dynamic and by 32.8 to 162.5 per cent in those of highly dynamic groups, has resulted in great expansion in net sown area (from 2.2 to 7.0 per cent in the former and from 19.8 to 24.5 per cent in the latter category of blocks) with the corresponding decreases in the culturable wasteland (from 5.5 to 10.2 per cent in the former and from 1.0 to 20.7 per cent in the latter), and in the forested area (from 0.5 to 5.5 per cent in the former and from 6.5 to 22.8 per cent in the latter category of blocks). The fallowland has decreased only in one block in the former category, but in three blocks (from 0.4 to 3.0%) in the latter category of blocks.

As regards the decennial variations in overall volume of change, we see almost the same trend as was perceived in the individual major uses of land. There were 23 blocks in the stable category and the remaining 8 were in the moderately dynamic category in the first decade. In the second and third decades, the number of the blocks in the stable category respectively decreased to 18 and 13, and that of the moderately dynamic category increased to 9 and 14. One block (Nautanwa) shifted to the category of dynamic change in the second decade and three (Mihipurwa, Sirsia and Khadda) to the category of highly dynamic change. This latter number became 4 (Nautanwa, Laxmipur, Pharenda and Khadda) in the third category. This shows that during the latter decades, the efforts to step up agricultural colonisation by more rapid expansion of net sown area, more intensified use of cultivated land by providing better irrigation facilities, use of chemical fertilisers and HYV of seeds etc. and by rapid change in the socio-economic agronomic structure, have been more effective.

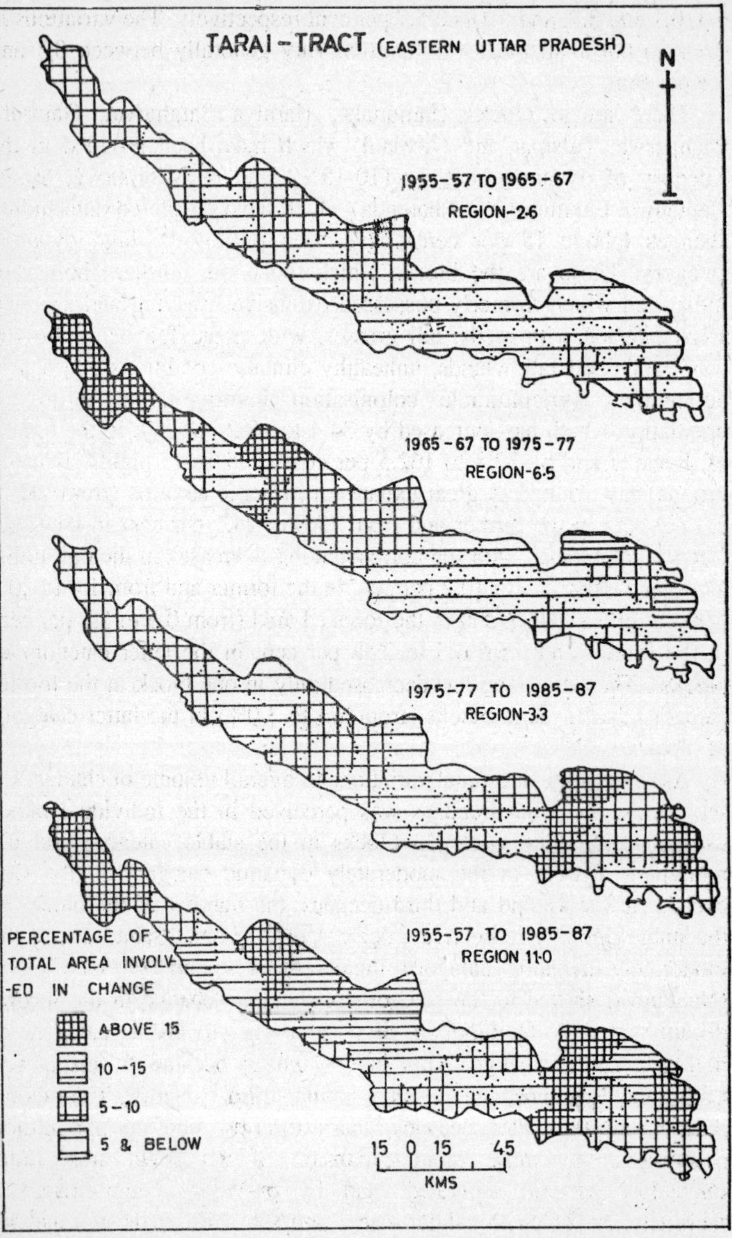

FIG. 3.12: Overall Volume of Change

TABLE 3.7: Overall Volume of Change

Change category (% of the total area involved in change)		No. of blocks			
		1955-57 to 1965-67	1965-67 to 1975-77	1975-77 to 1985-87	1955-57 to 1985-87
Highly dynamic	(Above 15)	—	3	4	6
Dynamic	(10-15)	—	1	—	6
Moderately Dynamic	(5-10)	8	9	14	13
Stable	(5 & below)	23	18	13	6

NOTES

1. Directorate of Economics and Statistics, Department of Agriculture and Cooperation, Ministry of Agriculture, Govt. of India (April 1988), Indian Agriculture (22nd edition), p. 224.
2. *Ibid.*

FOUR

INTENSITY OF CROPPING

THE INTENSITY of cropping which is the total cropped area as a percentage of the net sown area, refers to the number of crops raised on a field during an agricultural area. For example, if only one crop is grown on a field in a year (either as a kharif or as a rabi crop), the index of cropping is 100 per cent, and if two crops are grown in a year (as kharif and as rabi crops), the intensity index will be 200 per cent. If three crops are procured in a year (as kharif, rabi and zaid crops), the intensity of cropping will be 300 per cent. This index, thus, provides a measure of land-use efficiency which may be defined as the extent to which the net sown area is cropped or resown. Therefore, the higher the index of the intensity of cropping, the higher is the land-use efficiency, and the lower the index, the lower is the land-use efficiency and less utilised or under-utilised the net sown area.[1]

The intensity index for the study area is 149.4 per cent (i.e. 49.4 per cent of the net sown area is cropped more than once), as compared to 145.6 per cent in the State of Uttar Pradesh and 125.0 per cent in the country as a whole.[2] This shows that the region enjoys a moderate position as regards the intensity of cropping and agricultural efficiency with respect to the State figures, but when the figures of the country are compared, it appears to have a better position.

Table 4.1 and Fig. 4.1 exhibit the spatial pattern of distribution of the intensity of cropping at the micro-level. It may be seen from them that the cropping intensity index varies from 119.8 per cent (Nebua Naurangia) to 177.5 per cent (Mithaura). There are 6 blocks (Pachperwa in Gonda district, Nichlaul, Mithaura, Maharajganj, Ghughli and Siswa in Gorakhpur district) in the high intensity index group (Above 160%) and 16 blocks (54.8% of the total) in the moderately high intensity index group (145-160%). There are 7 blocks in the intensity index group of 130-145 per cent and the remaining in that of 115-130 per cent.

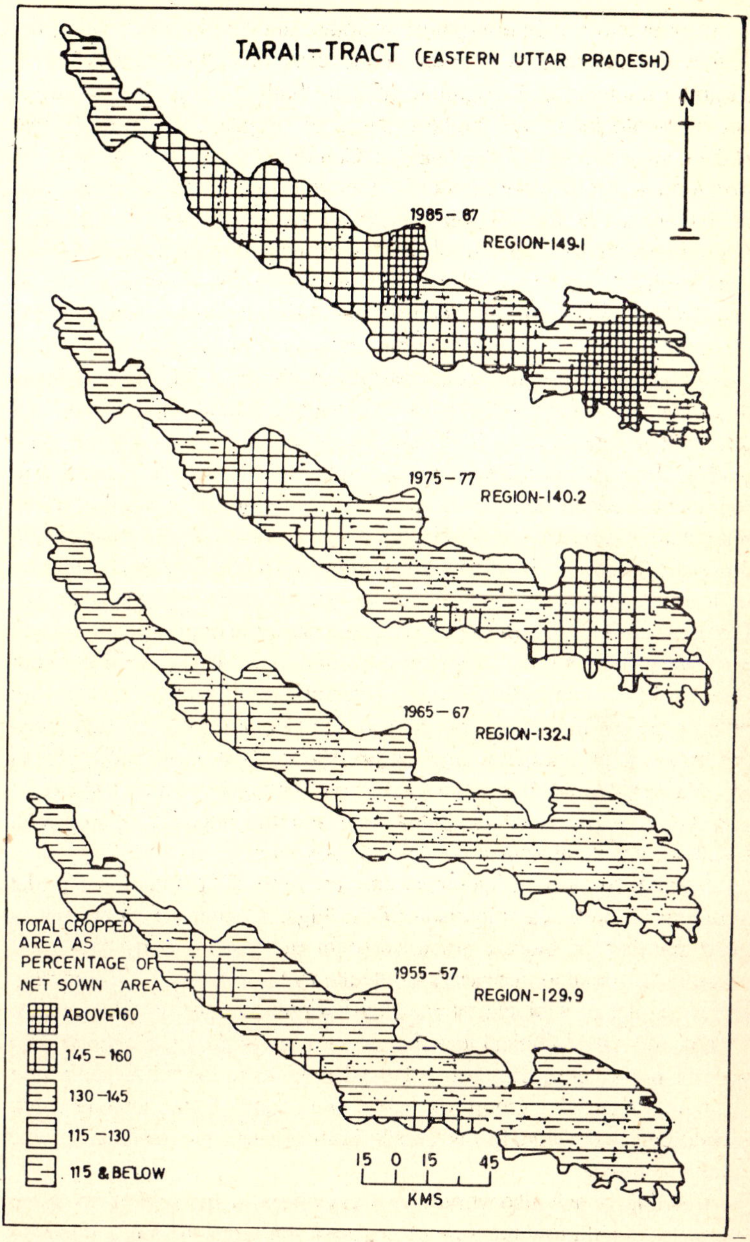

FIG. 4.1: Intensity of Cropping

Undoubtedly, irrigation plays an important role in determining the degree of intensity, as is seen in the case of certain blocks where cropping intensity and irrigation intensity both are high. For example, out of 6 blocks in the highest range of cropping intensity (above 160%), in 4 blocks of the Gandak Command Area, the intensity of irrigation is also of very high order (above 60%). In the same way, out of 16 blocks in the cropping intensity range of 145-160 per cent (moderately high), 2 are in the irrigation intensity range above 60 per cent (high), 4 and 5 respectively in that of 145-160 (moderately high) and 130-145 per cent (moderate). But this relationship between the cropping intensity and irrigation intensity does not hold good everywhere in the region. A comparison of the Figs. 2.6 and 4.1 shows that there are many blocks in the region, where the intensity of cropping is moderate to high, but the proportion of the net irrigated land is low to very low. For example, in the blocks Nawabganj, Jamunaha, Sirsia and Hariharpur Rani in Bahraich district, Haraiya Satgharwa, Tulsipur, Gainsari, and Pachperwa in Gonda district, the cropping intensity is moderately high to high, but the irrigation intensity is low and very low (below 30%). In Hariharpur Rani and Pachperwa, the irrigation intensity respectively is only 8.6 and 1.5 per cent (very low), but the cropping intensity is 159.1 and 161.4 per cent (high). In two blocks, the cropping intensity and the irrigation intensity both are in the low range, but in two, the proportions of irrigated area are higher than that of the area cropped more than once. These blocks are Nautanwa and Laxmipur. There is low positive co-efficient of correlation between net irrigated area and the intensity of cropping (+0.214), which also substantiates the above fact.

Another important factor is the pressure of population on the agricultural land. A comparison of the Figs. 2.3 and 4.1 shows that a good number of blocks, particularly in the eastern part, with high pressure of population possess moderate to high intensity of cropping. For example, in 4 blocks of the high cropping intensity index (above 160%) the physiological density is also of high order (above 550 persons per sq km). Similarly, of the 16 blocks in the moderately high cropping intensity group, 4 have high and 5 moderately high physiological density. In one block, both of these elements are of the same magnitude.

But this is not also very true everywhere in the region, as some blocks even with low pressure of population possess moderate intensity of cropping. For example, in two blocks of the high cropping intensity, the physiological density is moderately high (450-550) and in one moderate (350-450). Similarly, in 6 blocks in the moderately high cropping intensity, the physiological density is moderate (350-450) and in one low (250-350). One block in the low cropping intensity range (115-130), has high physiological densities (above 550) and one moderate (350-450). There, thus, exists a positive co-efficient of correlation between the population pressure and intensity of cropping, but it is again very low (+0.218).

A very important factor which appears responsible for moderately high intensity of cropping in this region is the presence of fertile soils with high contents of moisture in the sub-soils. These soils grow winter season crops even without irrigation. That is why even in spite of low proportion of irrigated land, the intensity of cropping in some of the blocks mentioned above is high.

It may be mentioned here that this does not rule out the necessity of extending irrigation facilities in this region. There are still large areas where double cropping can be extended efficiently, provided irrigation facilities are made available. Besides, irrigation is needed, for the required use of the HYV of seeds and chemical fertilizers and thus for boosting up of per hectare yield from the land already put to double cropping.

Table 4.1 and Fig. 4.1 exhibit the changing pattern of the spatial distribution of the intensity of cropping. It may be seen that high and moderately high intensity classes were occupying much smaller number of blocks in comparison to the low and very low intensity classes in 1955-57 and 1965-67. For example, high and moderately high classes embraced 3 and 2 blocks respectively in 1955-57 and 1965-67, as against 13 and 12 in the low and very low categories. The number in the high and moderately high categories became 11 in 1975-77 and 22 in 1985-87, as against 5 in low and very low categories in 1975-77 and 2 in low category in 1985-87.

Volume of Change in the Intensity of Cropping

The double-cropped area has undergone significant changes during the span of three decades. An additional 19.2 per cent of the net sown area has shifted to double-cropping in the region as a whole, but there are spatial variations within the region, for which the micro-regional variations in the geographical conditions are responsible.

About 87.1 per cent of the blocks have experienced increases ranging between 2.0 per cent (Pharenda) and 44.0 per cent (Maharajganj), while the remaining blocks have registered decreases ranging between 0.3 per cent (Barhni) and 27.3 per cent (Balrampur) the category of high increases (above 20%) embrace 12 blocks (increases vary between 21.6 per cent and 44.0 per cent), 3 of which belong to the western, 3 to the central and remaining to the eastern part in the Gorakhpur and Deoria districts. Most of these blocks, particularly in the Gandak Command Area have registered high increases in the proportion of net irrigated area due to the expanded networks of canals and other means. In the category of moderately high increases (10-20%), there are 8 blocks (4 belonging to western, 3 to central and 1 to the eastern part) where an increase in the irrigation facilities and pressure of population together with an improvement in the infrastructure have encouraged double-cropping. There are 6 blocks in the moderate (5-10%), 1 in the low (2-5%) and 1 in the very low (2% & below) category. In these blocks, either the double-cropped area was already high or slow development in the irrigation facilities, some limitations of soils, flood and water-logging hazards have impeded the growth of double-cropping.

Among 3 blocks witnessing decreases, one is in the high (Balrampur 27.3%) and one in the moderately (Nautanwa 6.4%) category. In one, the decreases have been quite negligible. These are the blocks where flooding, water-logging, insufficient irrigation facilities etc. have affected the intensity of cropping adversely.

If the decennial volume of changes are analysed (Table 4.2 and Fig. 4.2), we find that it is during the latter decades that the changes have been more phenomenal. During the first decade, of the 14 blocks experiencing increases and of the 17 blocks experiencing decreases, 10 and 12 were respectively in the low and very low categories. The number of the blocks experiencing increases became 26 during the second decade, of which only 6 were in the low and very low categories and the remaining were in the higher categories. Of the 5 blocks experiencing decreases, there were only 2 in these categories. During the third decade, of the 25 blocks registering increases, only 4 were in the low and very low categories and the remaining were in the moderate and higher categories, the highest number being in the moderately high category. Similarly, of the 6 blocks experiencing decreases, only 3 were in the low category and the remaining were in the higher categories.

TABLE 4.1: Intensity of Cropping

Intensity class	Total cropped area as % of net sown area	No. of blocks			
		1985-87	1975-77	1965-67	1955-57
High	Above 160	6	2	1	1
Moderately high	145-160	16	9	91	2
Moderate	130-145	7	15	17	15
Low	115-130	2	3	9	11
Very low	115 & below	—	2	3	2

TABLE 4.2: Volume of change in the Intensity of Cropping

Change category	% of the net sown area involved in change	No. of blocks			
		1955-57 to 1965-67	1965-67 to 1975-77	1975-77 to 1985-87	1955-57 to 1985-87
Increase					
High	Above 20	—	6	6	12
Moderately high	10-20	—	8	8	8
Moderate	5-10	4	6	7	6
Low	2-5	6	3	3	1
Very low	2 & below	4	3	1	1
Decrease					
High	Above 20	—	1	1	1
Moderately high	10-20	1	—	2	—
Moderate	5-10	4	2	—	1
Low	2-5	10	1	3	—
Very low	2 & below	2	1	—	1

The rapid changes in the latter decades appear due to the provision of better irrigation facilities, better farm technology and infrastructure, an increasing rate of growth of population and comparatively greater measures to combat recurring floods and water-logging.

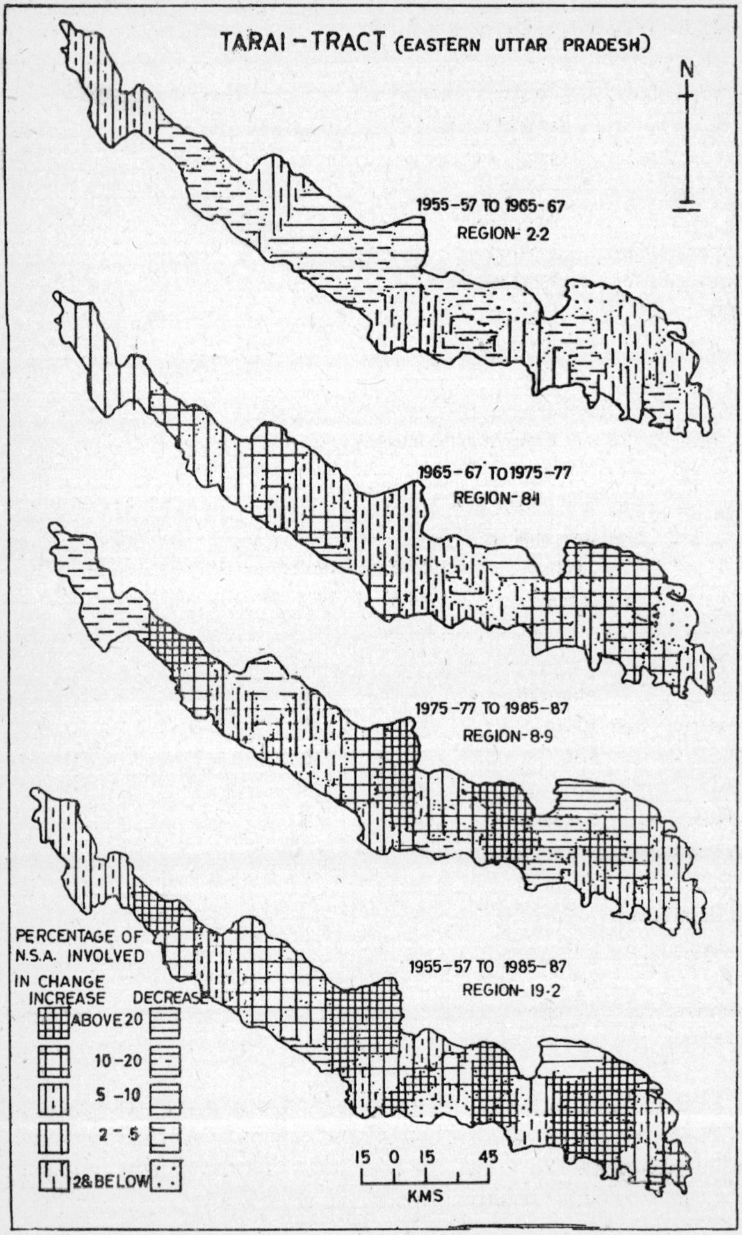

FIG. 4.2: Volume of Change in Intensity of Cropping

NOTES

1. Singh, J., (1976), *An Agricultural Geography of Haryana*, Vishal Publications, Kurukshetra, pp. 235-36.
2. Directorate of Economics and Statistics, Department of Agriculture Cooperation, Ministry of Agriculture, Govt. of India, (April 1988), Indian Agriculture in Brief (22nd edition), p. 224.

FIVE

CROPLAND USE AND CHANGING PATTERNS

THE CROPLAND use analysis of the study area reveals that kharif or summer monsoon crops which are usually sown in June-July are harvested in October-November, and rabi or winter season crops which are sown in October-November and harvested in March-April, together occupy almost all of the total cropped area with almost negligible area under Zaid (the intermediary crops occupying the soils from March to June) crops.

Table 5.2 and Fig. 5.1 show that the kharif crops account for 5,88,907 hectares or 84.1 per cent of the net and 56.4 per cent of the total cropped area, while the rabi crops claim 4,47,027 hectares or 63.8 per cent of the net and 42.8 per cent of the total cropped area of the region. This shows that the kharif crops are more important than the rabi crops, which is mainly due to the better suitability of the region for growing the former crops. As already discussed, the study area abounds in depressions and lowlying lands, the clayey soils of which grow nothing but paddy, particularly late paddy which sometimes occupies the field late in the winter season causing much delay in the sowing of the rabi crops.

The kharif crops dominate the agricultural landscape almost everywhere with some spatial variations in its predominance. In two blocks (Nebua Naurangia and Bishunpura, the eastern most blocks), their proportion is above 65 per cent, in 5 (Bhanwapur, Itwa, Barhni, Birdpur and Khadda, the former 4 in the middle and the latter 1 in the east), it is between 60 and 65 per cent, and in 11 (Mihipurwa, Nawabganj, Jamunaha, Sirsia, Tulsipur, Gainsari, Pachperwa, Naugarh, Nautanwa, Nichlaul and Siswa) between 55 and 60 per cent of the total cropped area. There are 12 blocks in the range of 50-55 per cent and the remaining 1 in the range of 50 per cent and below. This shows that in about 58 per cent of the blocks, the concentration vary from moderate to high and in the remaining 42 per cent, they are low to very low.

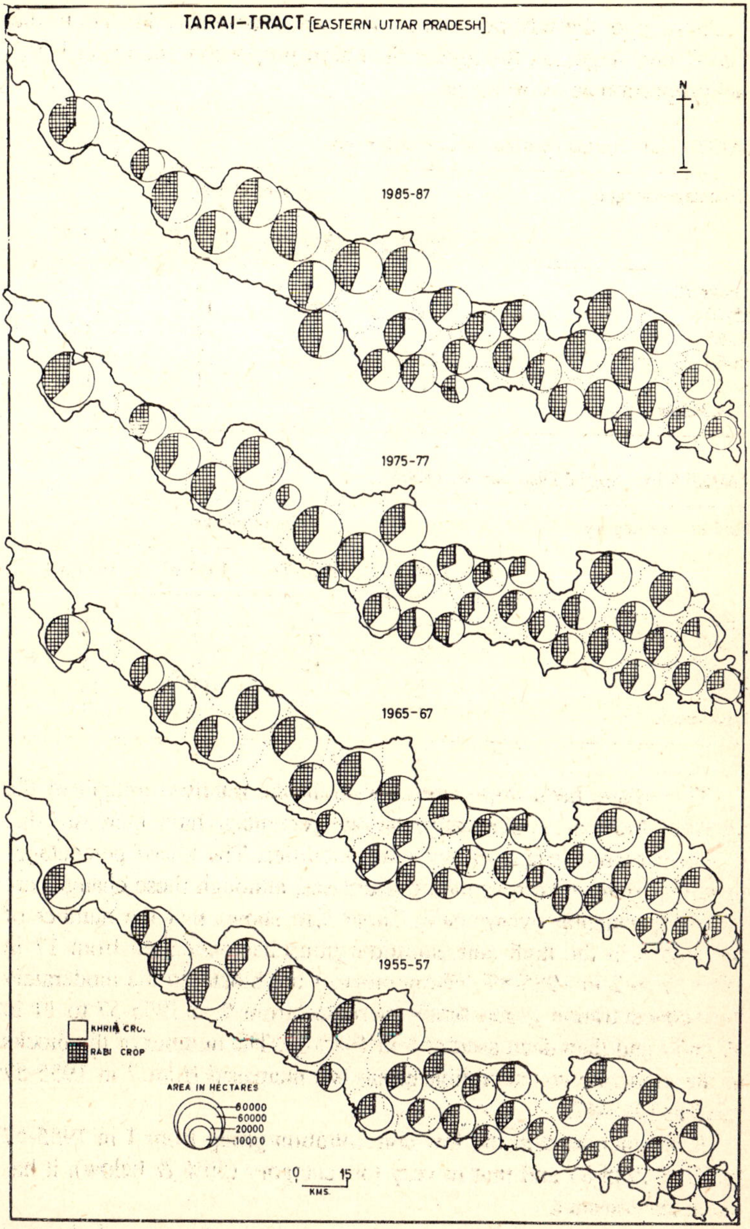

FIG. 5.1: Season-wise Cropland Use

As regards the rabi percentage, they are everywhere inverse to the kharif percentage, i.e. the higher the kharif proportion, the lower is the rabi proportion and *vise versa*.

TABLE 5.1a: Spatial Distribution of Kharif Crops

Percentage category	No. of blocks			
	1985-87	1975-77	1965-67	1955-57
Above 70	1	3	3	3
65-70	1	3	11	14
60-65	5	11	9	5
55-60	11	10	7	7
50-55	12	4	1	1
50 & below	1	—	—	1

TABLE 5.1b: Spatial Distribution of Rabi Crops

Percentage category	No. of blocks			
	1985-87	1975-77	1965-67	1955-57
Above 45	11	4	1	2
40-45	12	10	7	7
35-40	5	11	6	5
30-35	2	4	13	14
30 & below	1	2	4	3

There have been important changes in the relative strength of the kharif and rabi area. Generally, the rabi percentage have increased due to the increased irrigation and other facilities. The kharif percentages have experienced corresponding decreases, although these changes are not so spectacular everywhere. Table 5.1a shows that the number of the blocks in the high concentration group has decreased from 17 in 1955-57 to 2 in 1985-87. The number of the blocks in the moderately high concentration group firstly increased from 5 in 1955-57 to 11 in 1975-77 and then decreased to 5 in 1985-87. The number of the blocks in the moderate concentration group has increased from 7 in 1955-57 to 11 in 1985-87.

It has increased in the low concentration group from 1 in 1955-57 to 12 in 1985-87 and that in very low category (50% & below), it has remained constant.

There has been a corresponding increase in the case of the rabi crops (Table 5.1b). The number of the blocks in the high concentration group (above 45%) has increased from 2 (1955-57) to 11 (1985-87) and that in the moderately high (40-45%) group from 7 to 12 during this period. The number of blocks in the moderate concentration group (35-40%), firstly increased from 5 (1955-57) to 11 (1975-77) and then decreased to 5 (1985-87). Just reverse to this trend, we find that the number of the blocks in the low concentration group (30-35%) has decreased significantly, i.e. from 14 (1955-57) to 2 (1985-87). There exists only one block now in the very low concentration group. This number was 3 in 1955-57.

As regards the volume of change, i.e. the percentage of the total cropped area involved in change, the kharif area has experienced a decrease of 7.6 per cent during the span of three decades. During the decade 1955-57 to 1965-67, there was a slight increase and then a decrease was registered during the second and third decades. At micro-level, some variations are observed. For example, in all, 30 blocks have registered decreases ranging from 0.3 to 26.4 per cent. In 18 blocks, high decreases (above 6%) have been registered and in 3, they have been moderate (4-6%). The number of the blocks in the low and very low (2-4% and 2% & below) categories respectively is 5 and 4. Only one block (Uska) has registered an increase, that, too, very low. It may be seen from Fig. 5.2 that the higher decreases are particularly in those blocks where due to an increase in the irrigation facilities and some other factors, the area under rabi crops has increased largely.

A perusal of the Fig. 5.2 shows that the number of the blocks experiencing increases and decreases was almost the same during the decade 1955-57 to 1965-67, but the number of the blocks experiencing decreases rose to 24 and 30 during the second and third decade respectively. The number of the blocks experiencing increases has decreased to 7 and 3 respectively during these decades. This shows an increasing importance of the rabi crops in the cropping pattern of the study area.

LAND UTILISATION IN KHARIF SEASON

Table 5.2 shows that the foodgrains occupy a very prominent position among the kharif crops, as they account for 90.9 per cent of the kharif, 76.4 per cent of the net and 51.3 per cent of the total cropped area of the region. Paddy (Rice) is the only predominant foodgrain. Food-cum-commercial crops occupy 8.6 per cent of the kharif, 7.3 per cent of the net and 4.9 per cent of the total cropped area of the region. Non-food crops are quite negligible.

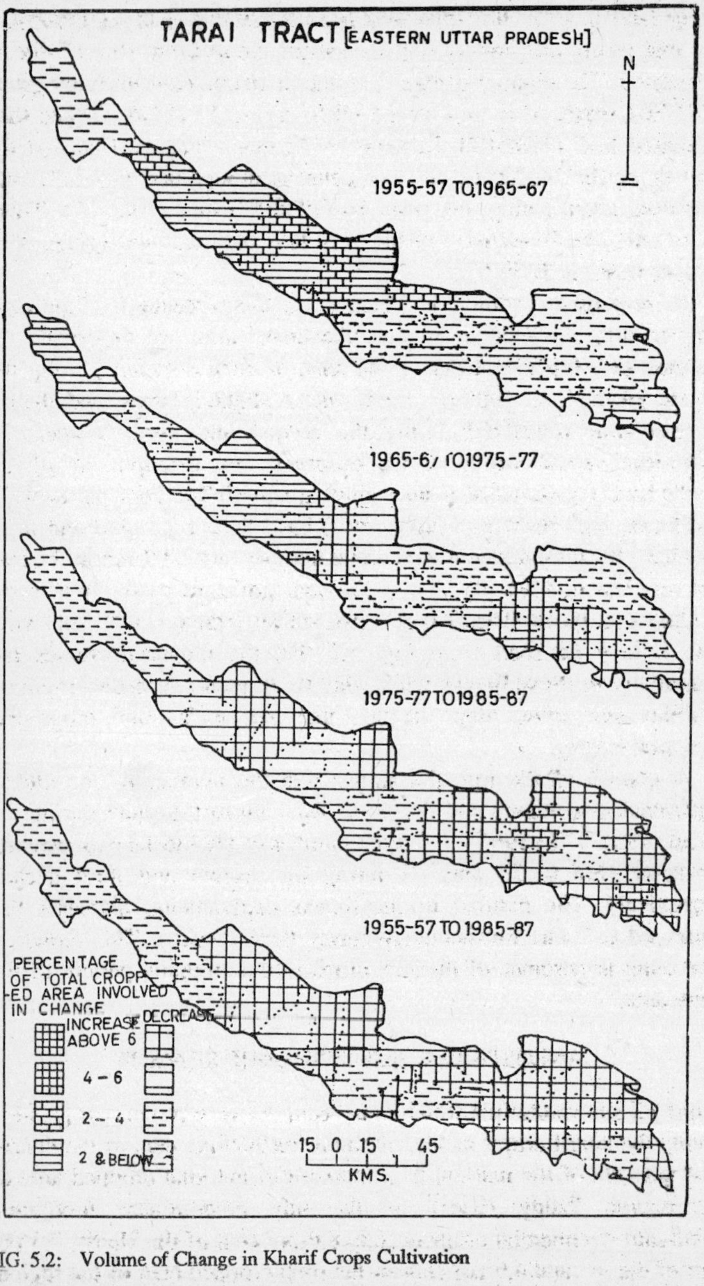

FIG. 5.2: Volume of Change in Kharif Crops Cultivation

TABLE 5.1c: Volume of Change in Kharif Area

Change category (% of the total cropped area involved in change)	No. of blocks			
	1955-57 to 1965-67	1965-67 to 1975-77	1975-77 to 1985-87	1955-57 to 1985-87
Increases				
Above 6	—	—	—	—
4-6	—	—	—	—
2-4	6	3	2	—
2 & below	9	4	1	1
Decreases				
Above 6	—	8	14	18
4-6	1	5	8	3
2-4	3	7	2	5
2 & below	12	4	4	4

The kharif foodgrains did occupy more important position in the past, as they claimed 60.2 per cent of the total cropped area in 1955-57 and 1965-67 and 56.3 per cent in 1975-77. The area under food-cum-commercial crops has witnessed an increase, which is due to an increase in the area of oil-seeds, sugarcane and vegetables. There is no significant change in the area under non-food crops.

LAND UTILISATION IN RABI SEASON

Table 5.3 shows the land utilisation in the rabi season, from which the predominance of the foodgrains is again established, as they account for 90.3 per cent of the rabi, 57.6 per cent of the net and 38.7 per cent of the total cropped area of the region. Wheat is the principal foodgrain. Food-cum-commercial crops are relegated to an insignificant position and non-food crops are quite negligible.

The area under rabi foodgrains has witnessed an increase to the tune of 5.8 per cent of the total cropped area, as they accounted for 32.9 per cent of the total cropped area in 1955-57. The area under food-cum-commercial crops, particularly that under oil-seeds has increased from 2.4 to 3.3 per cent of the total cropped area. No significant change is observed in the area of non-food crops.

TABLE 5.2: Land Utilisation in the Kharif Season

Crops	Area in hectares	% of the net sown area in the season	% of the net sown area of the region	% of the total cropped area of the region
Cereals				
Paddy	4,53,842	77.1	64.7	43.4
Maize	44,026	7.5	6.3	4.2
Kodon alone & mixed	14,310	2.4	2.0	1.4
Jowar	518	0.1	0.07	0.05
Bajra	203	0.03	0.02	0.02
Others	7,488	1.2	1.1	0.7
Total	5,20,387	88.3	74.2	49.8
Pulses				
Arhar	11,861	2.0	1.7	1.2
Urad & Moong	3,164	0.5	0.5	0.3
Others	227	0.03	0.03	0.02
Total	15,252	2.6	2.2	1.5
Total foodgrains	5,35,639	90.9	76.4	51.3
Food-cum-Commercial Crops				
Oil-seeds	4,133	0.7	0.6	0.4
Sugarcane	43,928	7.5	6.3	4.2
Vegetables	2,618	0.4	0.4	0.3
Total	50,679	8.6	7.3	4.9
Total foodcrops	5,86,318	99.5	83.7	56.2
Non-food crops	2,509	0.5	0.4	0.2
Grand Total	5,88,907	100.0	84.1	56.4

OVERALL CROPPING PATTERN

The overall cropping pattern of the study area (Appendix VIII) reveals that the gaps in the concentrations of the different sets of crops are very wide, as will be clear from the ensuing discussion.

Foodgrains

As already mentioned, the foodgrains occupy an excessively predominant position in the cropping pattern of the region, as 90.0 per cent of the total cropped area of the region is under these crops which include cereals and pulses. This may be attributed to their importance in sustaining the population and also to their climatic and edaphic preferences. In 14 blocks (Table 5.4a and Fig. 5.3), the proportions of the area under foodgrains are moderately high (93-97%) and in 1 high (above 97%). There are 7 blocks in the moderate range (89-93%), 3 in the low (85-89%) and 6 in the very low (85% & below) range. In no block, the share of these crops is below 65 per cent. In the blocks of the very low category, viz., Khadda, Nebua Naurangia and Bishunpura, the low percentages are due to the increasing importance of sugarcane which claims 22.4, 26.4 and 26.1 per cent of the total cropped area respectively, and in Tulsipur, gainsari and Nichlaul due to that of sugarcane and oil-seeds. In the blocks of the low category also (Mihipurwa, Ghughli and Siswa), sugarcane occupies important place.

TABLE 5.3: Land Utilisation in the Rabi Season

Crops	Area in hectares	% of the net sown area in the season	% of the net sown area of the region	% of the total cropped area of the region
Cereals				
Wheat	2,99,834	67.1	42.7	28.7
Wheat mixed with gram	2,106	0.5	0.3	0.2
Wheat mixed with barley	3,565	0.8	0.5	0.3
Barley	3,824	0.9	0.6	0.4
Barley mixed with gram	6,035	1.3	0.9	0.6
Others	1,281	0.2	0.2	0.1
Total	3,16,645	70.8	45.2	30.3
Pulses				
Gram	25,286	5.7	3.6	2.4
Pea	7,949	1.8	1.1	0.8
Masoor	48,764	10.9	7.0	4.7
Others	5,042	1.1	0.7	0.5
Total	87,041	19.5	12.4	8.4
Total foodgrains	4,03,686	90.3	57.6	38.7
Food-cum-Commercial Crops				
Fruits & vegetables	7,454	1.7	1.1	0.7
Oil-seeds	34,353	7.7	4.9	3.3
Total	41,807	9.4	6.0	4.0
Total foodcrops	4,45,493	99.7	63.6	42.7
Non-food crops	1,534	0.3	0.2	0.1
Grand Total	4,47,027	100.0	63.8	42.8

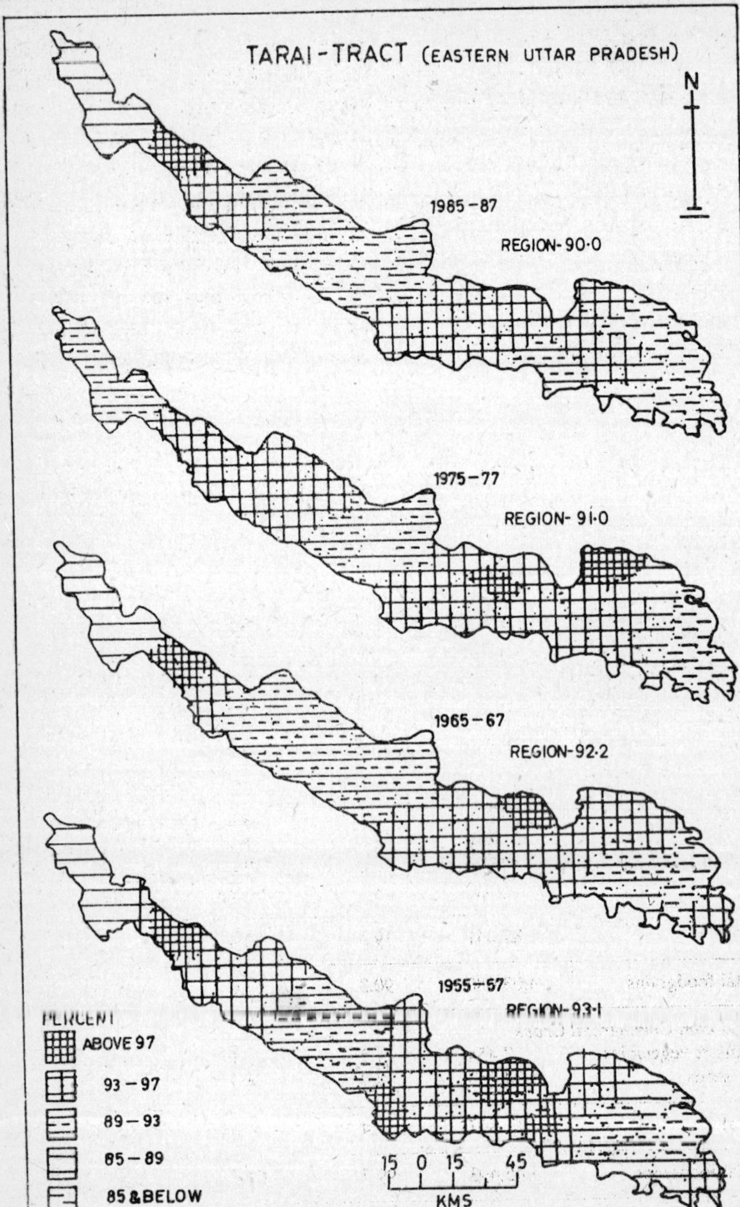

FIG. 5.3: Foodgrain Cropped Area

It may be noted here that kharif foodgrains are more important than the rabi foodgrains, as they claim 51.3 per cent of the total cropped area as against 38.7 per cent in the case of the latter.

TABLE 5.4a: Total Foodgrains as Percentage of Total Cropped Area

Percentage category	No. of blocks			
	1985-87	1975-77	1965-67	1955-57
Above 97	1	2	2	5
93-97	14	19	14	14
89-93	7	3	9	7
85-89	3	4	4	4
85 & below	6	3	2	1

The proportion of the area under foodgrains has registered a decrease, although the hectareage under them has increased with an increase in the cultivated area and in the intensity of cropping. Table 5.4a and Fig. 5.3 show that the number of the blocks in the high concentration group (above 97%) has decreased from 5 in 1955-57 to 1 in 1985-87. In the range of 93-97 per cent, the number has remained constant, except in 1975-77 when it was 19. In the concentration range of 89-93 per cent, there has not been any definite trend, for example, the number of the blocks increased from 7 in 1955-57 to 9 in 1965-67, then it decreased to 3 in 1975-77 and again increased to 7 in 1985-87. The number of the blocks in the low concentration range (85-89%) has remained almost stable, but that in the very low concentration range (85% & below), it has increased from 1 in 1955-57 to 6 in 1985-87. This substantiates the fact that there is a decreasing trend in the extent of the area under foodgrains, which is due to the increasing importance of the food-cum-commercial crops.

Almost the same trend may be observed in the spatial distribution of the volume of change. The region as a whole registered a decrease of 3.1 per cent. Within the region, only 5 blocks have experienced a minor increase (in the range 2-4% and below), while 26 blocks have experienced decreases, of which 3 are in high category (above 8%), 4 in the moderately high category (4-8%), 8 in the moderate category (2-4%), 7 in the low (1-2%) and 4 in the very low (1% & below) category. As regards the decennial changes, during the decade 1955-57 to 1965-67, 26 blocks experienced a decrease and only 5 an increase (Table 5.4b). These figures for the decade 1965-67 to 1975-77 respectively are 13 and 18 and for the decade 1975-77 to 1985-87, 10 and 21. This indicates that generally the proportion of hectareage under foodgrains has been decreasing slowly, although in some of the blocks, it has increased particularly more during the decade 1965-67 to 1975-77 (Table 5.4b and Fig. 5.4).

Cereals

Cereals constitute the major part of the foodgrains in the study area by claiming about 80.1 per cent of the total cropped area of the region. Kharif cereals are more important, as they claim 49.8 per cent of the total cropped area as against 30.3 per cent claimed by the rabi cereals.

TABLE 5.4b: Volume of Change in Total Foodgrains

Change category (% of the total cropped area involved in change)	No. of blocks			
	1955-57 to 1965-67	1965-67 to 1975-77	1975-77 to 1985-87	1955-57 to 1985-87
Increase				
Above 8	—	—	—	—
6-8	—	2	—	—
2-4	—	6	—	1
1-2	2	4	2	3
1 & Below	3	6	8	1
Decrease				
Above 8	—	2	—	3
4-8	1	—	4	4
2-4	6	2	8	8
1-2	4	7	5	7
1 & Below	15	2	3	4
No Change	—	—	1	—

Table 5.5a and Fig. 5.5 show that the concentration of the cereals is high (above 89%) in 5 blocks (Itwa, Bhanwapur, Naugarh, Birdpur and Laxmipur) and in 11 blocks (Khuniaon, Barhni, Uska, Jogia, Bansi, Nautanwa, Brijmanganj, Pharenda, Mithaura, Maharajganj, Ghughli), it is high (83-89%). It is moderate (77-83%) in 4 blocks (Nawabganj, Hariharpur Rani, Sirsia and Siswa). In most of these blocks the area under pulses, particularly that under the kharif pulses, is not very significant, which appears due to the extensive water-logging and

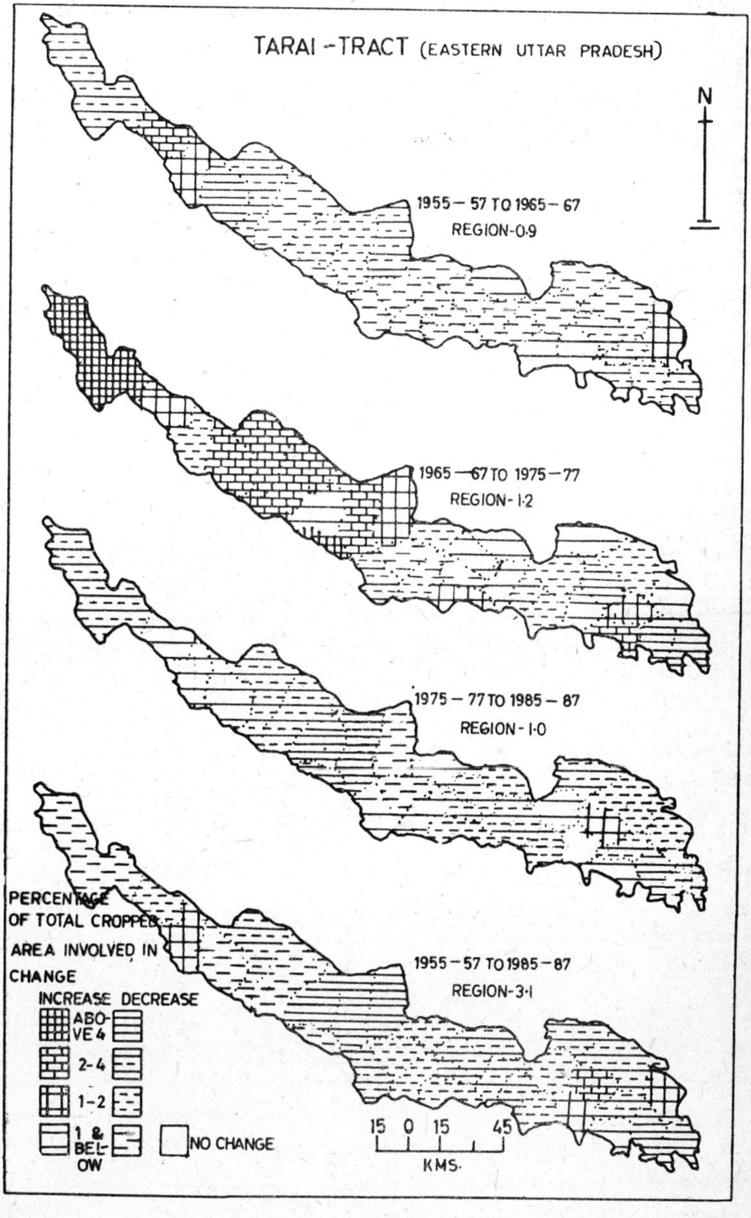

FIG. 5.4: Volume of Change in Total Foodgrains

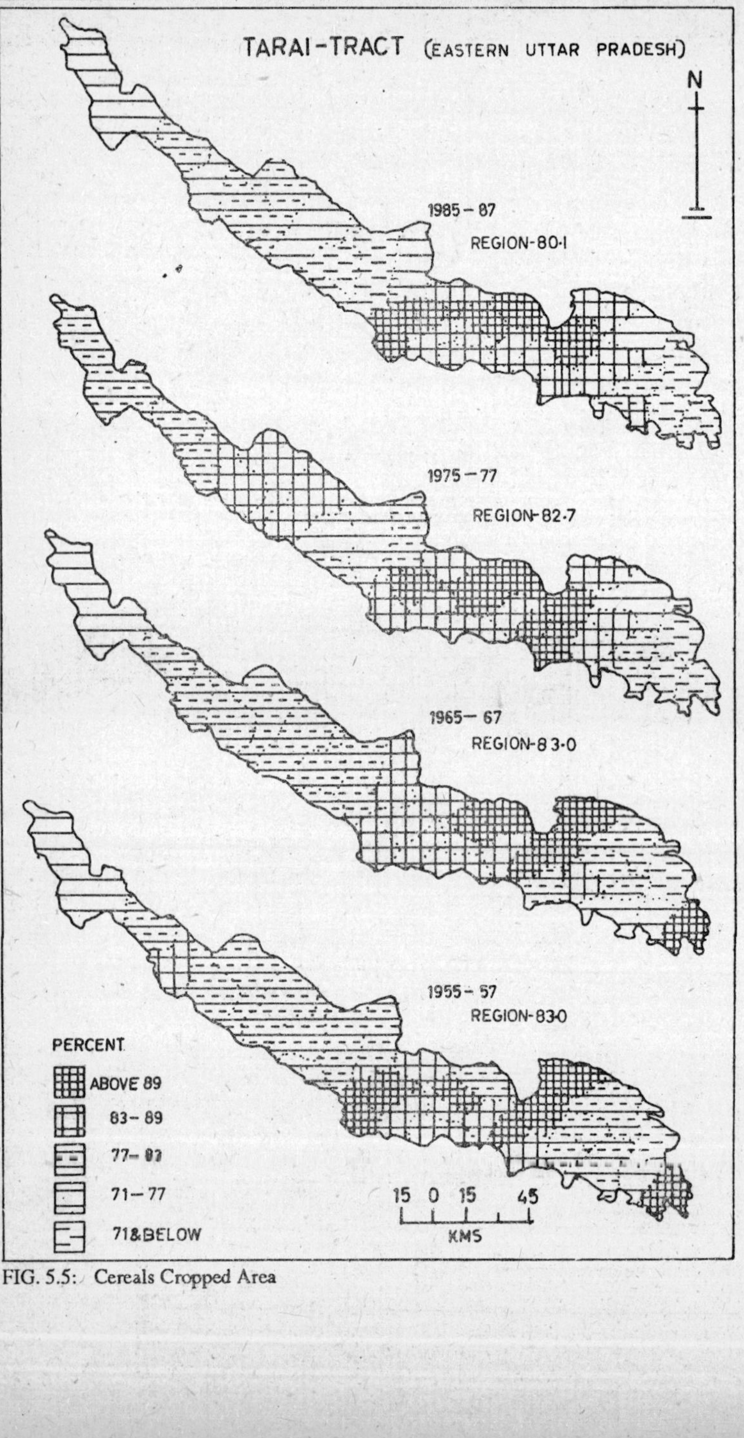

FIG. 5.5: Cereals Cropped Area

inundation causing vast devastation of kharif pulses. In 6 blocks (Mihipurwa, Jamunaha, Haraiya Satgharwa, Pachperwa, Balrampur and Nichlaul), the proportions of the area under cereals range between 71 and 77 per cent and in the remaining 5 (Tulsipur, Gainsari, Nebua Naurangia, Khadda and Bishunpura), it is in the range 71 per cent and below. In these blocks, the decreasing proportion is either due to the increasing proportion of the area under pulses or under sugarcane, which have encroached upon the area of cereals.

Table 5.5a and Fig. 5.5 further show that the number of the blocks in the high concentration group (above 89%) has slightly decreased, i.e. from 7 in 1955-57 to 5 in 1985-87, although the number of the blocks in the moderately high concentration group (83-89%) has increased from 7 in 1955-57 to 11 in 1985-87. There is a great reduction in the number of the blocks in the moderate concentration group (77-83%), i.e. from 13 in 1955-57 to 4 in 1985-87. The number of the blocks in the low concentration group (71-77%) has slightly increased, i.e. from 3 to 6, but that in the very low concentration group (71% and below), it has increased greatly, i.e. from 1 to 5 during this period.

TABLE 5.5a: **Area Under Cereals as Percentage of Total Cropped Area**

Percentage category	No. of blocks			
	1985-87	1975-77	1965-67	1955-57
Above 89	5	7	6	7
83-89	11	11	8	7
77-83	4	7	12	13
71-77	6	3	4	3
71 & below	5	3	1	1

The above discussion shows that there is generally a reduction in the proportion of the area under cereals, although this is not true in the case of all the blocks. This is simply on account of the fact that in certain blocks, the importance of cereals has increased and in some it has decreased due to an increase in the area of other crops, but the changes are not considerable.

As regards the volume of change, the region has undergone a change to the tune of 2.9 per cent during the span of three decades. At the micro-level (Table 5.5b and Fig. 5.6), we see that one-third blocks have registered an increase and two-thirds a decrease. There is no block in the range of high increase (above 8%), but there are 4 (Ghughli, Birdpur, Maharajganj and Mithaura) in the range moderately high (4-8%) increase. These are the blocks in which this magnitude of change in the area under cereals appears due to an increase in the extent of net sown area and in the intensity of cropping together with the pressure of population. In the remaining 6 blocks, the increases are from moderate to very low. Among the blocks experiencing decreases, more than 50 per cent are in the moderate range (2-4%) and a little less than one-third in the moderately high to high range. In 4 blocks, the changes are low to very low. The significant decreases are in those blocks in which there have been corresponding increases in the area under sugarcane and pulses etc.

TABLE 5.5b: Volume of Change in Cereals

Change category (% of the total cropped area involved in change)	No. of blocks			
	1955-57 to 1965-67	1965-67 to 1975-77	1975-77 to 1985-87	1955-57 to 1985-87
Increase				
Above 8	—	1	—	—
4-8	—	7	—	4
2-4	4	4	4	1
1-2	3	2	4	2
1 & Below	8	5	1	3
Decrease				
Above 8	1	1	1	4
4-8	2	3	12	1
2-4	7	3	3	12
1-2	4	1	4	2
1 & Below	2	4	2	2

As regards the decennial changes, the decade 1955-57 to 1965-67 witnessed increases and decreases in 15 and 16 blocks respectively. These numbers became 19 and 12 during the decade 1965-67 to 1975-77 and 9 and 22 during 1975-77 to 1985-87. This shows that during the second decade, the increases in the area under cereals were considerable but in the third decade, the decreases were more significant.

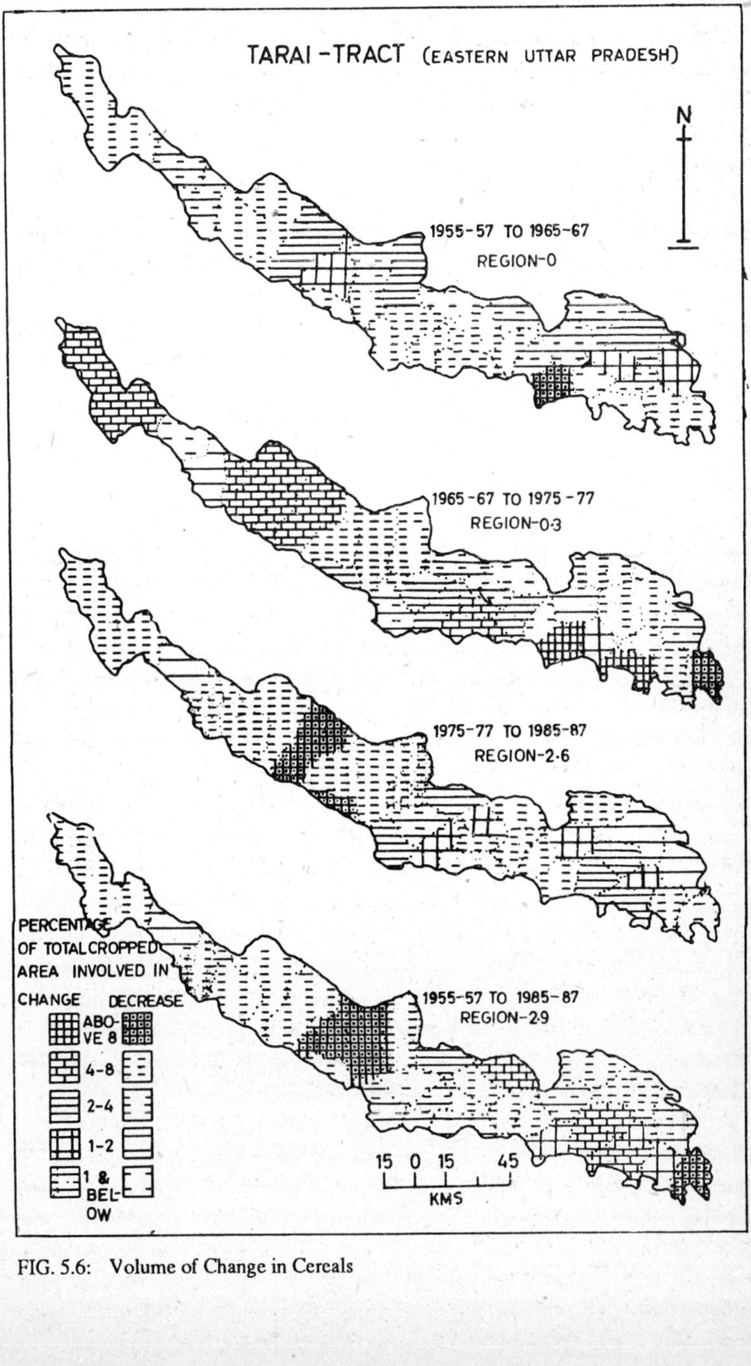

FIG. 5.6: Volume of Change in Cereals

As already mentioned, rice, wheat, maize, wheat mixed with gram, wheat mixed with barley, barley, kodon etc. are the main cereals which have been discussed in some details in the forthcoming pages.

Rice (Paddy or *Oryza Sativa*)

Rice is the principal crop not only in the kharif season but in the overall cropping pattern of the region also. It greatly dominates the agricultural landscape of the region and has been a staple food of the majority of the people. It occupies an area of 4,53,842 hectares, i.e. 77.1 per cent of the kharif, 64.7 per cent of the net and 43.4 per cent of the total cropped area of the region and thus ranks first, while the share of this crop in the total cropped area of the State of Uttar Pradesh and that the country as a whole respectively is 21.2 and 23.2 per cent (1986-87).[1] This predominance of rice in the study area is due to the predominance of clayey, clayey loam or heavy loam soils which are well suited to the production of this crop, the high water-table and abundance of moisture in the sub-soil, the lowlying lands that favour impounding of water which is beneficial for the aquatic crops, and hot and humid climate with moderately high rainfall in the region.

The spatial distribution of the area under rice as exhibited in Table 5.6a and Fig. 5.7 shows that the concentration of this crop is not uniform throughout the region. It varies from 28.2 per cent (Haraiya Satgharwa) to 58.7 per cent (Birdpur), albeit the areas of low concentration are not so extensive. In 28 blocks, there is moderate to high concentration of area under rice. High concentration (55-65% and above) is found in 4 blocks (Naugarh, Barhni, Birdpur and Nautanwa). There are 11 blocks (Nawabganj, Pachperwa, Khuniaon, Bhanwapur, Itwa, Laxmipur, Brijmanganj, Mithaura, Maharajganj, Ghughli and Siswa) in the moderately high (45-55%) concentration group. In this area of moderately high to high concentration is situated the 'rice bowl' of the region, which is famous for the distinguished varieties of rice such as 'kala namak'. The ecological conditions favouring rice culture are found in greater magnitude here.

The blocks with moderate concentration (35-45%) number 13. They are: Jamunaha, Sirsia, Tulsipur, Gainsari and Balrampur in the west, Uska, Jogia and Bansi in the middle, and Pharenda, Nichlaul, Khadda, Nebua Naurangia and Bishunpura in the east. The blocks with low concentration (25-35%) number 3 (Mihipurwa, Hariharpur Rani and Haraiya Satgharwa, all in the west). In these blocks, low concentration of rice appears due to the predominance of light sandy loam soils in the riverine area or in the sloppy uplands in the north.

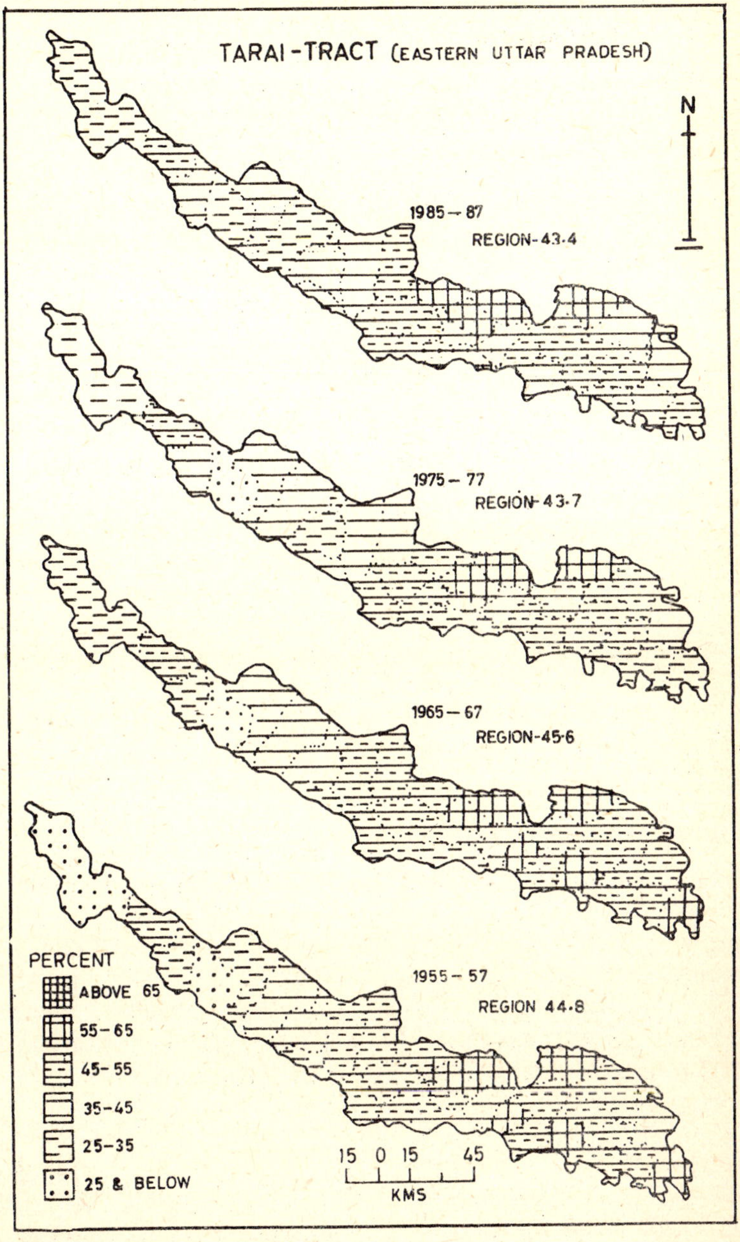

FIG. 5.7: Rice Cropped Area

TABLE 5.6a: Rice Cropped Area as a Percentage of Total Cropped Area

Percentage Category	No. of blocks			
	1985-87	1975-77	1965-67	1955-57
Above 65	—	1	1	1
55-65	4	2	5	6
45-55	11	13	15	13
35-45	13	8	5	4
25-35	3	6	4	5
25 & Below	—	1	1	2

TABLE 5.6b: Volume of Change in Rice Cultivation

Change category (% of the total cropped area involved in change)	No. of blocks			
	1955-57 to 1965-67	1965-67 to 1975-77	1975-77 to 1985-87	1955-57 to 1985-87
Increase				
Above 4	2	5	4	8
2-4	7	3	4	3
1-2	7	3	5	1
1 & Below	7	1	4	2
Decrease				
Above 4	1	5	8	13
2-4	—	6	1	1
1-2	1	3	4	—
1 & Below	6	5	1	3

Several varieties of rice are grown in the region, which may be classified into two broad categories on the basis of their peculiarities of cultivation:

Early paddy

It is known as *Bhadain* or *Kunwari*, as it is harvested in the month of *Bhadrapad* or *Ashwin*. It is also known as *Usahn*, as it is generally sown broadcast.

Late Paddy

It is locally called *Jarhan*, as it is transplanted. It is also known as *Aghani*, as it is harvested in the month of *Aghan* or (*Margashirsh*).

The early paddy which is sown by broadcasting, occupies generally the well drained loamy soils which do not retain water for long. It is generally followed by pea, gram or *masoor* in the rabi season. Major area of rice is under this variety in the region.

The late paddy for which transplantation method is adopted, thrives best on a clayey soil and it requires a constant supply of water till at least the end of October. The late paddy is sown in the nursery beds in July and transplantation is done in the well puddled fields with seedlings about 30 to 40 days old and 10 to 12 inches high. The transplanted fields do not require the same amount of weeding as the broadcast fields and give a better yield. It occupies comparatively less area which is mainly confined to those blocks which have high concentrations of lowlying areas with clayey or clayey loam soils.

Table 5.6a and Fig. 5.7 show that rice has occupied very predominant position in the past also. In 1955-57, 1965-67 and 1975-77, the blocks having moderate to high concentrations numbered 24, 26 and 24 respectively, and those having low and very low concentrations have almost remained the same. Only a few blocks have changed their categories.

In the study area, the hectareage under rice has risen from 3,80,700 (1955-57) to 4,53,842 (1985-87), showing a percentage increase of 19.2, which appears due to the reclamation of waterlogged areas and intensification of cultivated land, introduction of new varieties of rice and more demand of it to feed the increasing population. Within the region, there are spatial variations, some blocks have registered increases and some decreases.

The analysis of the volume of change (Table 5.6b and Fig. 5.8) indicates that the region has registered minor decrease, i.e. to the tune of 1.1 per cent of the total cropped area during the span of three decades. It is quite interesting to note that about 45 per cent of the blocks have witnessed increases and the remaining the decreases. In both the cases, the number of the blocks in the higher category is the maximum (11 towards increase and 14 towards decrease side). The high category increases are generally in those blocks where formerly, the concentration of the hectareage under rice was low or very low and with an increase in the net sown area, the area under rice has increased. The high category decrease (above 4%), on the other hand, are in those blocks which already had moderately high to high concentrations and with an increase in the importance of other crops such as sugarcane etc. (as in the case of Nebua Naurangia, Bishunpura, Maharajganj and Gainsari etc.), the area under rice has decreased.

There is one block in the increase range of 1-2 per cent and 2 and 3 respectively in the increase and decrease range of 1 per cent and below.

It may further be seen from Table 5.6b that in the decade 1955-57 to 1965-67, 23 blocks witnessed increases and the remaining 8 witnessed decreases, but this number decreased to 12 in the case of increase and increased to 19 in the case of decreases, showing a decreasing trend in the proportion of the rice hectareage, during the decade 1965-67 to 1975-77. During the decade 1975-77 to 1985-87, the position again appears to have been slightly changed, as the number of the blocks witnessing increases has increased to 17 and that of witnessing decreases has decreased to 14. This shows that with an increase in the intensity of cropping, the hectareage under this crop has also increased.

Wheat *(Triticum Sativum)*

Wheat is the chief rabi cereal and it ranks second in the cropping pattern of the region. It occupies 2,99,834 hectares of area or 67.1 per cent of the rabi, 42.7 per cent of the net sown and 28.7 per cent of the total cropped area of the region and now forms quite an important item in human diet in many parts. In the State of Uttar Pradesh and in the country as a whole, the share of this crop in the total cropped area respectively is 33.1 and 12.9.[2]

Although the Tarai climatic conditions may not be considered very ideal for wheat cultivation everywhere, its superior adaptability to environmental conditions being a winter crop, the prevalence of clayey loam to sandy loam soils with sufficient sub-soil moisture together with the irrigation facilities being available in many parts, and an improvement in the drainage conditions, are some of the factors which have provided sufficient incentive for the extension of hectareage under this crop.

The spatial distribution of wheat (Table 5.7a and Fig. 5.9) hectareage exhibits very interesting facts. Firstly, there are considerable spatial variations (21.6% in Sirsia to 40.8% in Bansi). Secondly, the concentration of wheat distribution is inverse to that or rice, as the number of the blocks with significant concentration is much higher in the case of rice in comparison to wheat, which indicates the preponderance of the former in the study area.

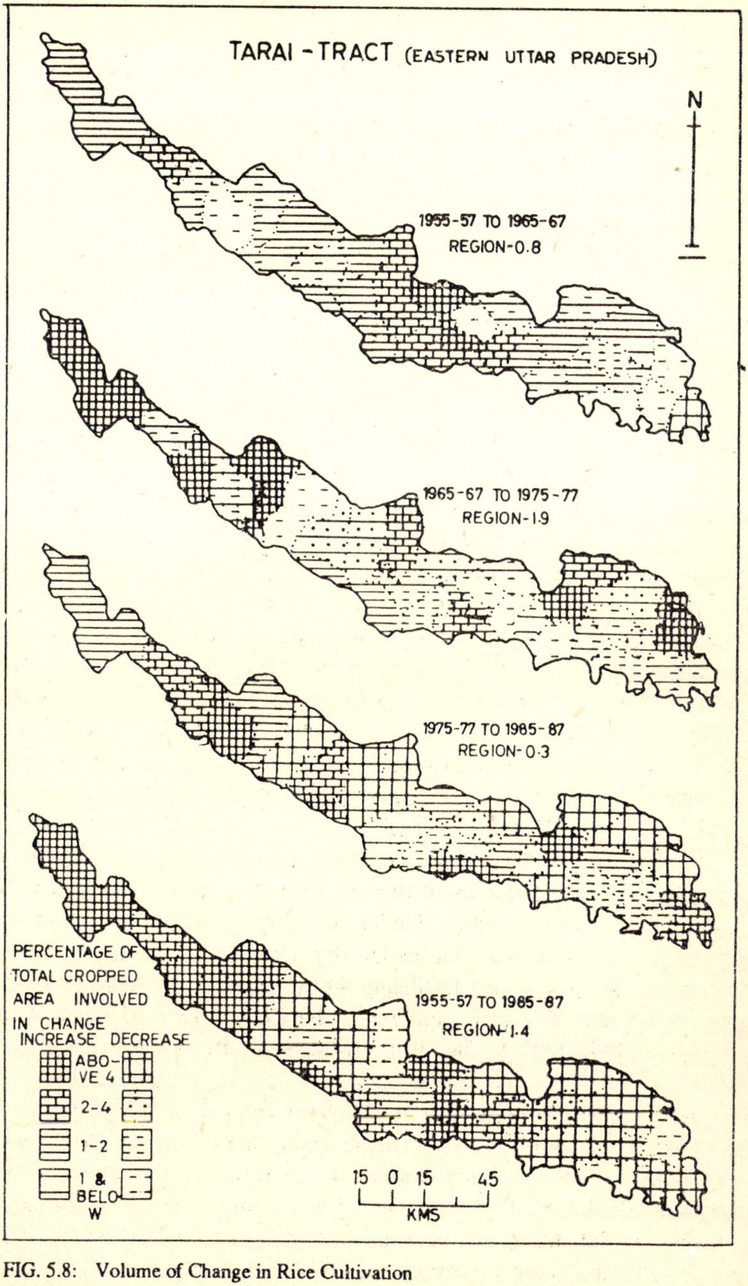

FIG. 5.8: Volume of Change in Rice Cultivation

It may be seen from Table 5.7a and Fig. 5.9 that in 12 blocks (Khuniaon, Naugarh, Jogia, Bansi, Laxmipur, Brijmanganj, Pharenda, Nichlaul, Mithaura, Maharajganj, Ghughli and Siswa, all belonging to the middle and eastern parts), the proportion of the total cropped area occupied by wheat is more than 30 per cent. A good number of these blocks possess adequate facilities of irrigation and favourable soils for wheat cultivation. In 10 blocks (Hariharpur Rani, Jamunaha, Haraiya Satgharwa, Balrampur in the west; Itwa, Bhanwapur, Barhni, Birdpur and Uska in the middle; and Nautanwa in the east), the proportion varies between 25 and 30. Most of these blocks have moderate proportions of loamy soils and moderate facilities of irrigation or moderate amount of moisture in the sub-soil. The blocks in the range 20-25 per cent number 9 (Mihipurwa, Nawabganj, Sirsia, Tulsipur, Gainsari and Pachperwa in the west; and Khadda, Nebua Naurangia and Bishunpura in the east). Most of these blocks either do not possess adequate facilities of irrigation or late paddy which is more popular here, seldom offers an opportunity for growing second crop.

It may be mentioned here that wheat is grown both as a dry as well as an irrigated crop. In the riverine and in some other tracts, particularly in the middle and western parts where the facilities of irrigation are inadequate but there is enough moisture in the sub-soil for growing this crop, it is raised without irrigation, although the yields are generally poor. But wherever a winter shower is received, good yields are obtained. As an irrigated crop, its cultivation is widely distributed in the eastern part, particularly in the Gandak Command Area.

Sowing of wheat is done generally in the month of November (although now it continues in the month of December also) and it is harvested in March-April. Under irrigated cultivation, wheat is irrigated 3 to 4 times, but under dry cultivation, no irrigation is provided. The practice of fertilising wheat is becoming more popular, but uncertainty of water supply is a handicap. HYV of wheat and enhanced application of chemical fertilizers require adequate supply of water.

The hectareage under wheat has risen from 87,506 (1955-57) to 2,99,834 (1985-87), i.e. by 343.1 per cent, which is much more than that of rice. This appears due to an increase in the intensity of irrigation, reclamation of waterlogged areas, improvement in drainage conditions and increasing importance of this crop in the diet of the people and due to some other factors. This crop has eclipsed a number

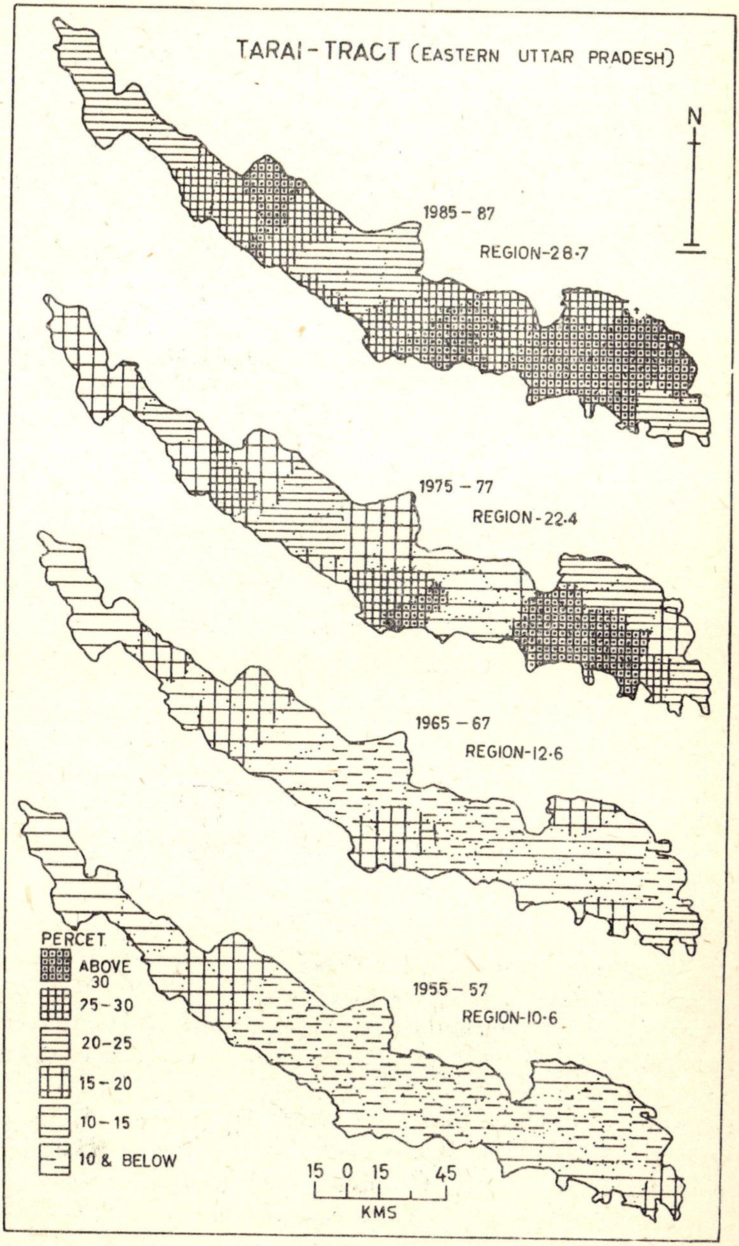

FIG. 5.9: Wheat Cropped Area

of other crops. There are remarkable variations in the percentage increases, some blocks have experienced increases more than 400 per cent.

Table 5.7a and Fig. 5.9 show that in 1955-57, there was no block in the range above 15-20 per cent. There were only 3 blocks in the range 15-20 per cent, while 13 blocks were in the range 10-15 per cent and 15 in the lowest concentration range (10% & below). In 1965-67, this situation improved slightly, as one block stepped up in the range 20-25 per cent and there were 9 and 14 blocks respectively in the range 15-20 and 10-15 per cent. There remained only 7 blocks in the lowest range. In 1975-77, a hike in the wheat hectareage is observed, as there is no block now in the lowest concentration range and in the ranges 10-15 and 15-20 per cent, the number of the blocks is decreased to 5 and 8 respectively. In the range 20-25 per cent and above, the number is now 18. In 1985-87, all the blocks are in the range 20-25 per cent and above. This shows that wheat cultivation has expanded greatly during the latter two decades, which may be ascribed mainly to an increase in the intensity of irrigation and some other factors already narrated.

TABLE 5.7a: Wheat Cropped Area as Precentage of Total Cropped Area

Percentage category	No. of blocks			
	1985-87	1975-77	1965-67	1955-57
Above 30	12	8	—	—
25-30	10	4	—	—
20-25	9	6	1	—
15-20	—	8	9	3
10-15	—	5	14	13
10 & Below	—	—	7	15

The volume of change which shows the change in the relative strength of the area under wheat, discerns that 18.5 per cent of the total cropped area has been added to the wheat hectareage (2.4 per cent in the first, 9.8 per cent in the second and 2.6 per cent in the third decade). On micro-level, wide disparities are observed (Table 5.8b and Fig 5.10). In 15 blocks (Khuniaon, Barhni, Birdpur, Naugarh, Uska, Jogia and Bansi in the central part; and Laxmipur, Brijmanganj, Pharenda, Nichlaul, Mithaura, Maharajganj, Ghughli and Siswa in the eastern part) high increases (to the tune of above 20%) have been registered, which appear mainly due to an increase in the irrigation facilities, particularly in the Gandak Command Area. In 6 blocks

(Haraiya Satgharwa, Pachperwa and Gainsari in the west, Itwa and Bhanwapur in the central and Khadda in the east), moderately high (15-20%) and in 8 (Mihipurwa, Nawabganj, Jamunaha, Hariharpur Rani, Tulsipur and Balrampur in the western, and Nautanwa and Nebua Naurangia in the eastern part), moderate (10-15%) changes have been registered. In most of these blocks, there has been moderate to moderately high increases in the intensity of irrigation and there do occur extensive areas with loamy or clayey loam soils. In 2 blocks, low to very low increases have been experienced, which appears due to the predominance of other crops or inadequate supply of water for irrigation and the inferior soils.

As regards the decennial changes, it may be seen that in the first decade, all the 27 blocks experiencing increases and 4 blocks experiencing decreases were in the low and very low categories of change, while in the second decade, the number of the blocks in these categories decreased to 18 and that in the higher categories, it became 13, but in the third decade, the number in the lower categories again rose to 24 and in the moderate to high categories decreased to seven. This indicates that in the third decade, there has been a registered growth in the wheat hectareage.

TABLE 5.7b: Volume of Change in Wheat Cultivation

Change category (% of the total cropped area involved in change)	No. of blocks			
	1955-57 to 1965-67	1965-67 to 1975-77	1975-77 to 1985-87	1955-57 to 1985-87
Increase				
Above 20	—	3	2	15
15-20	—	6	3	6
10-15	—	4	1	8
5-10	4	13	12	1
5 & Below	23	5	12	1
Decrease				
5 & Below	4	—	1	—

Maize *(Zea mays)*

Maize occupies an area of 44,026 hectares, i.e. 7.5 per cent of the kharif, 6.3 per cent of net and 4.2 per cent of the total cropped area of the region and thus stands second after paddy in the kharif and fourth after paddy, wheat and masoor in the overall cropping pattern of the region.

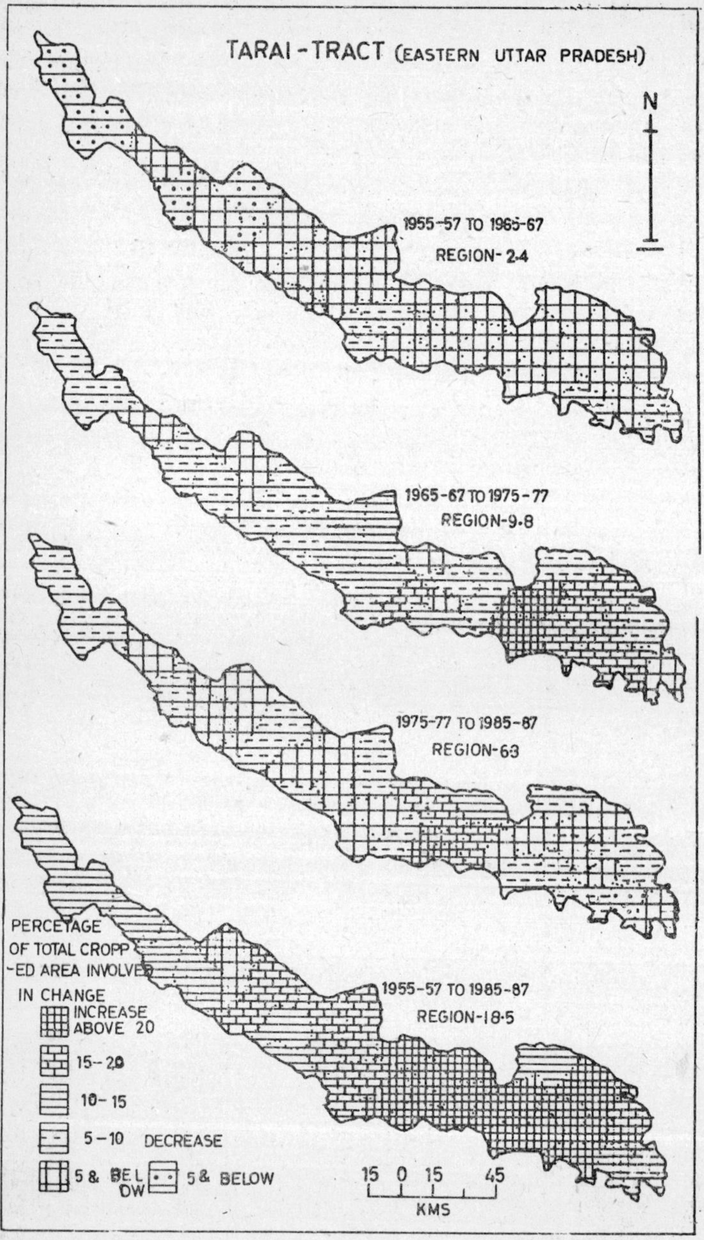

FIG. 5.10: Volume of Change in Wheat Cultivation

Maize flourishes on well drained loamy soils with plentiful moisture well distributed throughout the growing season. A long break in the rainfall or excessive rainfall and floods damage the crop. Water-logging is equally harmful and thus the crop is never sown in lowlying fields.

All these factors affect the spatial distribution of maize, which is not even. The share of maize in the total cropped area varies from nil to 19.9 per cent. Table 5.8a and Fig. 5.11 show that there are 4 blocks (Mihipurwa, Hariharpur Rani, Sirsia and Haraiya Satgharwa, all in the west) in the range 10-20 per cent and 1 in the range 5-10 per cent (Jamunaha, in the west). In all these blocks which belong to the western part of the region, the well drained and light soils best suited to the cultivation of maize occupy considerable area and the rainfall is also moderate. In 3 blocks (Nawabganj, Tulsipur and Balrampur), the proportion varies between 2 and 5 per cent. They also possess in greater or lesser degree the aforesaid favourable conditions. In the remaining 23 blocks very low concentrations (2% & below) are found. In these blocks, the conditions favouring the growth of rice are more in abundance and those best suited for the cultivation of maize are less prevalent. Flooding and water-logging, predominance of lowlying areas with heavy clayey loam soils, preferably being used for rice cultivation and comparatively heavier rainfall in the eastern part, are some of the adverse factors discouraging the cultivation of this crop.

It may be mentioned here that inter-culture is the chief feature of maize cultivation. Several vegetables, various kinds of pulses such as *urad*, *moong* and *arhar*, cereals such as *kodon* and *kokun* and some oil-seeds are sown with it.

The area under maize has decreased from 58,340 hectares (1955-57) to 44,026 hectares (1985-87), showing a percentage decrease of 24.5, but within the region, some parts have experienced increases and some decreases.

Table 5.8a and Fig. 5.11 show that there exists now no block in the range 20-30 per cent and above, while there were 2 to 3 blocks in these ranges in the previous years. In the range 10-20 per cent, the number of the blocks has almost remained static, but it has decreased from 5(1955-57) to 1(1985-87) in the range 5-10 per cent and from 7(1955-57) to 3(1985-87) in the range 2-5 per cent. The number of the blocks in the very low concentration range (2% & below) has increased from 14(1955-57) to 23(1985-87). This indicates a declining trend in the hectareage under maize.

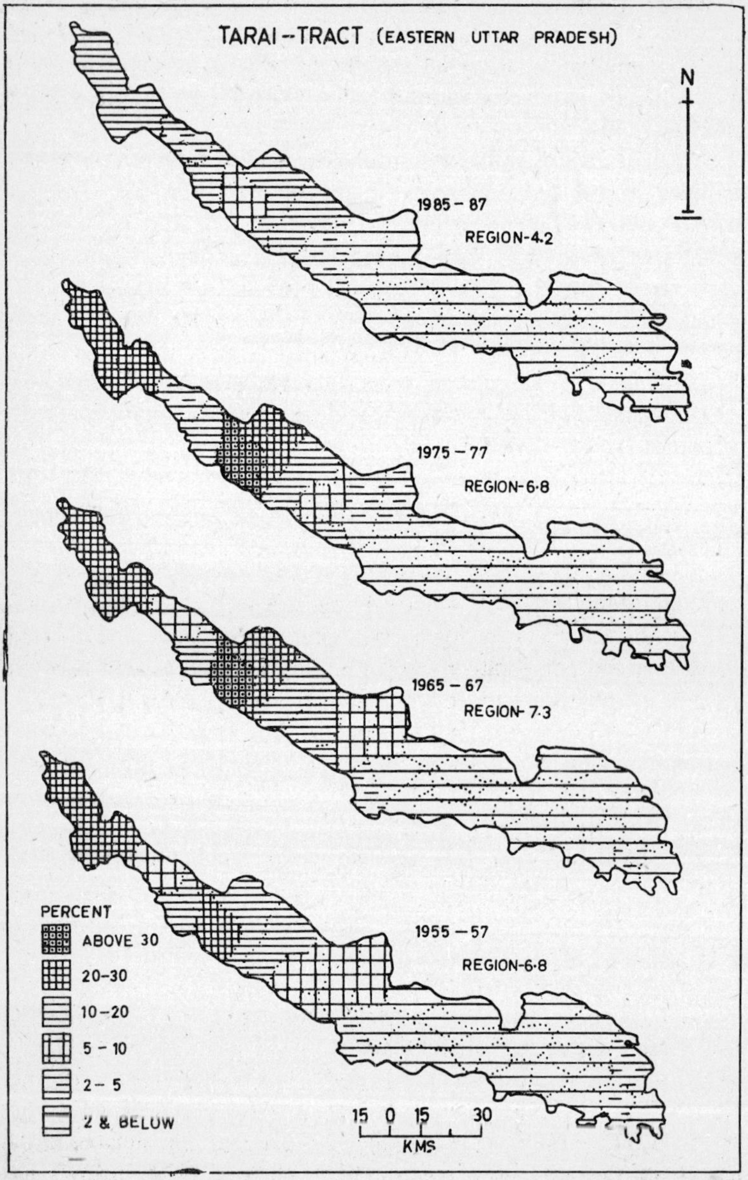

FIG. 5.11: Maize Cropped Area

TABLE 5.8a: Maize Cropped Area as Percentage of Total Cropped Area

Precentage category	No. of blocks			
	1985-87	1975-77	1965-67	1955-57
Above 30	—	1	1	1
20-30	—	2	2	1
10-20	4	3	4	3
5-10	1	1	3	5
2-5	3	6	3	7
2 & Below	23	18	18	14

TABLE 5.8b: Volume of Change in Maize Cultivation

Change category (% of the total cropped area involved in change)	No. of blocks			
	1955-57 to 1965-67	1965-67 to 1975-77	1975-77 to 1985-87	1955-57 to 1985-87
Increase				
Above 4	—	1	—	1
2-4	2	1	1	—
1-2	4	1	—	—
1 & Below	5	8	1	1
Decrease				
Above 4	—	—	5	8
2-4	—	7	2	2
1-2	4	4	8	11
1 & Below	10	9	12	8
No Change	6	—	2	—

As regards the volume of change (Table 5.8b and Fig. 5.12), the region has registered a decrease to the tune of 2.6 per cent of the total cropped area. Only 2 blocks have registered increases and the remaining 29 decreases. Of the 2 blocks experiencing increases, one (Haraiya Satgharwa) is in the range above 4 per cent and one (Bishunpura) in the range 1 per cent and below, while among the blocks experiencing decreases, 8 are in the range above 4 per cent and 2 in the range 2-4 per cent. The highest number of blocks, i.e. 11 is in the range 1-2 per cent and the remaining 8 in the range 1 per cent and below. This indicates that decreases are more prominent.

As regards the decennial changes, very minor variations are seen in the first and second decades. The number of the blocks experiencing increases remained 11 in both the decades, but the number of the blocks experiencing decreases increased from 14 to 20. It is during the third decade that the number of the blocks experiencing increases decreased to 2, while that of experiencing decreases rose to 27 and there was no change in 2 blocks. This again shows that the decreases in the proportion of the area under maize are more prominent during the third decade. This appears due to the low yields of this crop in comparison to rice and also due to the reluctance of the rural masses in accepting it as a staple food.

Small Millets

Kodon

Kodon is the next important kharif cereal which accounts for an area of 14,310 hectares, i.e. 2.4 per cent of the kharif, 2.0 per cent of the net sown and 1.4 per cent of the total cropped area of the region. Thus it ranks third among the kharif cereals and fifth in the overall cropping pattern of the region.

It is generally sown with pulses such as *arhar* and flourishes well when the rainfall is scanty. It may be grown on variety of soils, although light sandy loam soils suit it the most. The spatial distribution of this crop well substantiates this fact. Table 5.9a and Fig. 5.13 show that there are 2 blocks in the concentration range 4-8 per cent (Khuniaon and Itwa) and 4 (Pachperwa, Bhanwapur, Uska and Jogia) in the range 2-4 per cent. In the very low concentration range (2% and below), there are all the remaining 25 blocks. This shows that the blocks with the extensive riverine tracts possessing light sandy soils have higher concentration. Since per hectare yield of this crop is very low, its proportions are very low in most of the blocks of the eastern part. Here rice and sugarcane are preferred to this crop of poor yield.

It may further be seen from Table 5.9a and Fig. 5.13 that the number of the blocks in the high concentration range (above 8%) and that in the moderately high concentration range (4-8%) respectively has decreased from 7(1955-57) to zero (1985-87) and from 10(1955-57) to 2(1985-87). These figures for the range 2-4 per cent are 10 and 4 respectively. It is in the low concentration range (2% and below) that the number has increased from 4(1955-57) to 25(1985-87). This also shows that the hectareage under this crop has decreased immensely, i.e. from 44,060 in 1955-57 to 14,310 in 1985-87, i.e. by 67.5 per cent.

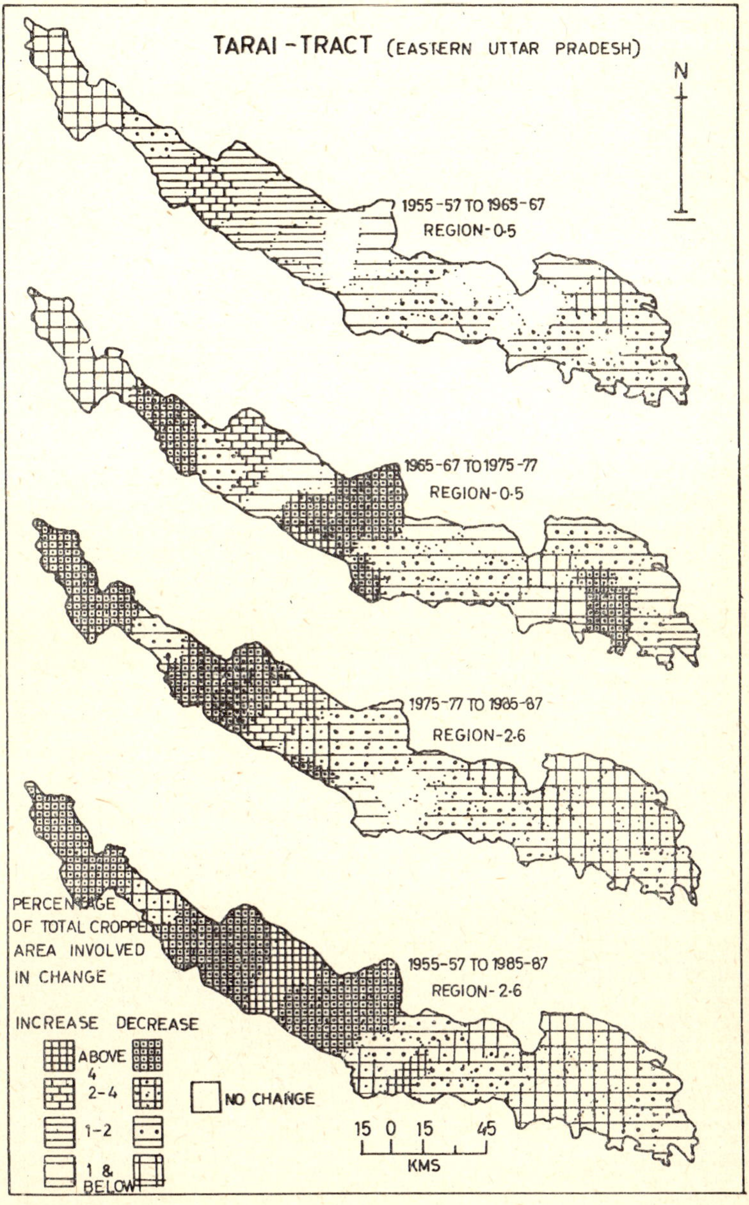

FIG. 5.12: Volume of Change in Maize Cultivaiton

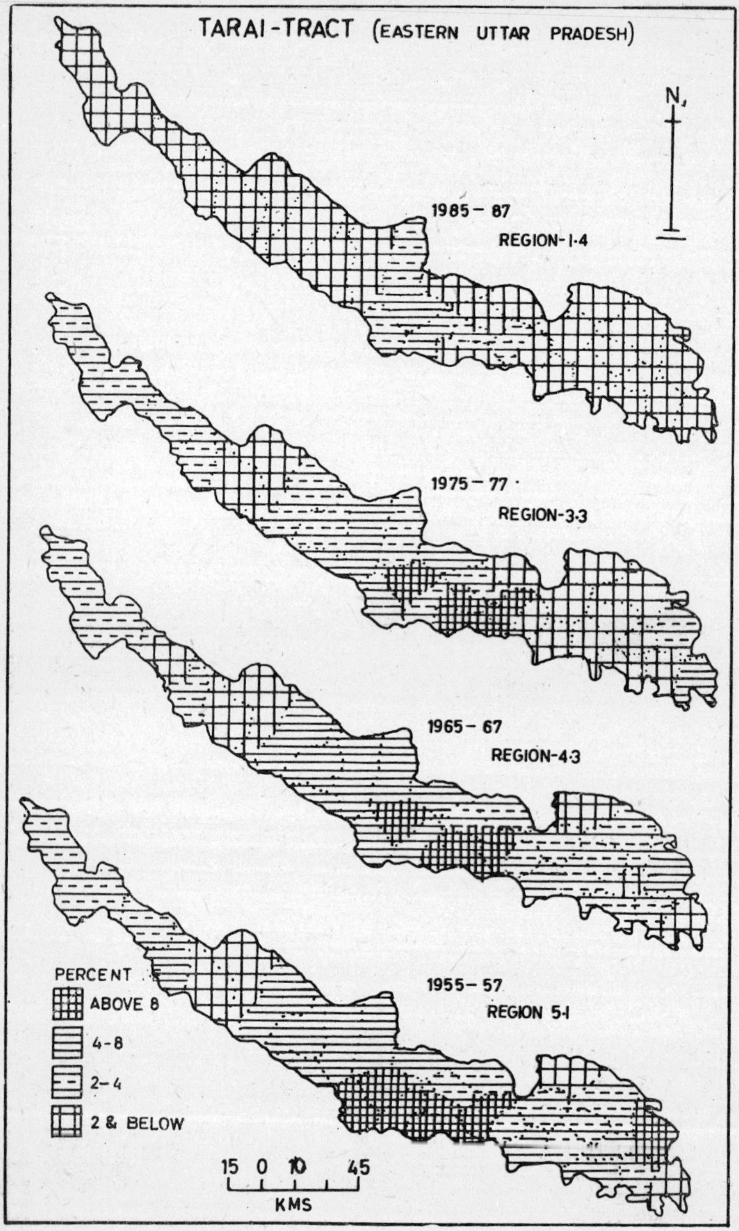

FIG. 5.13: Kodon Cropped Area

The volume of change (Table 5.9b and Fig. 5.14) in the area occupied by this crop amounts to 3.7 per cent (–0.8% in the first, –1.0% in the second and -1.8% in the third decade). The blocks experiencing decreases number 30 and that experiencing increase is only one. Of the 30 blocks experiencing decreases, 10 and 14 respectively are in the range above 4 per cent and 2.4 per cent. There are only 4 and 2 blocks respectively in the range 1-2 per cent and 1 per cent and below. This shows that there are considerable decreases in the hectareage under this crop which is an inferior cereal and has very low yields.

The trend in the *kodon* cultivated area has not been uniform throughout the study period. For example, the number of the blocks experiencing increases during the first decade was 4 which rose to 8 during the second decade, but it again decreased to 4 during the third decade. The number of the blocks experiencing decreases respectively was 27, 23 and 17 during these decades. The number of the blocks in the higher categories of decreases has been rising, which shows a general reduction in the area under this crop, which may be ascribed to its low yields and people's less liking for it as a foodcrop in comparison to crops such as rice and maize.

Other Minor Cereals

Among other minor cereals are mainly those of rabi, viz. barley, wheat mixed with barley, barley mixed with gram and wheat mixed with gram. *Jwar* and *Bajra* are quite insignificant.

Barley (Hardeun Vulgara)

Barley is grown on an area of 3,824 hectares, i.e. 0.8 per cent of the rabi, 0.5 per cent of the net sown and 0.4 per cent of the total cropped area of the region. Within the region, its share varies from almost negligible to 1.6 per cent. There are only 2 blocks (Table 5.10a and Fig. 5.15) in the concentration range 1-2 per cent and all the remaining 29 blocks are in the low concentration range (1% & below). This shows the insignificance of this crop which now rarely forms the diet of the poor only and which is sown on light poor soils not so suitable for the superior rabi crops.

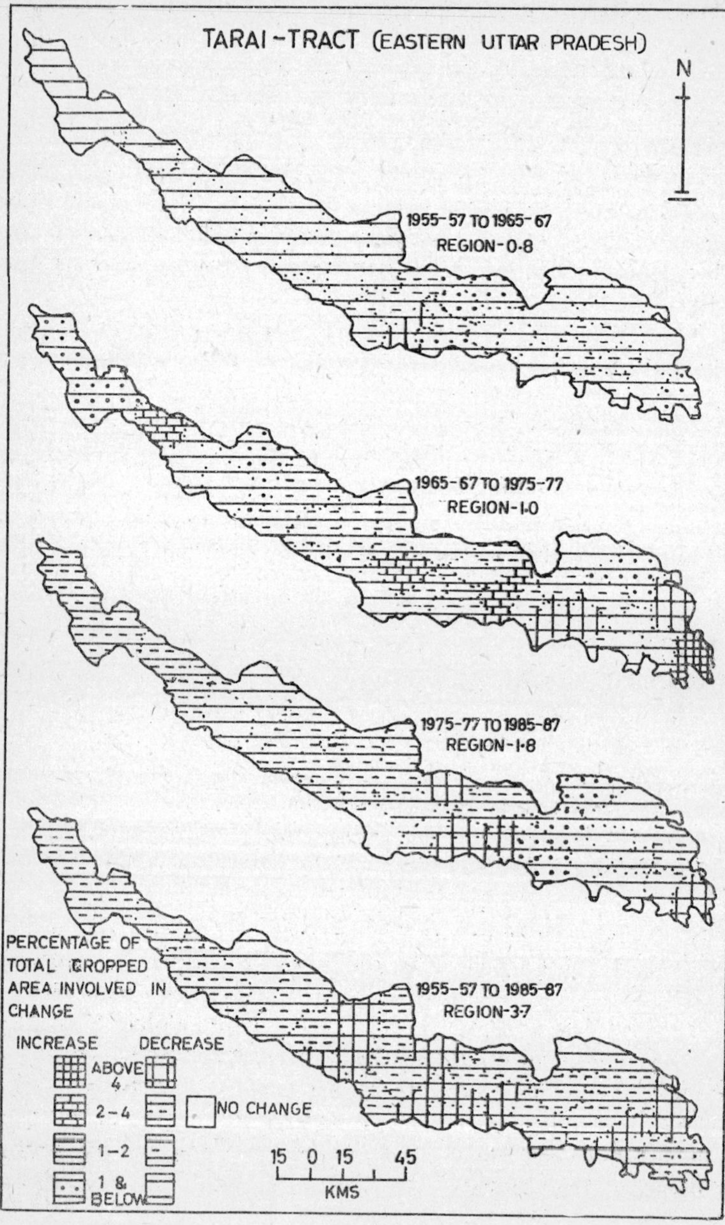

FIG. 5.14: Volume of Change in *Kodon* Cultivation

TABLE 5.9a: Kodon Cropped Area as Percentage of Total Cropped Area

Percentage category	No. of blocks			
	1985-87	1975-77	1965-67	1955-57
Above 8	—	4	3	7
4-8	2	5	12	10
2-4	4	9	10	10
2 & Below	25	13	6	4

TABLE 5.9b: Volume of Change in Kodon Cultivation

Change category (% of the total cropped area involved in change)	No. of blocks			
	1955-57 to 1965-67	1965-67 to 1975-77	1975-77 to 1985-87	1955-57 to 1985-87
Increase				
Above 4	—	1	—	—
2-4	—	3	—	—
1-2	—	1	—	—
1 & Below	4	3	4	1
Decrease				
Above 4	1	2	5	10
2-4	3	4	7	14
1-2	5	10	3	4
1 & Below	18	7	11	2
No change	—	—	1	—

Table 5.10a and Fig. 5.15 further show that significant decreases have been registered in the hectareage under this crop (from 66,213 hectares in 1955-57 to 3,824 hectares in 1985-87, showing a decrease by 94.2%). There was no block in the low concentration range (1% & below) in 1955-57, but the number in this category rose to 29 in 1985-87. The number of the blocks in the range 1-2 per cent rose from 1 in 1955-57 to 8 in 1975-77, but again decreased to 2 in 1985-87, as 6 of these blocks stepped down to the lower range. In the range 2-3 per cent, the number of the blocks almost remained static till 1975-77 and than became zero in 1985-87. In the range 3-9 per cent and above, there have occurred very significant reductions, i.e. from 23 blocks in 1955-57 to 6 in 1975-77 and then zero in 1985-87.

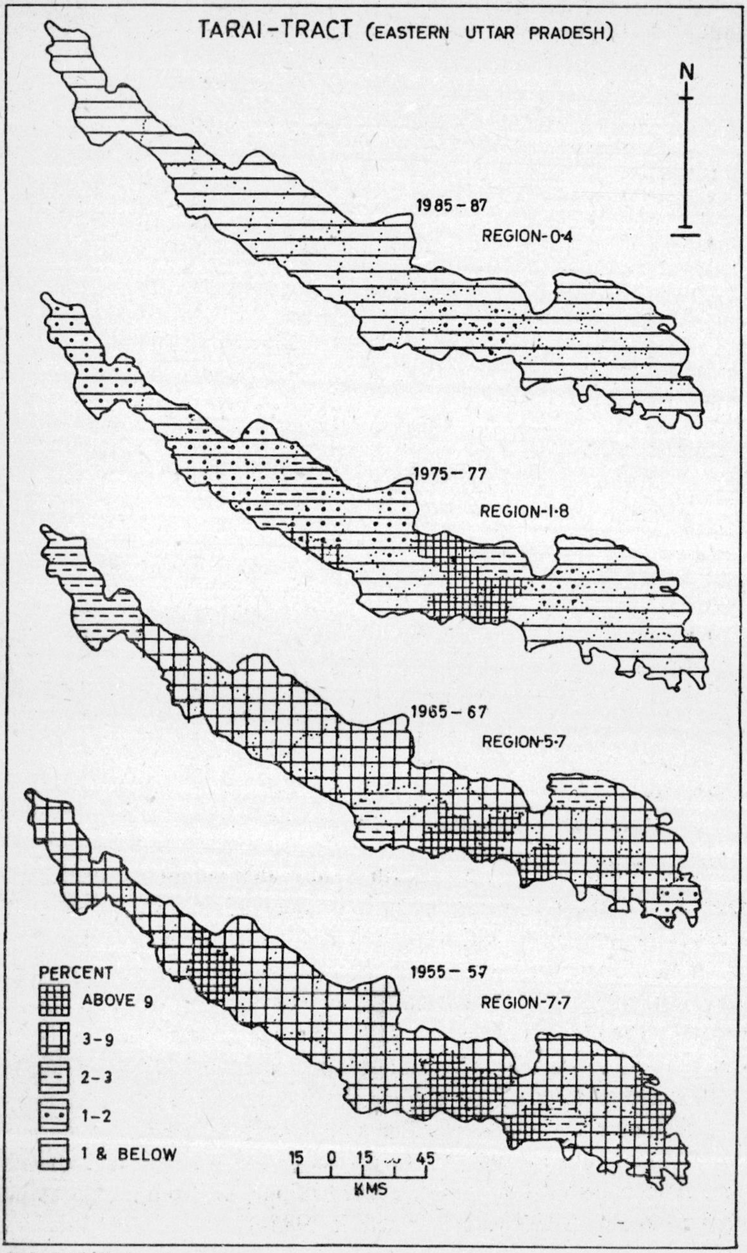

FIG. 5.15: Barley Cropped Area

TABLE 5.10a: Barley Cropped Area as Percentage of Total Cropped Area

Percentage category	No. of blocks			
	1985-87	1975-77	1965-67	1955-57
Above 9	—	5	4	7
3-9	—	1	22	21
2-3	—	2	3	2
1-2	2	8	1	1
1 & Below	29	15	1	—

TABLE 5.10b: Volume of Change in Barley Cultivation

Change category (% of the total cropped area involved in change)	No. of blocks			
	1955-57 to 1965-67	1965-67 to 1975-77	1975-77 to 1985-87	1955-57 to 1985-87
Increase				
2-5	1	—	—	—
2 & below	1	1	1	—
Decrease				
Above 10	1	—	1	6
5-10	6	13	2	17
2-5	4	8	3	7
2 & below	16	9	14	1
No Change	3	—	—	—

The volume of change (Table 5.10b and Fig. 5.16) amounts to 7.3 per cent during the span of three decades (-2.0% during the first, –3.9 during the second and –1.4 per cent during the third decade). As many as 6 blocks have experienced decrease to the tune of more than 10 per cent and 17 between 5 and 10 per cent. There are 7 blocks in the range of 2-5 per cent and 1 in the range of 2 per cent and below. The decreases are more prominent during the second decade than during the first and the third decades, as the number of the blocks experiencing decreases in the range 5-10 per cent and above rose from 6(1955-57 to 1965-67) to 13(1965-67 to 1975-77). It again decreased to 2 during the third decade. The same trend may be observed in the moderate range of 2 to 5 per cent. In the low range (2% & below), the number of the blocks decreased from 16 in the first decade to 9 in the second decade, although it again increased to 24 in the third decade.

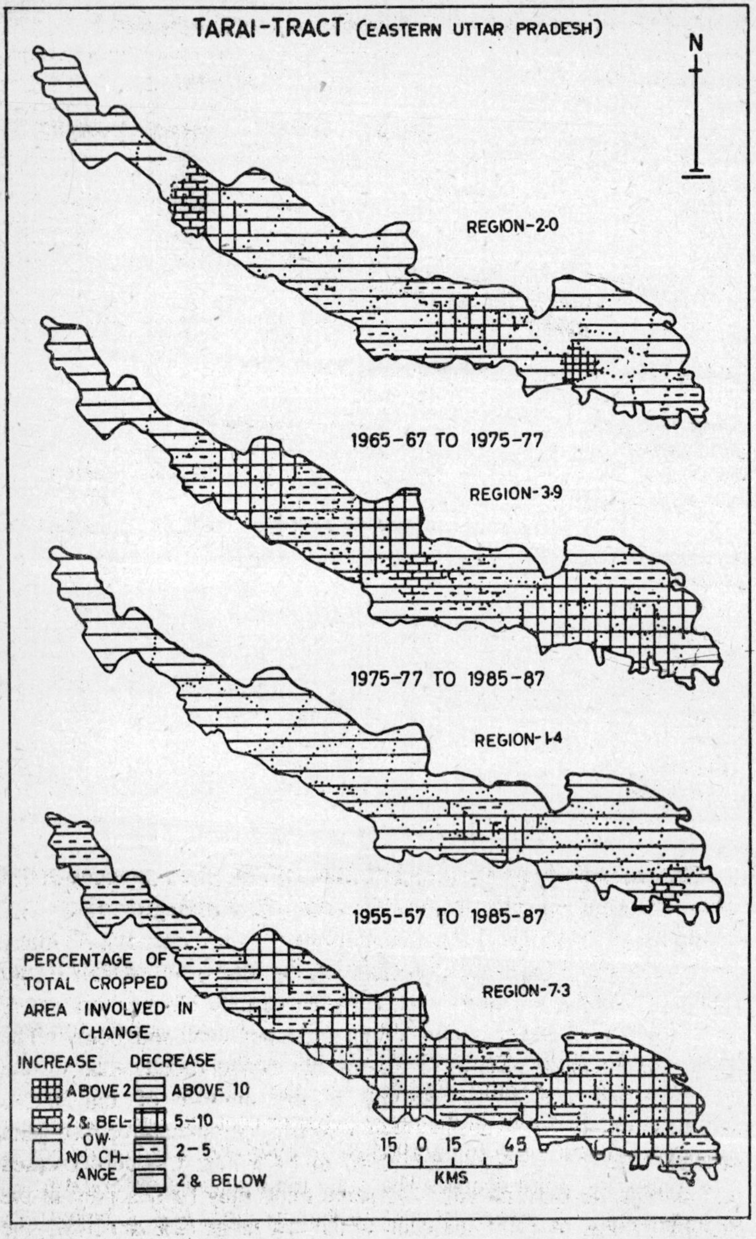

FIG. 5.16: Volume of Change in Barley Cultivation

This shows that the decreases are more significant during the second decade than that of the first and the third decades, which is due to a phenomenal increase in the area of wheat during this decade.

Wheat Mixed with Barley

It covers an area of 3,565 hectares or 0.8 per cent of the rabi, 0.5 per cent of the net sown and 0.3 per cent of the total cropped area of the region and thus is quite an unimportant crop. The climatic requirements of the crop are more or less the same as in the case of wheat, although it grows better on light soils as compared to wheat. Generally inferior soils with inadequate supply of water are devoted to its cultivation.

The proportion of the area under this crop varies (Table 5.11a and Fig. 5.17) from almost negligible to 2.3 per cent of the total cropped area. In 27 blocks, the proportions are very low (1% & below). It is only in 3 blocks that they vary from 1 to 2 per cent. Only 1 block is in the range 2-3 per cent. It is towards west that the concentration is a bit more, which is because of the more prevalence of light soils, less rainfall and inadequate supply of water for irrigation.

Table 5.11a and Fig. 5.17 show that the distributional pattern of this crop was quite different in the previous years. In 1955-57, the hectareage under this crop was 48,951, which has decreased to 3,565 in 1985-87, i.e. by 92.7 per cent during the span of three decades. In 1955-57, there were 11 blocks in the high concentration range (above 7%) as against 1 in 1975-77 and zero in 1985-87. In the moderately high concentration range (3-7%), the number of the blocks decreased from 14 in 1955-57 to 5 in 1975-77 and nil in 1985-87. In the range 2-3 per cent, not very significant changes are observed. Quite phenomenal changes are seen in the lowest range (1% & below), where the number of the blocks has increased from 4 in 1955-57 to 27 in 1985-87. The number of the blocks in the range 1-2 per cent was zero in 1955-57, two in 1965-67, 9 in 1975-77 and 3 in 1985-87. This discussion reveals that there is a general reduction in the hectareage under this crop.

The volume of change (Table 5.11b and Fig. 5.18) in the region is to the tune of 5.4 per cent (–1.1% during the first, –3.0 per cent during the second and –1.3 per cent during the third decade). No block has experienced an increase. In 7 blocks, the decreases are high (above 8%), in 16 moderate (4-8%), in 3 low (2-4%) and in 4 very low (2% & below). In one block, the position has remained constant. This shows

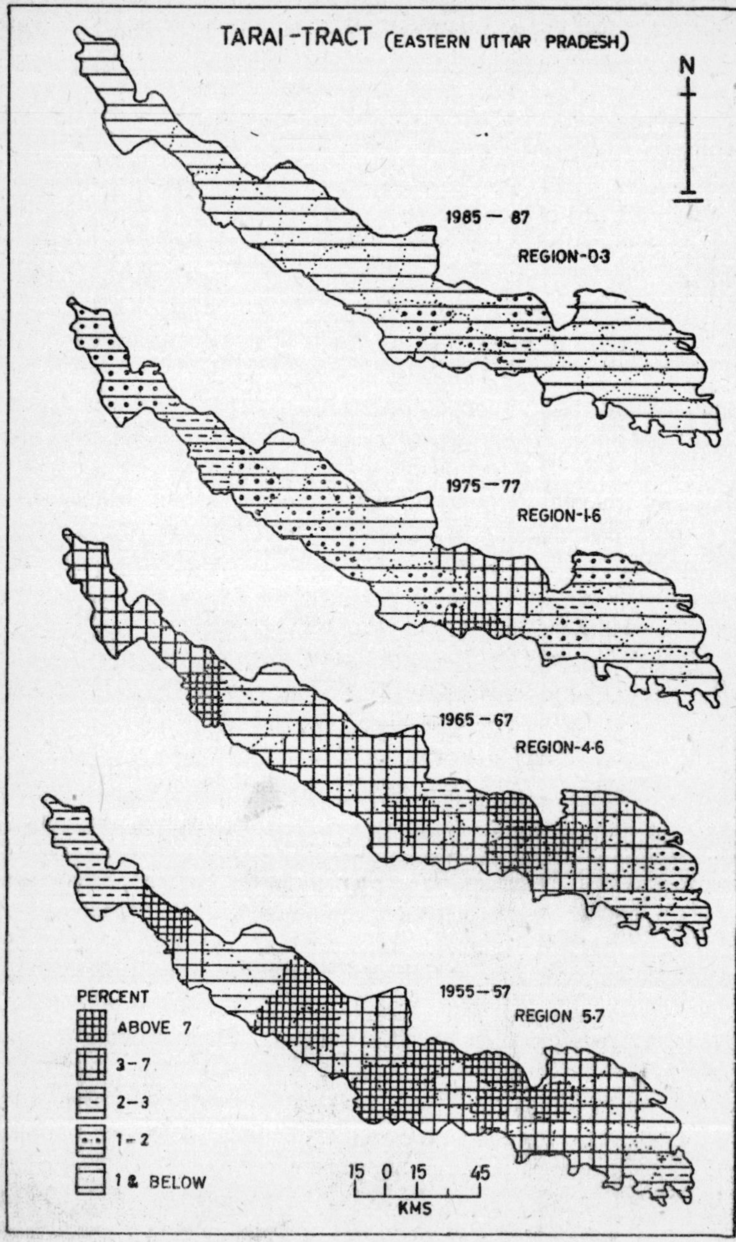

FIG. 5.17: Wheat Mixed with Barley Cropped Area

that this crop has also undergone a great reduction in its hectareage due to the increasing importance of wheat, pulses and some vegetables.

TABLE 5.11a: Wheat Mixed with Barkey Cropped Area as Percentage of Total Cropped Area

Percentage category	No. of blocks			
	1985-87	1975-77	1965-67	1955-57
Above 7	—	1	6	11
3-7	—	5	17	14
2-3	1	3	4	2
1-2	3	9	2	—
1 & below	27	13	2	4

TABLE 5.11b: Volume of Change in Wheat Mixed with Barley Cultivation

Change category (% of the total cropped area involved in change)	No. of blocks			
	1955-57 to 1965-67	1965-67 to 1975-77	1975-77 to 1985-87	1955-57 to 1985-87
Increase				
Above 8	—	—	—	—
4-8	1	—	—	—
2-4	—	—	—	—
2 & below	7	6	1	—
Decrease				
Above 8	—	1	—	7
4-8	2	9	3	16
2-4	7	9	4	3
2 & below	13	6	22	4
No change	1	—	1	1

Among other insignificant cereals are wheat mixed with gram and barley mixed with gram. Wheat mixed with gram occupies an area of 2,106 hectares (0.2% of the total cropped area). There exists no hectareage under this crop in 9 blocks and in the remaining blocks also it is quite negligible.

Barley mixed with gram occupies an area of 6,035 hectares, i.e. 1.4 per cent of the rabi, 0.9 per cent of the net sown and 0.6 per cent of the total cropped area of the region. Within the region, its share in the total cropped area varies from almost negligible to 2.1 per cent. In 18 blocks, its proportion is below 0.5 per cent, in 9 between 0.5 and 1.0 per cent, in 3 between 1.0 and 1.5 and in only 1 above 2.0 per cent. The distributional pattern of this crop exhibits almost the same pattern as exhibited by wheat mixed with barley.

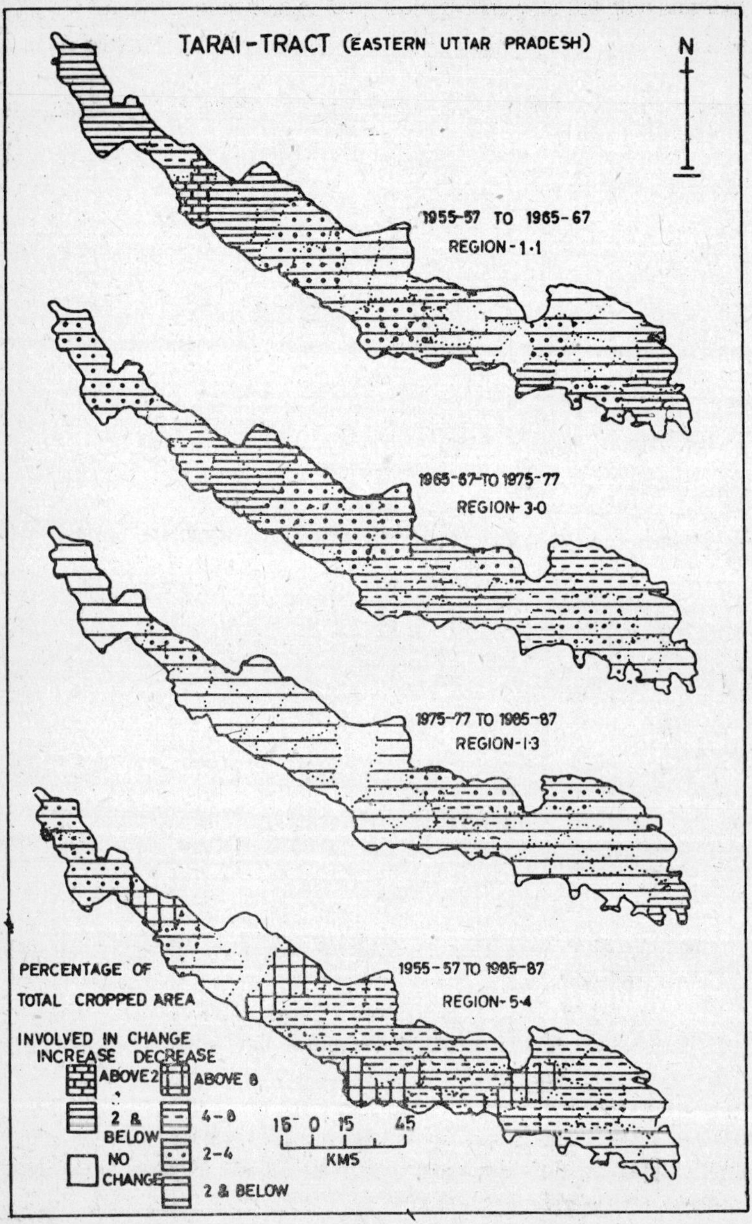

FIG. 5.18: Volume of Change in Wheat Mixed with Barley Cultivation

The climatic requirements of this crop are the same as those of wheat mixed with gram. It grows well in *khadar* areas. Clayey loam to loamy soils are very suitable for its cultivation. It is grown in dry conditions. It forms a part of the diet of the people of the poor classes only and that, too, very rarely. The area under this crop has also decreased greatly, which is due to the preference being given to wheat in place of this crop.

PULSES

The region grows a variety of pulses. *Arhar, moong* and *urad* are the main kharif and gram, peas and *masoor* are the main rabi pulses. The kharif and rabi pulses account for 1.5 and 8.4 per cent of the total cropped area respectively. Together they contribute 9.9 per cent of the total cropped area of the region.

It may be noted here that pulses constitute a very important element in the cropping pattern of the region. They are a necessity in the rural economy of the region, as they not only supply the major portion of the protein requirement of the rural communities, but also possess root-nodules and the bacteria which fix various amounts of atmospheric nitrogen and improve the soil fertility. They are cheap but good sources of vegetable protein and are a good substitute for the paucity of costly animal protein in the farmers' food.[3] They contain enough carbohydrates and starch and also small amount of fat. They thus provide nutritious food not only for human beings, but also for livestock. They not only play an important role in increasing the nitrogen contents in the soil, but also help in increasing the soil fecundity and reducing evaporation and soil erosion. The well developed root-system of pulses increases the aeration of soils also. Their green manures are also beneficial. Ecologically, the pulse crops generally do well with low moisture on light soils. They comparatively require less care.

Table 5.12a and Fig. 5.19 exhibit the spatial distribution of the area under pulses, which reveal considerable micro-level disparities (the proportions varying from 2.4% in Bishunpura to 18.7% in Jamunaha). In 5 blocks (Nawabganj, Jamunaha, Haraiya Satgharwa, Tulsipur and Balrampur, all in the west), its concentration is high (above 15%) and in 6 (Mihipurwa, Hariharpur Rani, Sirsia, Pachperwa, Gainsari and Nautanwa, 5 belonging to the western and one to the eastern part), it is moderately high (10-15%). As many as 12 blocks have moderate concentration (5-10%) and 8 have low concentration (5% & below). It is thus clear that high or moderately high concentrations are generally towards west where the ecological conditions favouring the cultivation of pulses are found better.

Table 5.12a and Fig. 5.19 show that the area under pulses has increased slightly, i.e. from 85,987 hectares in 1955-57 to 1,02,293 hectares (by 18.9%), but at micro-level, a mixed trend is seen, i.e. in some blocks, the hectareage has increased and in some it has decreased. An increase in the number of the blocks in the low concentration range (from 3 in 1955-57 to 8 in 1985-87) and the corresponding decrease in the moderately high concentration range (from 15 in 1955-57 to 6 in 1985-87) reveal that the proportions of the area under pulses have decreased in these blocks, but an increase in the high concentration range (from 1 in the 1955-57 to 5 in 1985-87) shows that the proportion in these blocks has increased.

The volume of change (Table 5.12b and Fig. 5.20) in the region as a whole is to the tune of 0.2 per cent only. In all, 19 blocks have registered decreases ranging from low (1% & below) to high (above 4%). Only 12 blocks have registered increases. Of the 19 blocks experiencing decreases, the number is almost uniform in different ranges, i.e. above 4 per cent, 2-4 per cent, 1-2 per cent and 1 per cent and below, while among the blocks experiencing increases, 7 are in the range 2-4 per cent and above.

As mentioned earlier, the region grows a variety of pulses. It would be worthwhile to discuss the spatial distribution of some important pulses.

Kharif Pulses

The kharif pulses which comprise of *arhar, urad* and *moong*, account for 15,252 hectares of area, i.e. 2.6 per cent of the kharif, 2.2 per cent of the net sown and 1.5 per cent of the total cropped area of region. They are, thus, quite insignificant when compared to the rabi crops. Among them, *arhar* is by far the most important, claiming 11,861 hectares of area, i.e. 2.00 per cent of the kharif, 1.7 per cent of the net sown and 1.1 per cent of the total cropped area of the region. It is sown mixed with *kodon, saman* and maize also. If these areas are also taken into consideration, the hectareage under *arhar* would be much higher than this. *Urad* and *moong* together claim only 3,164 hectares, i.e. 0.5 per cent of the kharif, 0.5 per cent of the net sown and 0.3 per cent of the total cropped area of the region.

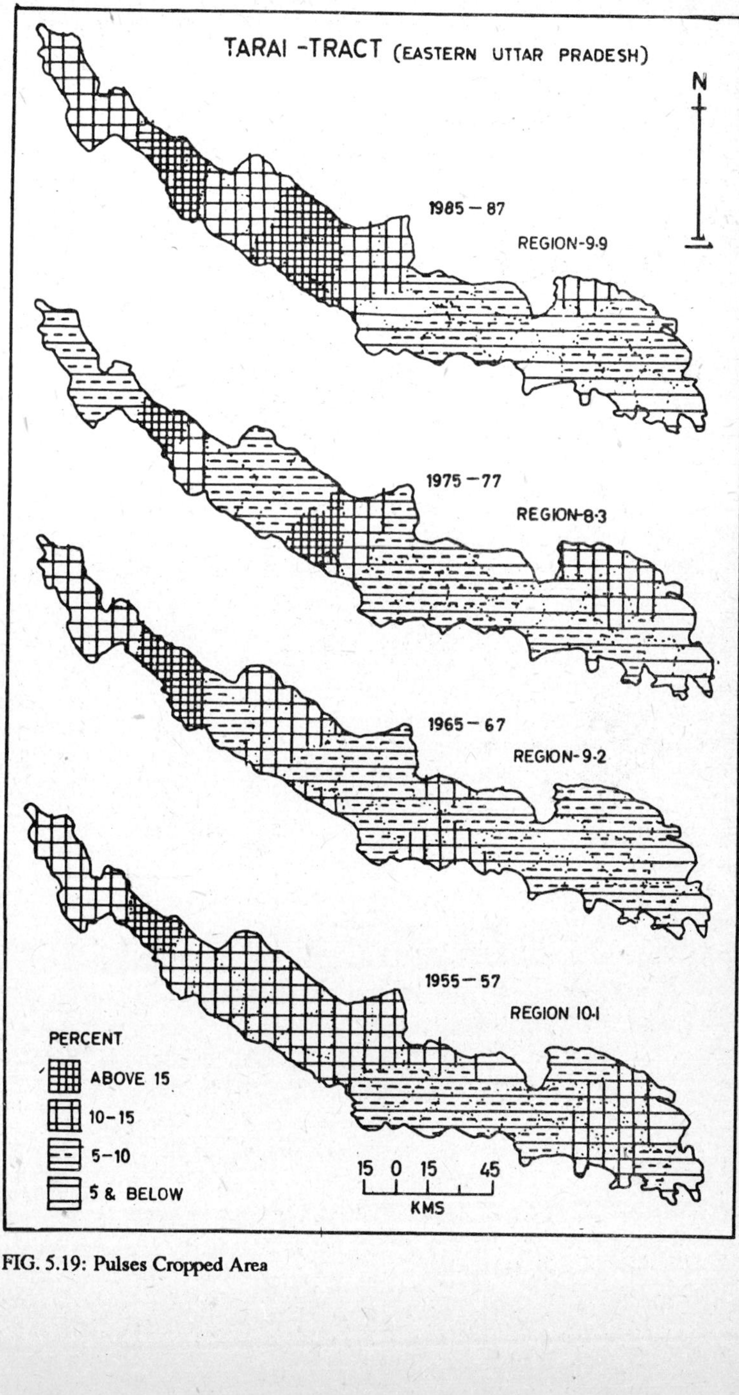

FIG. 5.19: Pulses Cropped Area

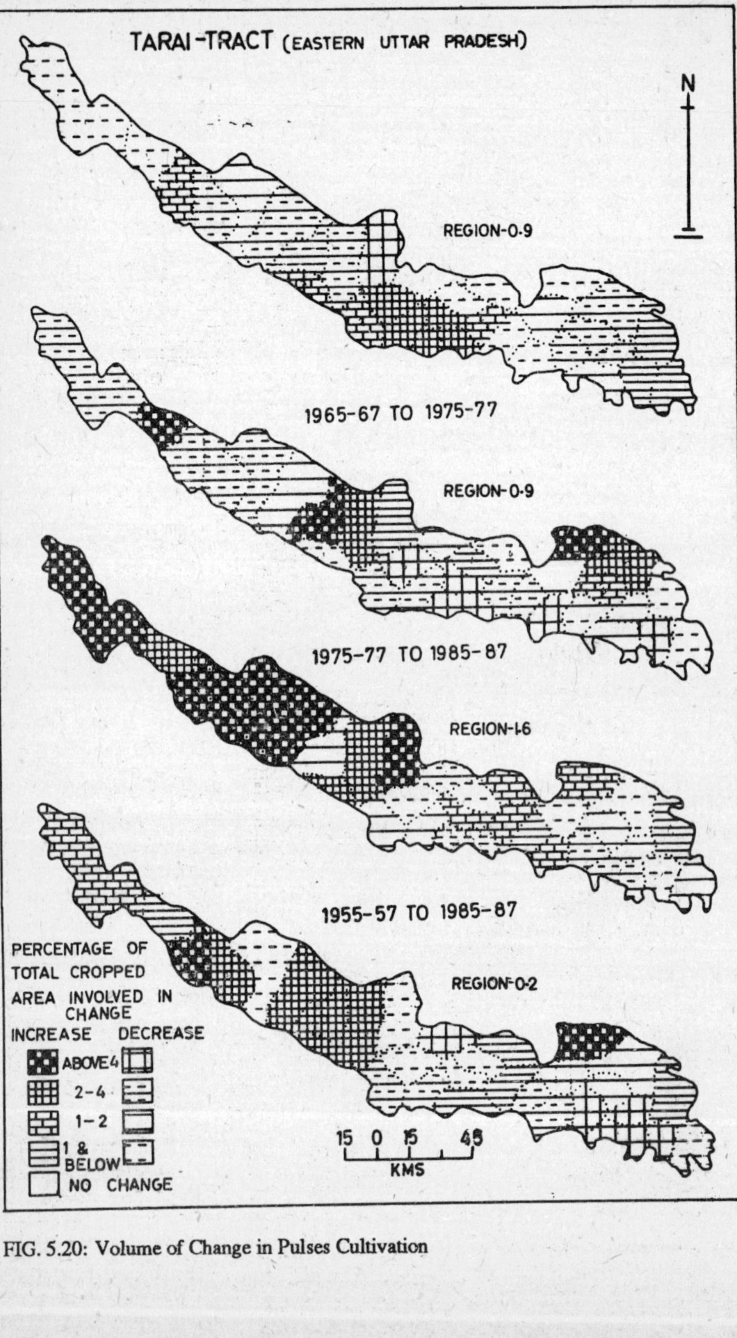

FIG. 5.20: Volume of Change in Pulses Cultivation

TABLE 5.12a: Pulses Cropped Area as Percentage of Total Cropped Area

Percentage category	No. of blocks			
	1985-87	1975-77	1965-67	1955-57
Above 15	5	2	2	1
10-15	6	5	8	15
5-10	12	17	17	12
5 & below	8	7	4	3

TABLE 5.12b: Volume of Change in Total Pulses

Change category (% of the total cropped area involved in change)	No. of blocks			
	1955-57 to 1965-67	1965-67 to 1975-77	1975-77 to 1985-87	1955-57 to 1985-87
Increase				
Above 4	—	2	6	2
2-4	4	2	3	5
1-2	4	1	5	1
1 & below	3	2	2	14
Decrease				
Above 4	1	5	—	5
2-4	8	8	3	5
1-2	3	5	6	4
1 & below	7	5	6	5
No Change	1	1	—	—

The spatial distribution of the kharif pulses within the region reveals that in 15 blocks, the share of these pulses in the total cropped area is below 1 per cent. In 10, it is between 1 and 2 per cent and only in 6 (all of which belong to the western part where the hectareage under arhar is higher), it is above 2 per cent. In Haraiya Satgharwa and Tulsipur, the northern parts of which have high concentration of *arhar* cultivation, the proportion is above 4 per cent. The blocks with low concentration generally belong to the eastern part. The area under kharif pulses has decreased slightly i.e. from 16,048 hectares in 1955-57 to 15,252 hectares in 1985-87, and their share in the total cropped area from 1.9 per cent to 1.5 per cent.

Table 5.13a shows that the number of the blocks in the low concentration category has increased from 10 (1955-57) to 15

(1985-87), but in the high concentration category, there is a reduction from 11 to 6. The number of the blocks in the moderate concentration category has remained static almost. This indicates a decreasing trend in the hectareage under kharif pulses. This may further be substantiated by the analysis of the volume of change (Table 5.13b). The region as a whole has experienced a decrease of 0.4 per cent (0.3 per cent decrease in the first, 0.1 increase in the second, and 0.2 per cent decrease in third decade). As many as 16 blocks have experienced decreases, almost equal in all the ranges of 2-4, 1-2 and 1 per cent and below. Fifteen blocks have experienced increases, of which 9 are in low category (1% & below), 5 in the moderate (1-2%) and only 1 in the range 2-4 per cent.

Table 5.13b reveals that the number of the blocks experiencing increases decreased from 11 (during 1955-57 to 1965-67) to 6 (during 1965-67 to 1975-77) and again increased to 18 (during 1975-77 to 1985-87). On the other hand, the number of blocks, experiencing decreases increased from 20 to 25 and then decreased to 13 respectively. This shows that increases were more significant during the third and first decade and decreases during the second decade.

Rabi Pulses

The rabi pulses which comprise of *masoor* (lentil), pea and gram etc., occupy more important position than the kharif pulses, as they cover 87,041 hectares of land, i.e. 19.5 per cent of the rabi, 12.4 per cent of the net sown and 8.4 per cent of the total cropped area of the region. The spatial distribution shows micro-regional variations. In 10 blocks (Mihipurwa, Nawabganj, Jamunaha, Hariharpur Rani, Sirsia, Haraiya Satgharwa, Tulsipur, Gainsari, Pachperwa and Balrampur, all belonging to the Bahraich and Gonda districts in the west), there is high concentration (above 10%). In 7 blocks (Khuniaon, Uska, Jogia, Bansi, Nautanwa, Nichlaul and Mithaura, belonging to the central and eastern parts), their concentration is moderate (5-10%) and in the remaining 14, it is low (5% & below). The higher concentration towards west is due to the predominance of light loamy soils and comparatively drier conditions which are suitable for the rabi pulses such as masoor, gram etc. These pulses are sown in the fields vacated after early paddy is harvested.

TABLE 5.13a: **Area Under Kharif Pulses as Percentage of Total Cropped Area**

Percentage category	No. of blocks			
	1985-87	1975-77	1965-67	1955-57
Above 2	6	5	7	11
1-2	10	9	12	10
1 & below	15	17	12	10

TABLE 5.13b: **Volume of Change in Kharif Pulses Cultivation**

Change category (% of the total cropped area involved in change)	No. of blocks			
	1955-57 to 1965-67	1965-67 to 1975-77	1975-77 to 1985-87	1955-57 to 1985-87
Increase				
Above 4	—	1	—	—
2-4	—	—	1	1
1-2	—	1	3	5
1 & below	11	4	14	9
Decrease				
Above 4	—	—	1	—
2-4	—	2	—	5
1-2	2	4	1	5
1 & below	16	17	9	6
No Change	2	2	2	—

Table 5.14a shows that in 1955-57, there were 6 blocks in the high concentration category, which number decreased to 3 in 1975-77, but it again increased to 10 in 1985-87. The number of the blocks in the moderate category, remained the same till 1975-77, but it decreased to 7 in 1985-87. The number of the blocks in the low category, has increased from 6 in 1955-57 to 14 in 1985-87.

TABLE 5.14a: **Area Under Rabi Pulses as Percentage of Total Cropped Area**

Percentage category	No. of blocks			
	1985-87	1975-77	1965-67	1955-57
Above 10	10	3	5	6
5-10	7	19	21	19
5 & below	14	9	5	6

TABLE 5.14b: Volume of Change in Rabi Pulses Cultivation

Change category (% of the total cropped area involved in change)	No. of blocks			
	1955-57 to 1965-67	1965-67 to 1975-77	1975-77 to 1985-87	1955-57 to 1985-87
Increase				
Above 4	—	1	7	2
2-4	5	4	2	4
1-2	3	2	3	2
1 & below	3	4	3	5
Decrease				
Above 4	1	4	1	2
2-4	5	7	—	6
1-2	5	5	8	4
1 & below	8	4	7	6
No change	1	—	—	—

The volume of change (Table 5.14b) for the region is to the tune of 0.2 per cent, but there are spatial variations. For example, 13 blocks have registered increases (2 above 4%, 4 from 2 to 4%, 2 from 1 to 2% and 5 from 1% & below). The number of the blocks experiencing decreases is 18 (2 above 4%, 6 from 2 to 4%, 4 from 1 to 2% and 6 from 1% & below). As regards the decennial changes, the position in the first and second decades almost remained the same, as the number of the blocks registering increases and decreases remained static, but during the third decade, the number of the blocks registering increases increased to 15 and that of registering decreases decreased to 16. This indicates a slight rise in the hectareage under rabi pulses.

Masoor (Lentil)

Masoor is by far the most important crop not only among the rabi but also among all the pulses grown in the region, as it covers an area of 48,764 hectares or 10.9 per cent of the rabi, 7.0 per cent of the net sown and 4.7 per cent of the total cropped area. It ranks third among the foodgrains grown in the region.

Although, *masoor* preferably requires fertile soils with considerable moisture, it is generally sown on a variety of soils, viz., from light loamy soils to clayey loam or clayey soils. There is, however, a usual practice to sow it in the fields vacated by paddy. It is grown as dry and unirrigated crop with the help of available moisture in the sub-soil. The crop receives no further attention by way of weeding and inter-culture. The per hectare yield is also low.

As regards the distributional pattern of the area under *masoor*, it may be seen from Table 5.15 and Fig. 5.21 that its share in the total cropped area is above 4 per cent in 13 blocks, 10 of which belong to the western part. It is from 2 to 4 per cent in 4 blocks which belong to the eastern part and in 5 blocks, it varies from 1 to 2 per cent. It is very low (1% & below) in the remaining 9 blocks which belong to the middle and the adjoining eastern part. This anomalous distribution of the crop shows the wider range of adaptability of this crop.

Table 5.15 shows that the importance of this crop during the decade 1975-77 to 1985-87 has remained almost the same, as the number of the blocks in the higher and lower categories has remained almost static.

Gram

Gram which is rarely used as pulse, is the next important pulse crop of the rabi season, claiming 25,286 hectares or 5.7 per cent of the rabi, 3.6 per cent of the net sown and 2.4 per cent of the total cropped area of the region. It also has higher concentration towards west, as almost all the blocks (Nawabganj, Jamunaha, Tulsipur, Gainsari, Pachperwa, except Uska) in the high concentration range (above 4 per cent) hail from this part. Similarly, moderately high (2-4%) concentration is found in the western and middle parts, while the blocks with low concentration are confined generally to the eastern part, as of the 10 blocks in this category, 8 belong to this part. The increasing importance of this crop during the decade 1975-77 to 1985-87 is evidenced by the fact that the number of the blocks in the higher concentration range has increased from 6 to 14 and that in the moderate and low ranges has decreased from 25 to 17.

Pea

Pea such as gram is also rarely used as a pulse in the region. The immature pods are used as fresh vegetables or are roasted. Dry grains are used for stock-feeding or turned into flour. Well drained heavy soils such as loams or clayey loam with lime content are more suitable for it. It is often grown on the fields occupied by rice and maize etc. in

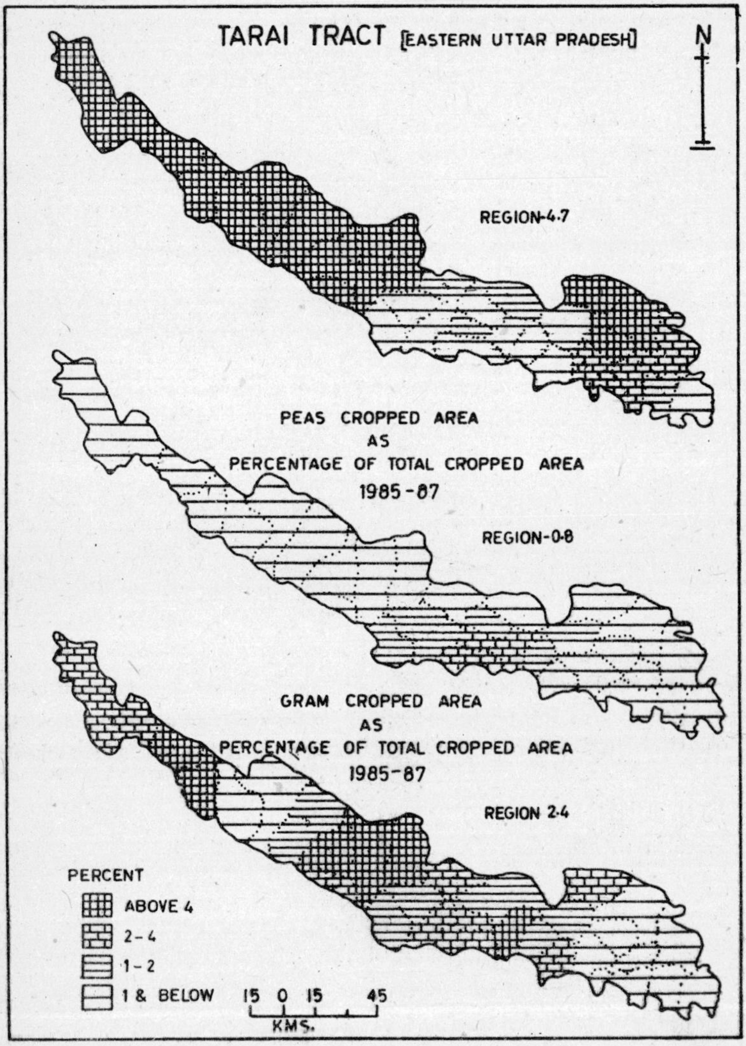

FIG. 5.21: Lentil, Pea and Gram Cropped Area

TABLE 5.15: Gram, Peas and Masoor Cropped Area

Percentage category	No. of block					
	Masoor		Gram		Peas	
	1985-87	1975-77	1985-87	1975-77	1985-87	1975-77
Above 4	13	5	6	2	—	3
2-4	4	10	8	4	3	11
1-2	5	5	7	13	5	11
1 & below	9	11	10	12	23	6

the kharif season. An area of 7,949 hectares or 1.8 per cent of the rabi, 1.1 per cent of the net sown and 0.8 per cent of the total cropped area is under this crop. This is thus quite an insignificant crop. The micro-level spatial distribution (Table 5.15 and Fig. 5.21) does not decipher any specific pattern. It is amorphous, as the concentration is low (1% & below) in the western and eastern parts both, it is moderate (1-2%) and moderately high (2-4%) mostly in the central part. The decreasing trend is the hectareage under pea is clear from the fact that the number of the blocks in the low concentration range has increased from 6 (1975-77) to 23 (1985-87) and that under moderate and moderately high ranges has decreased from 11 each to 5 and 3 respectively during this period. There exists no block now in the high concentration range while this number in 1975-77 was three.

FOOD-CUM-COMMERCIAL CROPS

They consist of fruits and vegetables, oil-seeds and sugarcane and together cover 50,679 hectares of land, i.e., 8.6 per cent of the kharif, 7.3 per cent of the net sown and 4.9 per cent of the total cropped area in the kharif season; and 41,807 hectares, i.e., 9.4 per cent of the rabi, 6.0 per cent of the net sown and 4.0 per cent of the total cropped area of the region in the rabi season. In all, they claim 8.9 per cent of the total cropped area of the region, which is almost the same as that of pulses as a whole. Generally, their concentration, as it may be seen in the case of some individual crops, is higher towards east.

Their hectareage has increased from 50,967 (1955-57) to 92,486 (1985-87), i.e. by 81.5 per cent. The volume of change in the region as whole is to the tune of 3.0 per cent during the span of three decades (1.1 during the first, 1.5 during the second and 0.4 per cent during the third decade). There are micro-level variations which may be seen in the case of some important crops.

Sugarcane

Among the food-cum-commercial crops, sugarcane ranks first, as it occupies an area of 43,928 hectares or 7.5 per cent of the kharif, 6.3 per cent of the net sown and 4.2 per cent of the total cropped area of the region. This proportion is lower than that of the State of Uttar Pradesh, where 9.8 per cent of the net sown and 6.8 per cent of total cropped area is devoted to the cultivation of sugarcane, but it is higher than that of the country, where 2.2 per cent of the net sown and 1.7 per cent (1986-87) (4) of the total cropped area is occupied by this crop.

Sugarcane flourished well on heavy loam with good drainage and slightly acidic to alkaline soils. Its moisture requirement is considerable (1250-1500mm). It requires careful culture, heavy manuring and adequate irrigation facilities and heavy demand on labour. All these agro-climatic and socio-economic conditions and competition from other crops affect the spatial distribution of this crop within the region (Table 5.16a and Fig. 5.22).

Three easternmost blocks (Nebua Naurangia, Khadda and Bishunpura) have very high (above 20%) concentration. In 5 blocks (Tulsipur and Gainsari in the west, and Nichlaul, Ghughli and Siswa in the east) the proportion varies from 5 to 10 per cent, 7, from 2 to 5 per cent, and in 16, it is 2 per cent and below.

From the above discussion, it emerges that higher concentrations are generally towards east where the conditions for its cultivation are comparatively better. Besides, the nearby location of sugar factories has also been responsible for higher percentage under sugarcane in certain blocks. There has been an increase by 76.6 per cent in the area under sugarcane during the period of 30 years (from 24,859 hectares to 43,928 hectares), some blocks have experienced increases and some decreases.

The number of the blocks in the low concentration range (Table 5.16a and Fig. 5.22) has decreased from 19 (1955-57) to 16 (1985-87) and that in the moderate (2-5%), it first increased from 6 (1955-57) to 12 (1975-77) and then decreased to 7 (1985-87). The number of the blocks in the range 5-10 per cent has almost remained constant, although, there is first reduction from 5 (1955-57) to 2 (1975-77). In the range above 20 per cent, the number has increased from 1 to 3 during this period.

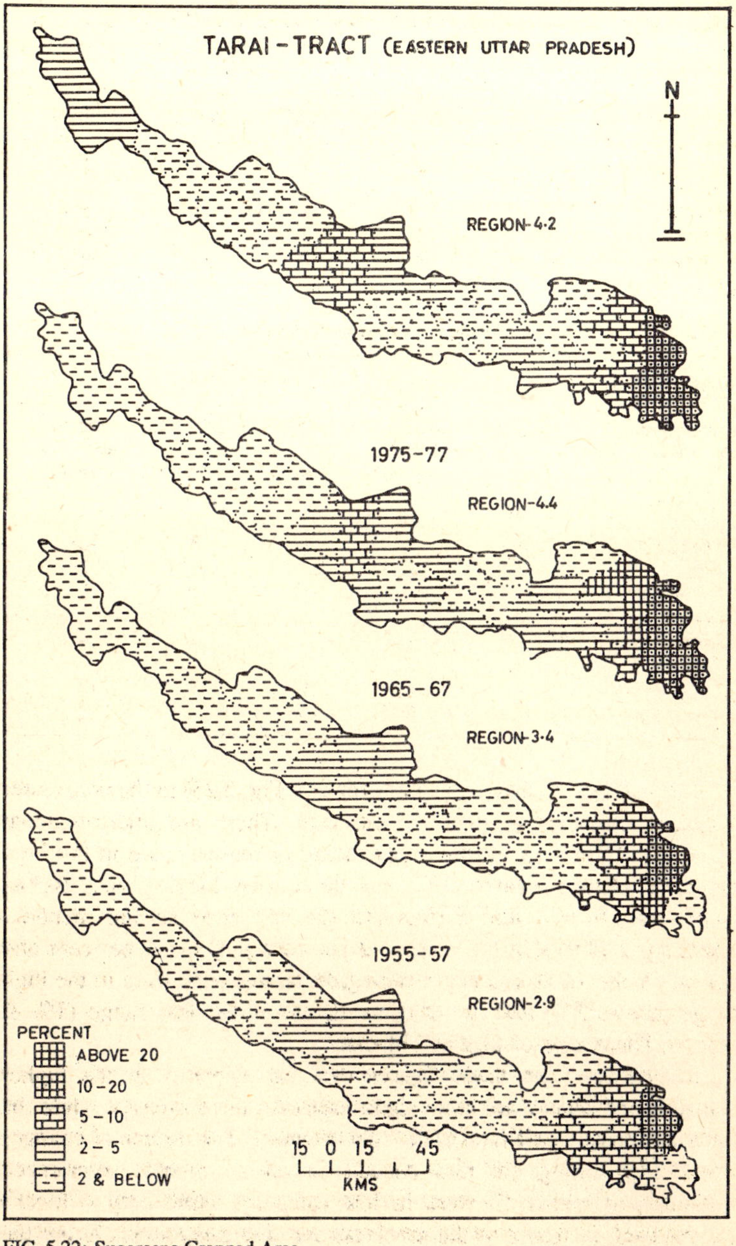

FIG. 5.22: Sugarcane Cropped Area

TABLE 5.16a: Sugarcane Cropped Area as Percentage of Total Cropped Area

Percentage category	No. of blocks			
	1985-87	1975-77	1965-67	1955-57
Above 20	3	3	1	1
10-20	—	2	1	—
5-10	5	2	4	5
2-5	7	12	7	6
2 & below	16	12	18	19

TABLE 5.16b: Volume of Change in Sugarcane Cultivation

Change category (% of the total cropped area involved in change)	No. of blocks			
	1955-57 to 1965-67	1965-67 to 1975-77	1975-77 to 1985-87	1955-57 to 1985-87
Increase				
Above 2	1	3	4	8
1-2	3	5	1	2
1 & below	13	6	5	3
Decrease				
Above 2	—	2	2	4
1-2	2	1	5	—
1 & below	4	9	12	12
No Change	8	5	2	2

The volume of change (Table 5.16b and Fig. 5.23) in the area under sugarcane amounts to plus 1.3 per cent. There are micro-regional variations. Of the 13 blocks experiencing increases, 8 are in the high range of change (above 2%). Of these, two blocks, viz., Nebua Naurangia (14.4%) and Bishunpura (24.5%) need special mention. There are 2 blocks in the range 1-2 per cent and 3 in 1 per cent and below. Of the 16 blocks experiencing decreases, only 4 are in the high range (above 2%) and as many as 12 are in the low range (1% & below). There is no change in 2 blocks.

It emerges from these figures that the increases in the higher categories of change are more significant than the decreases, while the latter are more significant in the low category. The decennial changes reveal that during the first decade, in all 17 blocks experienced increases, of which 13 were in low category, while only 6 blocks experienced decreases in the moderate and low categories. During the second decade, the number of the blocks experiencing increases decreased to 14, but in the range 1-2 per cent and above, it rose from 4 to 8, showing significant increase in the importance of this crop. The number of the blocks experiencing decreases, increased from 6 to 12, the larger increases are in low range of change, showing that the increases have been minor. During the third decade, the number of the blocks experiencing increases again decreased to 10, 50 per cent of which are in the range 1 per cent and below and 50 per cent above it, showing an arrested growth in the hectareage under this crop. There is a corresponding further rise to 19 in the number of the blocks experiencing decreases, of which 12 are in the low range. This again shows a lesser magnitude of change. The number of the blocks with no change respectively was 8, 5, and 2 during the first, second and the third decade.

It may be concluded from the above discussion that the hectareage under sugarcane increased greatly in those blocks where the facilities of irrigation have increased substantially and where sugar factories have been established. In the blocks where there already existed sugar factories, the proportion of the area under this crop was higher in the past also. In other blocks, minor changes have occurred.

Oil-seeds

The area devoted to oil-seeds accounts for 38,486 hectares, i.e. 5.3 per cent of the net and 3.7 per cent of the total cropped area of the region. The importance of oil-seeds in the subsistence agricultural economy of the region is next to foodgrains. They supply an essential requirement in the nutrition of the farm economy. They also serve as a source of plant food and constitute a valuable group of cash crops.

The spatial distribution of the area under oil-seeds (Table 5.17a and Fig. 5.24) decipher that the zones of high (6-8%) and moderately high (4-6%) concentrations are in the western districts of Bahraich and Gonda (Nebua Naurangia is the only exception). There are 15 blocks in the moderate range (2-4%), only 3 of which (Jamunaha, Hariharpur Rani and Balrampur) belong to the western part. Of the 9 blocks in the low concentration range (2% & below), only Nawabganj hails from the western part. The rest come from the middle and eastern parts. It may be noted here that too much moisture is harmful to the rabi oil-seeds and loamy sand, preferably alluvial soils are suited to their growth. This may explain, to some extent, the higher concentration of these crops towards west.

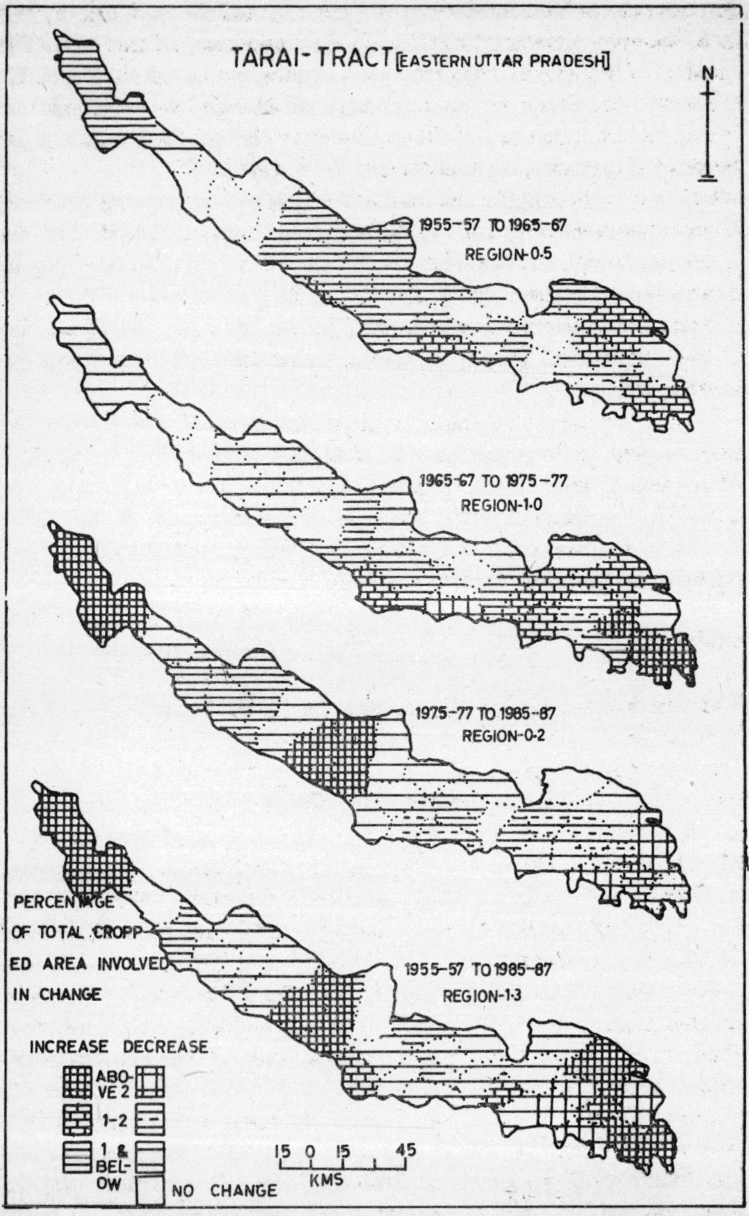

FIG. 5.23: Volume of Change in Sugarcane Cultivation

The hectareage under oil-seeds has increased from 24,899 (1955-57) to 38,486 (1985-87), showing a percentage increase by 62.6 per cent. Table 5.17a and Fig. 5.24, show that the number of the blocks in the range 6-8 per cent has increased from 1 (1955-57) to 4 (1985-87) and that in the range 2-4 per cent from 10 to 15. It has decreased in the range 2 per cent and below from 16 to 9 during this period. The number of the blocks in the range 4-6 per cent has remained almost static. This reveals an increasing trend which is evidenced by the volume of change also (Table 5.17b and Fig. 5.25), as there is an addition of 0.9 per cent in the area under oil-seeds. As many as 24 blocks have experienced increases (4, above 2%, 8, 1-2%, and 12, 1% & below). All the blocks in the high category of change belong to the western part, while of the 8 in the range 1-2 per cent, 4 belong to the middle and 3 to the eastern part. The distributional pattern of the blocks in the low category of change is irregular. Of the 6 blocks experiencing decreases, 4 are in the low and one each in the moderate and high category of change. The position of one block has remained unchanged.

TABLE 5.17a: Oil-seeds Cropped Area as Percentage of Total Cropped Area

Percentage category	No. of blocks			
	1985-87	1975-77	1965-67	1955-57
Above 8	—	—	2	1
6-8	4	3	2	1
4-6	3	4	5	3
2-4	15	11	7	10
2 & below	9	13	15	16

TABLE 5.17b: Volume of Change in Oil-seeds Cultivation

Change category (% of the total cropped area involved in change)	No. of blocks			
	1955-57 to 1965-67	1965-67 to 1975-77	1975-77 to 1985-87	1955-57 to 1985-87
Increase				
Above 2	5	1	3	4
1-2	—	2	4	8
1 & below	14	12	10	12
Decrease				
Above 2	—	4	1	1
1-2	3	3	3	1
1 & below	7	9	8	4
No Change	2	—	2	1

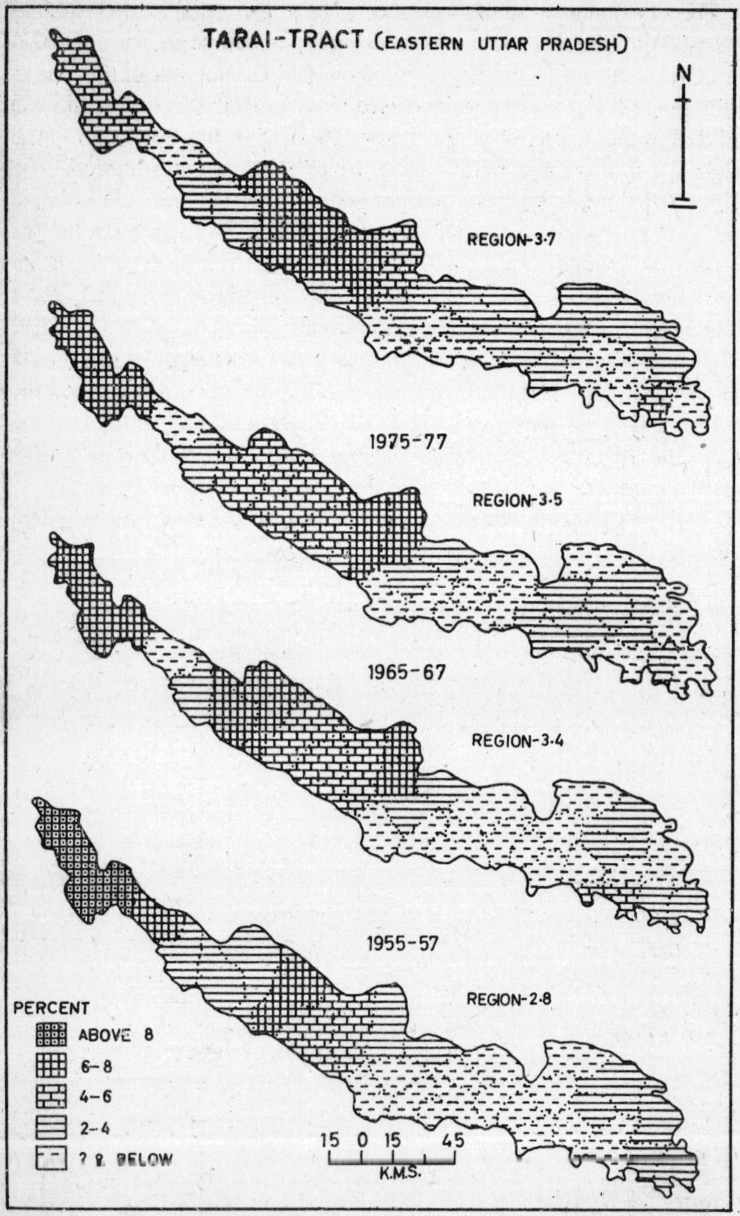

FIG. 5.24: Oil-seeds Cropped Area

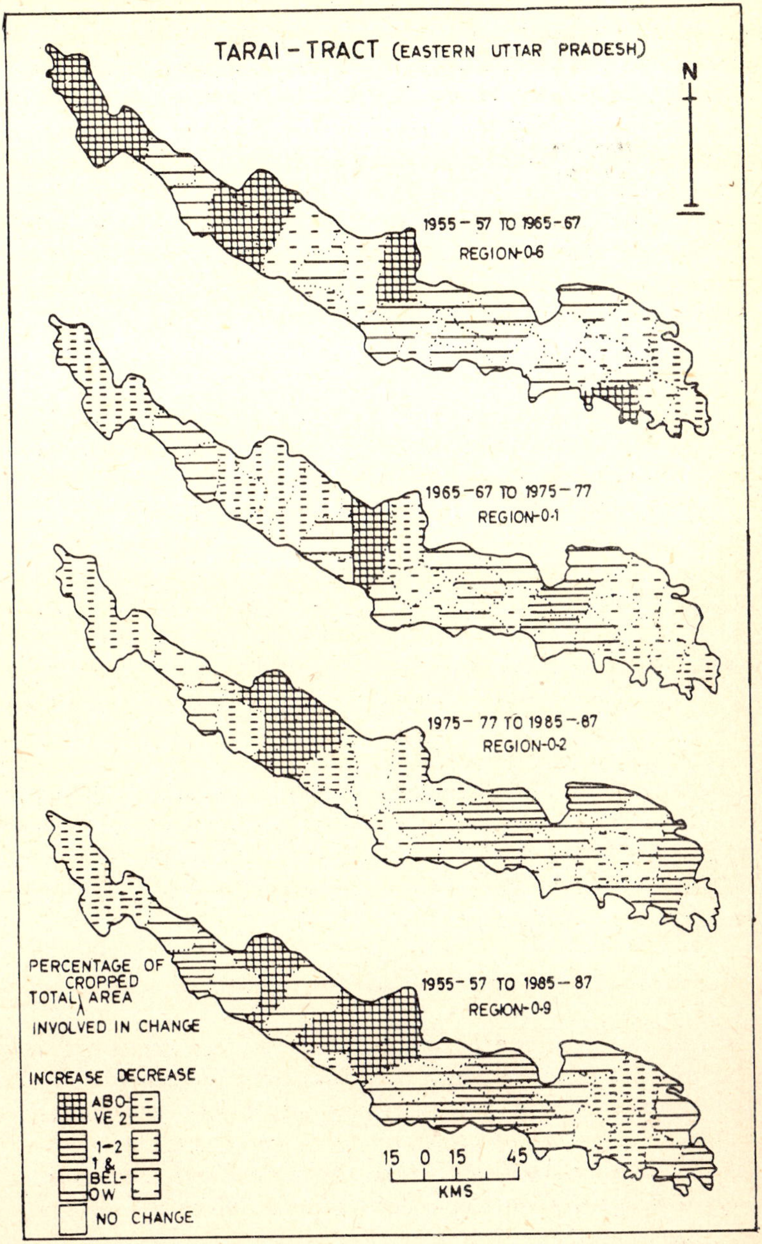

FIG. 5.25: Volume of Change in Oil-seeds Cultivation

Table 5.17b and Fig. 5.25 show that during the first decade, 19 blocks experienced increases and 10 decreases, majority of these being in the low categories of change. These figures decreased from 19 to 15 in the former and increased from 10 to 16 in the latter case during the second decade. Again, the number is larger in the low category of change. During the third decade, the number of the blocks experiencing increases, again, rose to 17 and that of experiencing decreases decreased to 12. In this case also, the larger number of the blocks is in the low category of change, showing that not very significant changes have occurred.

Fruits and Vegetables

The fruits and vegetables occupy only a very meagre and negligible area which accounts for only 1.0 per cent of the total cropped (kharif 0.3% and rabi 0.7%) area. The various types of vegetables grown for household consumption are beams, tubers, root vegetables such as carrots, potatoes, pumpkins, lady-finger, tomato, leafy vegetables such as cauliflower etc. Potato is most important. Mango, banana, jack-fruits, papayas etc. are the main fruits. The hectareage under them has increased a little, but that is quite insignificant.

Non-food Crops

The non-food crops which comprise of fibre crops, spices and fodder crops are also very insignificant, as only 0.6 per cent of the net and 0.3 per cent of the total cropped area is devoted to them. The hectareage under them has decreased from 5,183 (1955-57) to 4,043 (1985-87), but the volume of change is almost negligible.

THE RANKING OF CROPS

The regional dominance of various crops has been determined by comparing the relative strength of various crops by putting them in order of percentages of the total cropped area, as has been shown in Appendix IX. Taking the region as a whole, we find that rice, wheat and rabi pulses respectively are the first, second and third ranking crops, although there are some micro-regional variations in it (Figs. 5.26, 5.27 and 5.28).

The First Ranking Crops

Figure 5.26 shows that rice ranks first in 30 out of the 31 blocks of the region in 1985-87. It is only in blocks Pharenda that wheat takes its place. This shows the super dominance of rice in the region. The crop enjoyed the same privilege even in the past and no change worth-mentioning has occurred in its rank in the region, except Mihipurwa where maize ranked first in 1955-57 and 1965-67; Hariharpur Rani where maize ranked first in 1955-57, 1965-67 and 1975-77.

The Second Ranking Crops

Figure 5.27 shows the second ranking crops. It reveals that a few more crops step in when the second crops in order of dominance are considered. But it is wheat which dominates the agricultural landscape by snatching second rank in 28 blocks out of 31 blocks. In 1 block (Pharenda), it is rice and in 2 (Nebua Naurangia and Bishunpura), it is sugarcane which ranks second.

Figure 5.27 shows that there were 6 second ranking crops in 1955-57 and 1965-67, while 4 in 1975-77. Rice ranked second in 2 blocks in 1955-57 and in 1 in 1965-67. Wheat ranked second in 11 blocks in 1955-57, in 10 in 1965-67 and in 26 blocks in 1975-77. These blocks mostly belong to the central and eastern parts. Wheat mixed with barley occupied this position in 3 blocks (Itwa, Birdpur and Laxmipur) in 1955-57 and in 1 (Birdpur) in 1965-67. Maize held this position in 3 (Jamunaha, Sirsia and Haraiya Satgharwa), in 4 (Jamunaha, Sirsia, Haraiya Satgharwa and Tulsipur), and in 2 (Mihipurwa and Sirsia) blocks respectively in these years. Barley which is conspicuous by its absence on the second ranking crop map in 1985-87 and 1975-77 both, ranked second in 5 blocks (Gainsari, Naugarh, Jogia and Uska and Bansi) in 1955-57 and, in 3 blocks (Pachperwa, Jogia and Bansi) in 1965-67. Kodon covered only 1 block in 1975-77 and in other years it does not exist. Sugarcane occupied this position in one block (Khadda) in 1955-57 and in 2 each (Khadda and Bishunpura) in 1965-67 and 1975-77. The above analysis reveals that the number of the second ranking crops has decreased from 6 in 1955-57 to 3 in 1985-87 and the importance of wheat has increased, as the number of the blocks with wheat as a second ranking crop has increased from 11 to 28 (1955-57 to 1985-87).

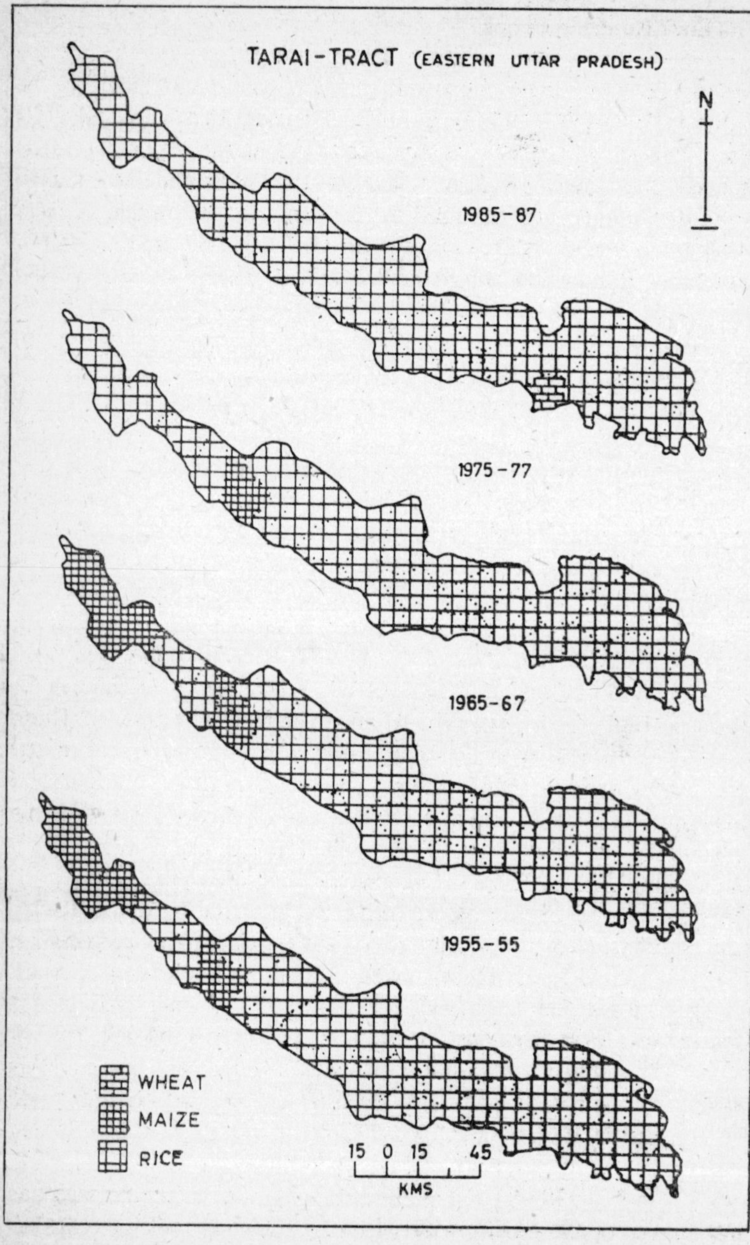

FIG. 5.26: First Ranking Crops

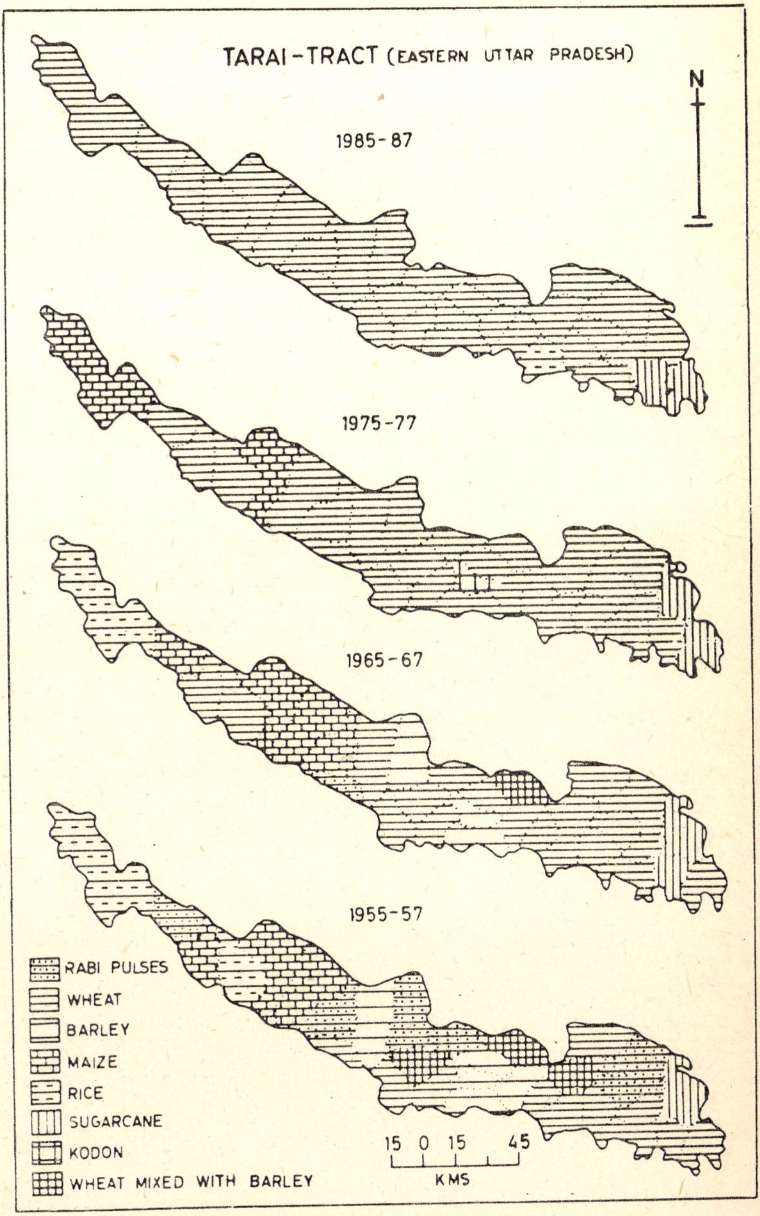

FIG. 5.27: Second Ranking Crops

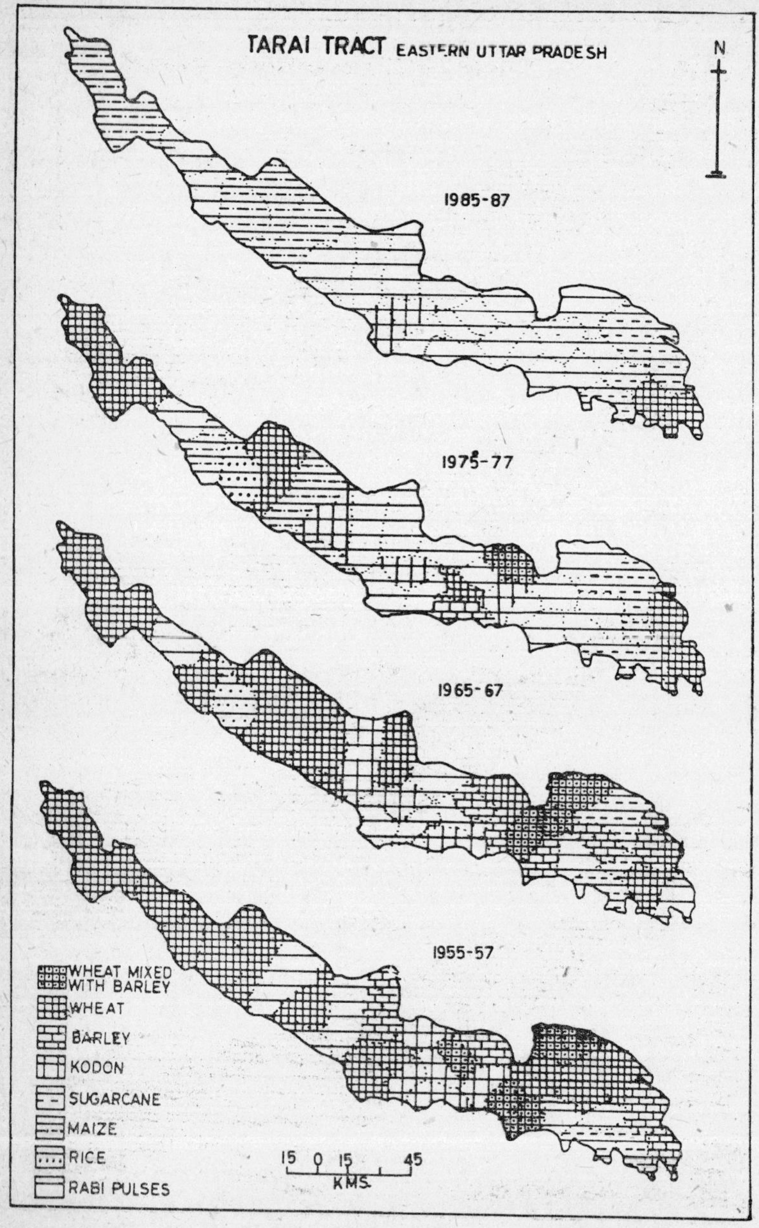

FIG. 5.28: Third Ranking Crops

The Third Ranking Crops

Figure 5.28 portrays the distributional pattern of the third ranking crops in the region, which shows that the number of the crops is now 5 (wheat, rabi pulses, maize, *kodon* and sugarcane). This number would have been still larger, if different rabi pulses had been taken individually, which could not be done due to the non-availability of the data of the individual crops for the years 1955-57 and 1965-67 and thus the rabi and kharif pulses have been taken as a whole. Rabi pulses emerge out as the most important third ranking crop by occupying 19 blocks which spread widely, although with higher concentration in the middle and eastern parts. Next come maize and sugarcane which occupy 4 blocks each (maize in Mihipurwa, Hariharpur Rani, Sirsia and Haraiya Satgharwa and sugarcane in Ghughli, Siswa, Khadda and Nichlaul, the former belonging to the western and the latter to the eastern part). Wheat and kodon also occupy 2 blocks each (wheat in Nebua Naurangia and Bishunpura and *kodon* in Itwa and Mankapur).

It may be seen from Fig. 5.28 that the relative importance of the crops has changed immensely in the cropping pattern of the region. Wheat ranked third in 10 blocks in 1955-57, in 9 in 1965-67 and in 5 in 1975-77, showing that the number of the blocks with this rank of wheat has been decreasing and it occupies this position only in 2 blocks. Wheat held that rank mostly in the blocks of the western part, where rice and maize were more important than this crop. Rabi pulses held this rank in 5 blocks in 1955-57, in 7 blocks in 1965-67 and in 13 blocks in 1975-77. These figures for kodon are 5, 3 and 3 respectively. Barley which does not appear on the third ranking crop map now occupied 5 blocks each in 1955-57 and 1965-67 and 1 block in 1975-77. Wheat mixed with barley held this position in 4 blocks in 1955-57, in 3 blocks in 1965-67 and in 1 block in 1975-77. Sugarcane occupied this rank respectively in 2, 3 and 4 blocks. Maize appeared only in 3 blocks in 1975-77 at this rank and rice in 1 block each in 1965-67 and 1975-77, but has disappeared in 1985-87.

It is clear from the above discussion that rabi pulses, maize and sugarcane are the main third ranking crops in the region, while in the previous years, wheat, kodon, barley, wheat mixed with barley occupied important place.

CROP COMBINATION REGIONS

A perusal of the distributional pattern and ranking of the crops as discussed at length in the preceding pages reveals that the crops are seldom grown in absolute isolation. Various geographical factors tend to confine these crop associations to rather restricted areas where they get requisite ecological conditions.[5] It is thus necessary to study and analyse the closely inter-related crop-combinations prevalent in different parts of the region.

It may be noted here that for delineating the crop-combinations, Doi's method which shows an improvement over that of Weaver's by substituting latter's $\Sigma d^2/N$ with Σd^2, i.e. the sum of squared differences, has been adopted.

The whole of the region taken as a unit falls under the two-crop combination, viz. R-W (Rice-Wheat). On the micro-level also, this combination prevails extensively with some spatial variations.

Table 5.18a and Fig. 5.29 show that on block-level, the following crop-combination regions emerge out, which have been discussed in the forthcoming pages.

Two-crop Combination Region

There are two two-crop combination regions which cover 19 blocks belonging mostly to the central and eastern parts. In one region, wheat ranks first and rice occupies the second position, while in the remaining 18, rice followed by wheat is the dominant crop. They together eclipse all other crops which are not so significant. Various ecological factors, viz. soil, climate and better irrigation facilities have favoured the large concentration of these two crops which are preferable due to their greater per hectare yield and more demand as foodgrain.

Three-crop Combination Region

There are three 3-crop combinations which occupy 8 blocks in all, 3 (Khadda, Nebua Naurangia and Bishunpura) belonging to the easternmost part and 5 (Jamunaha, Tulsipur, Gainsari, Pachperwa and Balrampur) belonging to the western part. The ecological conditions of these blocks, particularly, soil, sub-soil moisture and availability of irrigation facilities etc. lessen, to some extents, the exclusive predominance of the first and the second ranking crops. That is why they combine third crops also with them. Rice is the first ranking crop in all the blocks. In one region which occupies the western part, it is followed by wheat as the second and rabi pulses as the third crop. In one region which occupies only one block (Khadda), the second rank is held by wheat, but third one is held by sugarcane. There is another region which occupies 2 blocks (Nebua Naurangia and Bishunpura) where the second and third ranks respectively are occupied by sugarcane and wheat. Thus wheat occupies second rank in two regions and sugarcane in one. As regards the third rank, rabi pulses, sugarcane and wheat occupies this position in one region each respectively.

Four-crop Combination Region

There is only one four-crop combination region which occupies 4 riverine dominant blocks (Hariharpur Rani, Haraiya Satgharwa, Mihipurwa and Sirsia), all belonging to the western part, which are widely subjected to inundation in rainy season. Rice is the first ranking crop which is followed by wheat, maize and rabi pulses respectively as the second, third and fourth ranking crops.

TABLE 5.18a: Crop Combination Regions (1985-87)

No. of crops	Crop in order	Blocks
2	R-W	Nawabganj, Itwa, Bhanwapur, Khuniaon, Birdpur, Barhni, Naugarh, Jogia, Uska, Bansi, Nautanwa, Laxmipur, Brijmanganj, Nichlaul, Mithaura, Mahrajganj, Ghughli, Siswa
	W-R	Pharenda
3	R-W-RP	Jamunaha, Tulsipur, Gaisari, Pachperwa, Balrampur
	R-W-S	Khadda
	R-S-W	Nebua Naurangia, Bishunpura
	R-W-M-RP	Hariharpur Rani, Haraiya Satgharwa, Mihipurwa, Sirsia

Note: R-Rice, W-Wheat, M-Maize, RP-Rabi Pulses

Changes in the Crop Combinations

Figure 5.29 shows that the crop combinations in the region have undergone considerable changes which are due to the changes in the ecological conditions.

Tables 5.18a, b, c and d and Fig. 5.29 show that there were 4, 3 and 1 blocks in one-crop combination region respectively in 1955-57, 1965-67 and 1975-77, 3 of which (Naugarh, Birdpur and Nautanwa)

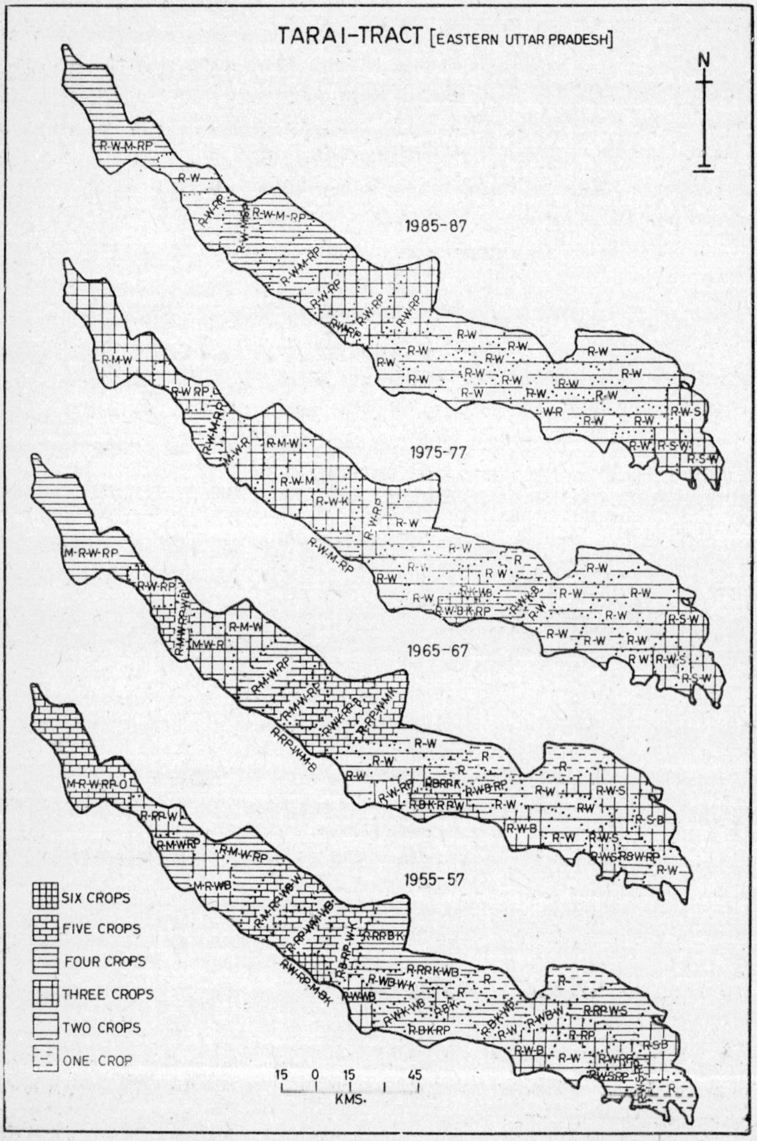

FIG. 5.29: Crop Combinations

are now in two-crop combinations and 1 (Bishunpura) in a three-crop combination. There exists no block in one-crop combination in 1985-87. There were 3, 8 and 15 blocks in two-crop combination regions respectively in 1955-57, 1965-67 and 1975-77, which number now is 19. This reveals the increasing dominance of rice and wheat, particularly of the latter over other crops, which appears due to an increase in the irrigation facilities and due to some other socio-economic factors.

In three-crop combination regions, there were 7, 10 and 10 blocks respectively in these years. This number remains the same (10) in 1985-87 also. The number of the blocks in the four-crop combination regions respectively was 11, 5 and 4 and it is 4 in 1985-87 also. The five-crop combination regions respectively had 5, 5 and 1 blocks respectively, while there exists no block now in this crop combination. As regards the six-crop combination, there was one block only in 1955-57 and after this, this combination disappeared from the agricultural landscape of the region.

It may be concluded from the foregoing analysis that the number of the crops in the crop combination regions has decreased. This shows that there is now greater emphasis on the major crops which suit the local ecological conditions and are in greater demand to feed the growing population.

TABLE 5.18b: Crop Combination Regions (1975-77)

No. of Crops	Crops in order	Blocks
1	R	Birdpur
2	R-W	Itwa, Bhanwapur, Khuniaon, Barhni, Nautanwa, Naugarh, Laxmipur, Brijmanganj, Pharenda, Nichlaul, Mithaura, Maharajganj, Ghughli, Siswa, Pachperwa
3	R-M-W	Mihipurwa, Sirsia, Gainsari
	R-W-RP	Nawabganj
	M-W-R	Hariharpur Rani
	R-W-M	Haraiya Satgharwa
	R-W-K	Tulsipur
	R-S-W	Khadda, Bishunpura
	R-W-S	Nebua Naurangia
4	R-W-M-RP	Jamunaha, Balrampur
	R-K-W-B	Jogia
	R-W-K-B	Uska
5	R-W-B-K-RP	Bansi

TABLE 5.18c: Crop Combination Regions (1965-67)

No. of crops	Crop in order	Blocks
1	R	Birdpur, Nautanwa, Naugarh
2	R-W	Itwa, Bhanwapur,Barhni, Laxmipur, Brijmanganj Mithaura, Maharajganj, Bishunpura
3	M-W-R	Hariharpur Rani
	R-M-W	Sirsia
	R-W-RP	Nawabganj, Khuniaon
	R-W-B	Pharenda
	R-W-S	Nichlaul, Ghughly, Siswa
	R-S-B	Khadda
4	M-R-W-RP	Mihipurwa
	R-M-W-RP	Haraiya Satgharwa, Tulsipur
	R-B-RP-K	Jogia
	R-W-B-RP	Uska
	R-S-W-RP	Nebua Naurangia
5	R-M-W-RP-WB	Jamunaha
	R-W-K-RP-B	Gainsari
	R-RP-W-M.K.	Pachperwa
	R-RP-W-M.B.	Balrampur
	R-B-K-RP-W	Bansi

OVERALL VOLUME OF CHANGE IN CROPPING PATTERN

The percentage of the total cropped area involved in change among different crops of the region has been compiled on a micro-level basis and mapped in Fig. 5.30. In the region as a whole, this percentage amounts to 22.5 for the total period of three decades (5.6% during the first, 12.1 per cent during the second and 8.8 per cent during the third decade). On examining the map, it is found that within the region, the percentage of the total cropped area involved in change varies from 18.9 (Nawabganj) to 46.3 (Bansi). These figures for the first decade are from 2.3 to 13.8 per cent, for the second decade from 5.9 to 38.2 per cent and for the third decade from 3.7 to 46.3 per cent. This reveals that the magnitude of change, particularly on the maximum side has been increasing, for which various physico-cultural factors are responsible.

Table 5.19 and Fig. 5.30 show that during the first decade, 30 blocks experienced low and very low changes, but during the second decade, there remained only 13 blocks in these categories, while 11 shifted to moderate (15-20%), 2 to moderately high (20-25%) and 5 to high (above 25%) category. During the third decade, 19 more blocks shifted to the high degree of change and there remained only 5 blocks in the moderately high and 2 in the moderate degree of change. If overall period of three decades is taken into consideration, we find that 12 blocks witnessed very slow rate of change (10% & below) and 5

TABLE 5.18d: Crop Combination (1955-57)

No. of Crops	Crops in order	Blocks
1	R	Birdpur, Naugarh, Nautanwa, Bishunpura
2	R-W	Brimanganj, Maharajganj
	R-RP	Milhawra
3	R-RP-W	Nawabganj
	R-W-WB	Bhanwapur
	R-B-K	Jogia
	R-WB-W	Laxmipur
	R-W-B	Pharenda
	R-W-RP	Siswa
	R-S-B	Khadda
4	R-M-W-RP	Jamunaha, Sirsia
	R-R-W-B	Hariharpur Rani
	R-RP-B-K	Pachperwa
	R-WB-W.K.	Itwa
	R-W-K-WB	Khuniaon
	R-RP-K-WB	Barhni
	R-B-K-WB	Uska
	R-RP-WB-S	Nichlaul
	R-W-RP	Ghughli, Nebua Naurangia
5	M-R-W-RP.O	Mihipurwa
	R-M-RP-WB-W	Haraiya Satgharwa
	R-RP-W-M-WB	Tulsipur
	R-B-RP-W-K	Gainsari
	R-B-K-W-RP	Bansi
6	R-W-RP-M-B-K	Balrampur

Note: R-Rice, W-Wheat, M-Maize, K-Kodon, B-Barley, WB-Wheat mixed with barley, RP-Rabi pulses, S-Sugarcane, O-Oil-seeds slow (10-15%). There has been lesser degree of shifting from one to the other crop here. In 9 blocks, the changes have been moderate (15-20%). In these blocks, the development in agricultural infrastructure has also been moderate. Three blocks (Balrampur, Mihipurwa and Uska) have witnessed moderately high changes (20-25%) and 2 (Jogia and Bansi) high (above 25%).

These blocks exhibit great shifting of the area covered by barley, *kodon*, wheat mixed with barley and such other crops which require lesser moisture and have lower yields to that covered by superior crops such as wheat, paddy etc., for which there is now greater demand. To some extents, the higher shifting may also be ascribed to the riverine characteristics of the area causing sometimes a great decrease in the kharif crops and a corresponding increase in the rabi crops.

It may be concluded that an increase in the irrigation facilities together with the preference of the cultivators to grow such crops as may feed the growing population and fetch them good income and some allied factors narrated above may be considered responsible for the dynamic changes in the cropping pattern in the study area. As observed earlier, the shift generally is from inferior grains to superior grains or from grains to cash crops such as sugarcane, oil-seeds etc. The region has great agricultural potential and still awaits more outstanding and dynamic swing.

TABLE 5.19: Over-all Volume of Change in Cropping Pattern

Change category (% of the total cropped area involved in change)	No. of blocks			
	1955-57 to 1965-67	1965-67 to 1975-77	1975-77 to 1985-87	1955-57 to 1985-87
Above 25	—	5	24	2
20-25	—	2	5	3
15-20	1	11	2	9
10-15	7	8	—	5
10 & below	23	5	—	12

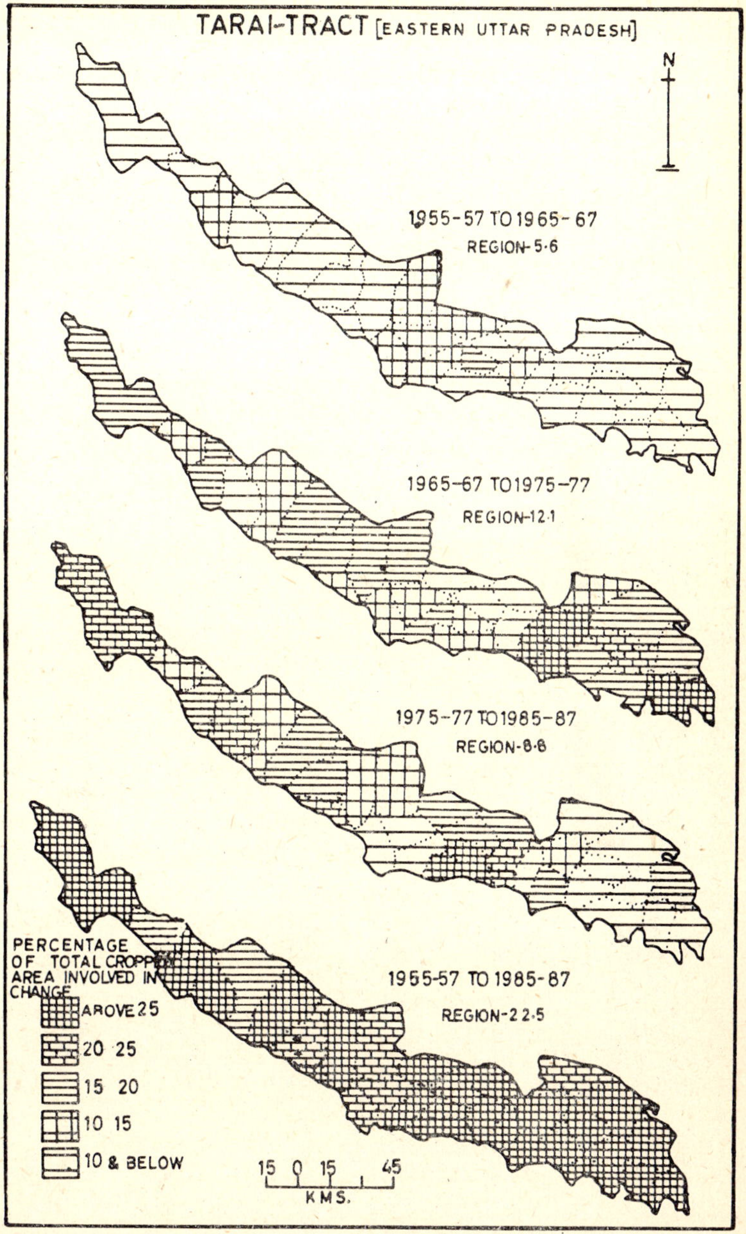

FIG. 5.30: Overall Volume of Change in Cropping Pattern

TABLE 5.20: Blockwise Changes in the Overall Cropping Pattern

Blocks	Percentage of total cropped area involved in change			
	1955-57 to 1965-67	1965-67 to 1975-77	1975-77 to 1985-87	1955-57 to 1985-87
1	2	3	4	5
1. Mihipurwa	7.5	15.7	22.0	31.5
2. Nawabganj	7.9	10.5	11.7	18.9
3. Jamunaha	11.5	18.1	16.8	27.4
4. Hariharpur Rani	10.0	9.0	19.7	27.3
5. Sirsia	5.7	14.6	15.2	18.2
6. Haraiya Satgharwa	9.1	17.0	17.6	27.3
7. Tulsipur	9.6	19.6	16.9	24.2
8. Gainsari	5.4	18.3	12.5	28.6
9. Pachperwa	10.5	17.3	12.1	23.9
10. Balrampur	9.1	13.9	24.6	31.1
11. Itwa	13.8	14.6	3.7	13.2
12. Bhanwapur	10.8	14.2	3.5	24.3
13. Khuniaon	10.3	17.4	6.9	27.3
14. Barhni	12.8	5.7	18.5	29.5
15. Birdpur	3.8	8.9	18.4	26.8
16. Naugarh	5.6	7.2	19.7	29.6
17. Jogia	15.1	11.4	26.7	39.0
18. Uska	12.0	5.9	22.0	29.7
19. Bansi	7.8	10.5	36.0	46.3
20. Nautanwa	3.4	12.0	9.5	21.6
21. Laxmipur	4.9	27.2	12.0	30.2
22. Brijmanganj	2.3	32.0	7.2	32.7
23. Pharenda	3.3	25.1	15.8	39.1
24. Nichlaul	8.6	19.4	8.4	25.5
25. Mithaura	6.0	22.8	4.8	28.5
26. Maharajganj	8.4	19.4	7.1	27.5
27. Ghughli	8.7	18.8	8.0	21.0
28. Siswa	6.4	20.4	6.7	28.5
29. Khadda	4.5	19.6	17.0	25.3
30. Nebua Naurangia	4.0	32.5	8.5	34.1
31. Bishunpura	8.8	38.7	8.1	40.0
Region	5.6	12.1	8.8	22.5

NOTES

1. Director of Economics and Statistics, Department of Agriculture and

Cooperation, Ministry of Agriculture, Govt. of India, (April, 1988), *Indian Agriculture in Brief* (22nd Edition), pp. 286-87.

2. *Ibid.*
3. Singh, J., (1974-75), *op. cit.*, p. 192.
4. Director of Economics and Statistics, Department of Agriculture and Cooperation, Ministry of Agriculture, Govt. of India, (April, 1988), *Ibid.*
5. Sharma, S.C., (1971), Cropping Pattern and Crop Combination Regions in Etawah District, *Deccan Geographer*, Vol. IX, No. 1, p. 85.

PART III

LAND-USE SURVEY OF SELECTED VILLAGES

SIX

LAND-USE SURVEY OF SELECTED VILLAGES

VILLAGE-DHAMPUR

Location

THE VILLAGE Dhampur which belongs to the northern gently rolling to nearly level lowlying tract, lies in the Balrampur tahsil of Gonda district at 27°4′ N latitude and 82°14′ E longitude. It is at a distance of about 58 km from the tahsil headquarters. During the rainy season, its accessibility is lessened by overflowing rivers and rivulets.

Climate

No climatic data are collected here. The nearest rainfall recording station is the tahsil headquarters, the rainfall data of which may be taken as the close approximation of the climatic conditions prevailing in the village. They have already been dealt with in detail in the chapter on climate. It may be mentioned that of the average annual rainfall of 1301.9 mm, 2.5 per cent is received during the cold, 6.5 per cent during the hot and 91.0 per cent during the rainy season. The co-efficient of annual variability of rainfall is 21.45 per cent and that of monthly variability is 56.2 per cent for June, 52.5 per cent for July, 50.8 per cent for August, 52.2 per cent for September and 94.3 per cent for October. The high variability in the months of June and October is of paramount importance from the agricultural point of view.

Soils

Towards north and west, the soils are sandy or light loam, towards south-west, they are light loam and clayey, towards east loamy and towards north-east sandy (Fig. 6.1). As regards the revenue classification, they may be grouped into 'goind' (those around the village site), loamy, No. 1, 2 and 3, clayey loam No. 1 and 2.

Land Classification

The land of the village may be classified into the following land capability classes*

Land Capability Class I

This is very good land with few limitations. This comprises of the 'goind' and loamy soils which occupy three blocks, one towards west, south and east of the main settlement site in the north; second in the middle-east and the third in the south. The soils are well drained, slope ranges up to 1 per cent or even more. The permeability is moderately rapid and water-holding capacity is moderately low. The soil reaction is mildly alkaline (pH 7.6). The fertility status is low in respect of organic carbon and phosphorous, and medium in respect of potash. The soils are best suited for the cultivation of vegetables, sugarcane, wheat, *arhar* etc.

Land Capability Class II

This is good land with minor limitations and occupies extensive areas towards north, east and south-west. This comprises of loamy 2 and 3 or clayey loam 1 and 2 soils. This is comparatively less fertile than that of the former and suffers either from minor problems of erosion or water-logging. Accordingly, it may be sub-divided as follows:

(i) *Land Capability Sub-Class IIe:* This is the land where erosional hazards are important and where the permeability of soil is rapid and the water-holding capacity is low. The other characteristics are more or less the same as above.

(ii) *Land Capability Sub-class IIw:* This suffers from the minor limitation of water-logging and comprises of clayey loam soils. The permeability is slow and water-holding capacity is high.

* The basis and criteria of land classification have already been discussed in the chapter on soils and given in Appendix I.

The soil reaction is mildly alkaline. The fertility status is low. The soils are suitable for growing paddy, maize, sugarcane, wheat, masoor etc.

Land Capability Sub-class IIIe: This is moderately good land with severe limitations of soil erosion. It occupies a narrow strip of land towards west and east and comprises of loam 3 and clayey loam 3, well drained soils, where the slope ranges between 1 and 3 per cent. The permeability is rapid and water-holding capacity is slow. Other characteristics are the the same as those of the previous categories. It is suitable for growing maize, arhar, groundnuts, sweet-potato and other cover crops.

Land Capability Sub-class IVe: This is fairly good land with very severe limitations of soil erosion. The drainage is good. This type of land is found along the rivulets in the west and east and comprises of sandy loam sloppy soils. The permeability is highly rapid and the water-holding capacity is very slow. Other features are similar to those of the previous categories, the difference is that of degree only. This is suitable for limited cultivation and tree plantation.

Land Not Suitable for Cultivation:

This consists of the following categories:

Land Capability Class V: This is the land with minor limitations other than erosion. This is suitable for 'boro' cultivation only, which is due to its being submerged with water.

Land Capability Class VI: This land is unsuitable for cultivation due to very severe erosion. This is found immediately on the banks of *nalas*, where this has been severely been eroded and converted into deep gullies. This is suitable for afforestation and should be devoted to social forestry.

Irrigation

Irrigation is rarely practised. Only a few farmers utilise water of rivulets for irrigation. Due to the hard strata below the surface, the sinking of tube-wells is not easily feasible. Government tube-wells and canals may be developed.

General Land-use Pattern

Table 6.1 and Fig. 6.1 depict the general land-use pattern of the village. It may be seen from them that 19.5 per cent of the total area of the village is not available for cultivation, as 14.8 per cent is under water and 4.7 per cent is occupied by settlements and path etc. Culturable wasteland occupies but only 0.7 per cent of the total area. The proportion under fallow-land is again very small, e.g. 0.9 per cent. The remaining 78.9 per cent is under plough.

TABLE 6.1: General Land-use (1987) Total area of the village — 241.72 hactares

Land-use	Area in hectares	% of the total areas
A. *Area not available for cultivation*	47.31	19.5
(i) Land under water	35.38	10.8
(ii) Settlements & paths etc.	11.93	4.7
B. *Other Uncultivated land excluding fallowland*	1.32	0.7
(i) Culturable wasteland	1.32	0.7
C. *Arable land*	193.09	79.8
(i) Fallow land	2.34	0.9
(ii) Net sown area	190.75	78.9
Double cropped area (as % of net sown area)	23.90	12.5

It may be mentioned here that the area under culturable wasteland has decreased and that under plough has increased. The changes in other categories are not very significant.

Cropland Use

Table 6.2 and Fig. 6.1 show the cropland use of the village. It is clear from them that 52.4 per cent of the total and 58.9 per cent of the net cropped area is devoted to the cultivation of kharif crops. Maize is the main kharif crop which accounts for 62.5 per cent of the kharif, 36.8 per cent of the net and 32.7 per cent of the total cropped area of the village. It is widely scattered in the village, as the loamy and clayey loam well drained soils which are extensively found are favourable for its growth. Rice is the next important crop which occupies 19.4 per cent of the kharif, 11.4 per cent of the net and 10.1 per cent of the total

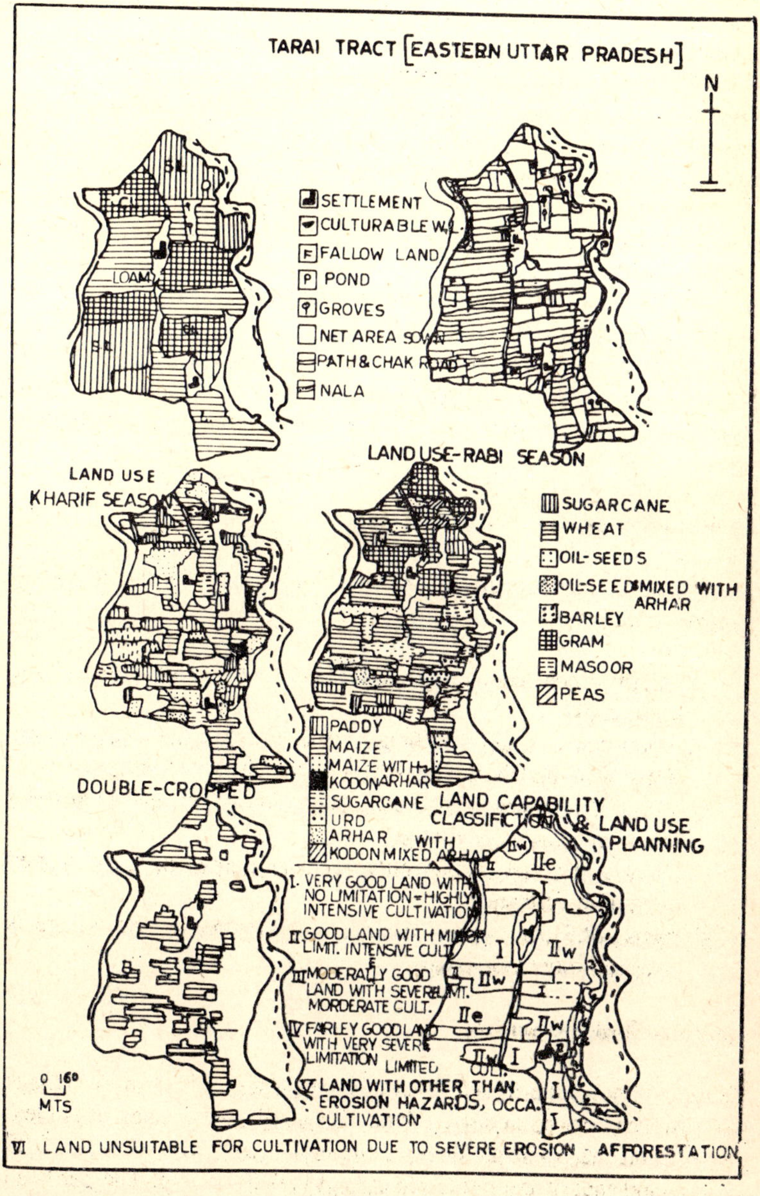

FIG. 6.1: Village Dhampur

cropped area of the village. It is also widely scattered with higher concentration towards south-west on the clayey loam soils. Sugarcane is the third important crop, occupying 8.9 per cent of the kharif, 5.2 per cent of the net and 4.8 per cent of the total cropped area of the village. The only important patch is towards west of the main settlement site in the north where it is grown on the loamy soils without irrigation. *kodon, arhar* etc. are other minor crops.

Rabi crops occupy 53.5 per cent of the net 47.6 per cent of the total cropped area of the village. Wheat is the dominant crop, claiming 68.0 per cent of the rabi, 36.4 per cent of the net and 32.4 per cent of the total cropped area of the village. It is extensively grown in the village with a higher concentration towards south of the main settlement site where loamy and clayey loam well drained soils with higher sub-soil moisture are found. Next important rabi crops are oil-seeds and pulses. The minor crops are wheat mixed with barley, barley mixed with gram etc.

General Features

A perusal of the cropland use of the village indicates the following special features:

1. Kharif crops occupy a slightly higher proportion than that of the rabi crops.
2. Foodgrains occupy 82.6 per cent of the total cropped area (46.4% in the Kharif and 36.2% in the rabi) of the village.
3. Food-cum-commercial crops occupy 16.3 per cent of the total cropped area of the village.
4. Non-food crops occupy only 1.1 per cent of the total cropped area of the village.
5. Only 12.5 per cent of the net cropped area of the village is cropped more than once.

Land-use and Population

Table 6.3 shows the totals of various categories of land in the village and also the per capita share of the villagers in these lands. It is clear from it that the per head share of the total area in the village is 0.39 hectare and that of the net cropped land is 0.30 hectare. The per head net cropped area in the kharif and rabi crops respectively is 0.18 and

0.16 hectare and thus the total cropped area comes to 0.34 hectare. This is the actual amount of land supporting one person in the village. It may be mentioned here that 97.3 per cent of the total workers of the village are engaged in primary rural occupation and thus depend directly on land for their livelihood.

TABLE 6.2: Cropland Use (1987)

Crops	Area in hectares	% of kharif/ rabi cropped area	% of net cropped area of the village	% of total cropped area of the village
1	2	3	4	5
Kharif	*112.44*	*100.0*	*58.9*	*52.4*
Cereals	*94.29*	*83.8*	*49.5*	*43.9*
Maize	70.28	62.5	36.8	32.7
Rice	21.72	19.3	11.4	10.1
Kodon alone & mixed	2.07	1.8	1.2	1.0
Others	0.22	0.2	0.1	0.1
Pulses	*5.41*	*4.8*	*2.8*	*2.5*
Arhar	3.83	3.1	1.8	1.6
Urad	1.88	1.7	1.0	0.9
Total foodgrains	*99.70*	*88.6*	*52.3*	*46.4*
Food-cum-commercial crops	*10.40*	*9.3*	*5.4*	*4.9*
Groundnuts	0.06	0.1	0.0	0.0
Sugarcane	10.04	8.9	5.2	4.8
Vegetables	0.30	0.3	0.2	0.1
Non-food crops	*2.34*	*2.1*	*1.2*	*1.1*
Fodder crops	2.34	2.1	1.1	1.1
Rabi	*102.21*	*100.0*	*53.5*	*47.6*
Cereals	*75.63*	*73.9*	*39.6*	*35.3*
Wheat	69.62	68.1	36.4	32.4
Wheat mixed with barley	3.91	3.8	2.1	1.9
Barley	2.10	2.0	1.1	1.0
Pulses	*2.03*	*2.0*	*1.1*	*0.9*
Gram	1.35	1.3	0.7	0.6
Peas	0.08	0.1	0.0	0.0
Masoor	0.60	0.6	0.3	0.3
Total foodgrains	*77.66*	*75.9*	*40.7*	*36.2*
Food-cum-commercial crops	*24.55*	*24.1*	*12.8*	*11.4*
Oil-seeds	24.45	23.9	12.8	11.4
Vegetables & spices	0.10	0.2	0.0	0.0

Net cropped area of the village 190.75 hectares.
Total cropped area of the village 214.65 hectares.

TABLE 6.3: Land-use and Population
[Total Population of the Village 625 (1987): *Area in hectare*]

	Total area of the village	Net cropped area of the village	Net cropped area in kharif	Net cropped area in rabi	Total cropped area of the village	Double cropped area of the village
	241.72	190.75	112.44	102.21	214.65	23.90
Land per head of population	0.39	0.30	0.18	0.16	0.34	0.04

Note: Population data of this and other sample villages were collected at the time of field study.

Carrying Capacity of Land*

With the present agricultural practices, one hectare of total cropped area is capable of supporting 3.3 persons (3.3 SNU), while the present burden is that 2.9 persons.

Land-use Planning

The following scheme for the optimum utilisation of the land in accordance with its capability is suggested (Fig. 6.1):

Land Capability Class I

Since this is the most fertile land in the village, highly intensive cultivation is recommended here with vegetables, sugarcane, wheat, maize and oil-seeds as the main crops. Addition of bulky organic manures is required for maintaining the fertility of the soils and obtaining high yields.

Land Capability Sub-class IIe

This is capable of intensive cultivation with maize, *arhar*, wheat, gram

* Carrying capacity of land denotes the number of people that a unit area of land can support and it can thus be used as a yard-stick for measuring the farming or agricultural efficiency. The best way for measuring the carrying capacity is to determine the calorie output available for ingestion per hectare of area and then to divide it by SNU (method described in Appendix X).

barley and oil-seeds as the main crops. The conservation practices recommended are:

1. Addition of bulky organic manures,
2. Growing of cover crops like groundnut and leguminous crops such as *urad, moong, arhar* etc.
3. Adoption of scientific crop rotation, viz.

    (i) Early paddy-peas-sugarcane,
    (ii) Medium paddy-wheat-*moong*,
    (iii) Maize-wheat-*arhar*.

Land Capability Sub-class IIw

This is also suitable for intensive cultivation with paddy, *masoor*, wheat, gram etc. as main crops. The conservation practices recommended are:

1. Proper provision of surface drainage,
2. Addition of bulky organic manures,
3. Adequate and timely fertiliser use, and
4. Scientific crop rotation, viz.

    (i) Late paddy-berseem/*masoor*,
    (ii) Medium paddy-wheat-moong

Land Capability Sub-class IIIe

Since this land is comparatively less fertile and the limitations of erosion are severe, moderate cultivation is recommended with maize, *arhar, urad* and other leguminous crops, and groundnut, sweet-potato and gram etc. as main cover crops. Contour bunding and cultivation of cover crops are specially recommended to check soil erosion. Other recommendations are more or less the same as in the case of sub-class IIe.

Land Capability Sub-class IVe

This land suffers very severely from soil erosion and thus this is suitable for limited cultivation and orchards. Inter-cultivation of leguminous crops and growing of orchards are recommended specially. As regards the measures for soil conservation, contour bunding, levelling and cultivation of cover crops are recommended.

Land Capability Class V

This is the land which has been eroded and lies submerged under water. This can be used for 'boro' paddy cultivation.

Land Capability Class VI

This is the land which has very severely been eroded and converted into gullies. Reclamative measures should be adopted for reclaiming this land to cultivation and afforestation.

VILLAGE KEWALPUR

Location

The village Kewalpur which belongs to the gently rolling to nearly level lowlying land of tahsil Balrampur of Gonda district, lies at 27°34′ N latitude and 82°32′ E longitude. It is situated along the N.E. Railway and is also connected by a metalled road. It is at a distance of about 37 km from the tahsil headquarters.

Climate

No climatic data are recorded here, but the climatic conditions already discussed in the case of previous village may be taken as the close approximation of these conditions of the village in question also.

Soils

The soils of the village belong to the lowlying gently rolling and moderately to imperfectly drained soil group, the texture of which vary from loam and clayey loam to silty clay loam (Fig. 6.2).

Land Classification

The land of the village may be grouped into the following capability classes (Fig. 6.2):

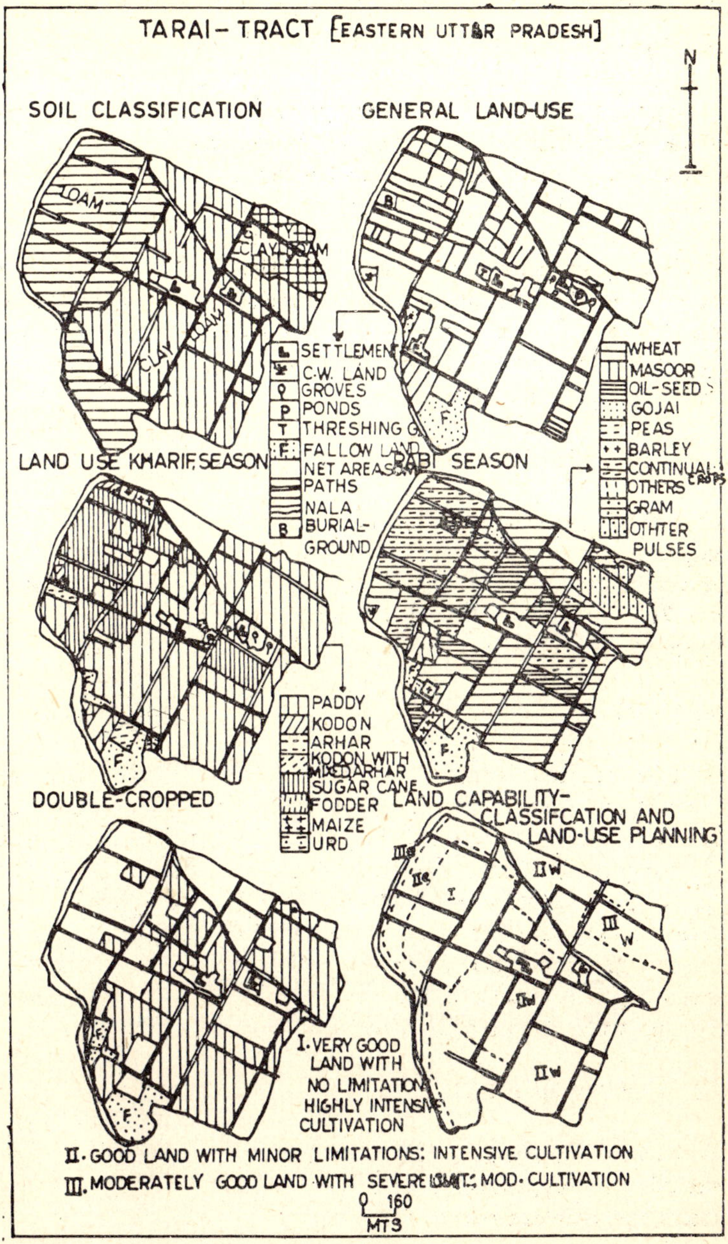

FIG. 6.2: **Village Kewalpur**

Land Capability Class I

This land which occupies an extensive area in the western part is without any limitation and possesses predominantly loamy fairly drained soils. The proportions of clay, silt and sand vary from 20-25 to 24.50, 20.75 to 21.25 and 59.25 to 59.8 per cent respectively. The slope ranges up to 1 per cent. The permeability is moderately rapid and water-holding capacity is moderately low. The soil reaction is mildly alkaline (pH 7.8). The natural fertility of the soils is low in respect of organic carbon and phosphorous and medium in respect of potash (0.3%, 13.4kg/ha and 123.2kg/ha respectively). The suitable crops are early paddy, sugarcane, wheat etc.

Land Capability Sub-class IIe

This type of land which suffers from minor limitation of soil erosion occupies a narrow strip towards west and south-west. The slope here varies from 1 to 3 per cent and the permeability is moderately rapid to rapid. The water-holding capacity is moderately low to low. The soil texture and other properties are the same as in the case of class I.

Land Capability Sub-class IIw

This type of land occupies an extensive area around the main settlement sites and possesses predominantly clayey loam fairly to imperfectly drained soils. The proportions of silt, clay and sand respectively are 23.3, 26.0 and 50.7 per cent, slope ranges up to 1 per cent. The permeability is slow and water-holding capacity is high. The soil reaction is mildly alkaline. The natural fertility is very low to low in respect of organic carbon (0.3%) and phosphorous (17.92 kg/ha) respectively. It is medium in respect of potash (134.40 kg/ha). Paddy, sugarcane, wheat, *masoor* and oil-seeds are the most suitable crops.

Land Capability Sub-class IIIe

A very narrow strip of land on the extreme west along the *nala* comes under this category, where the limitation of erosion is severe. In other respects, it resembles the land of IIe category.

Land Capability Sub-class IIIw

This consists of silty clay loam soils occupying the north-eastern corner of the village. It also resembles IIw sub-class in many respects. The different is that of degree only. Here the problem of water-logging is more severe.

Irrigation

The area under irrigation is almost negligible, as the means of irrigation do not exist and the sub-soil moisture is sufficiently high.

General Land-use Pattern

Table 6.4 and Fig. 6.2 show the general land-use of the village. It may be seen from them that 8.6 per cent of the total area is not available for cultivation. Settlements and paths claim about 5.0 and water 3.6 per cent of the total area. Groves claim only 1.3 per cent of the total area. Fallow land accounts for 3.1 per cent and net sown area for 86.1 per cent of the total area.

TABLE 6.4 General Land-use (1987)

Land-use	Area in hectares	% of the total area
A. *Area not available for cultivation*	9.09	8.6
(i) Water	3.81	3.6
(ii) Settlements, paths etc	5.28	5.0
B. *Other uncultivated land excluding fallowland*	2.31	2.2
(i) Culturable wasteland	0.95	0.9
(ii) Groves	1.36	1.3
C. *Arable land*	94.73	89.2
(i) Fallowland	3.36	3.1
(ii) Net sown area	91.37	86.1
Total	106.13	100.0
Double cropped area (as % of net sown area)	41.03	100.0
Total area of the village	106.13 hectares	

Cropland Use

Table 6.5 and Fig. 6.2 show the cropland use of the village. It is clear from them that kharif crops occupy 86.2 per cent of the net and 59.5 per cent of the total cropped area of the village. Paddy is the principal crop of this season, occupying 64.9 per cent of the kharif, 56.0 per cent of the net and 38.6 per cent of the total cropped area of the village. Its higher concentration is in the central and eastern parts. *Kodon* alone and mixed and maize are other insignificant cereals. Arhar and urad are the pulses which also occupy insignificant areas. Sugarcane is the principal food-cum-commercial crop occupying 30.6 per cent of the kharif, 26.3 per cent of the net and 18.2 per cent of the total cropped area.

Rabi crops claim 58.6 per cent of the net and 40.5 per cent of the total cropped area of the village. Wheat is the main crop of this season, which occupies 61.2 per cent of the rabi, 35.6 per cent of the net and 24.8 per cent of the total cropped area. Barley and wheat mixed with barley are other minor rabi cereals. Pulses occupy 17.2 per cent of the rabi, 10.1 per cent of the net and 6.9 per cent of the total cropped area. Masoor, peas, and gram are the pulse crops. Food-cum-commercial crops comprise of mainly oil-seeds which occupy 15.6 per cent of the rabi, 9.1 per cent of the net and 6.4 per cent of the total cropped area.

Double Cropping

Double-cropped area claim 38.7 per cent of the net cropped area and occupies the central and eastern parts.

Salient Features

A perusal of the cropland use of the village shows the following salient features of the cropping pattern of the village:

1. Kharif crops are more important than the rabi crops.
2. Foodgrains occupy only 74.8 per cent of the total cropped area (40.7% in kharif and 34.1% in rabi).
3. Food-cum-commercial crops occupy 24.6 per cent (18.2% in kharif and 6.4% in rabi) of the total cropped area. This shows that these crops are quite important in the cropping pattern of the village. This is mainly due to great importance of sugarcane in the kharif and oil-seeds in the rabi season.

Land-use and Population

Table 6.6 shows the totals of various categories of land in the village and also the per capita share of the villagers in these lands. It is clear from it that the per head share of the total area in the village is 0.48 hectare and that of the net cropped area is 0.41 hectare. The per head share in the net cropped area in the kharif and rabi crops respectively is 0.36 and 0.24 hectare and thus the per head total cropped area actually supporting one person in the village comes to 0.60 hectare. It may be mentioned here that 73.3 per cent of the total population of the village is enumerated as workers, of whom 62.9 per cent are cultivators and 30.9 per cent agricultural labourers.

TABLE 6.5: Cropland Uses (1987)

(Net cropped area of the village: 91.37 hectares
Total cropped area of the village: 132.40 hectares)

Crops	Area in hectares	% of kharif/ rabi cropped area	% of net cropped area of the village	% of total cropped area of the village
1	2	3	4	5
Kharif	*78.80*	*100.0*	*86.2*	*59.5*
Foodgrains	*54.0*	*68.4*	*59.0*	*40.7*
a. Cereals	52.80	66.9	57.7	39.8
Paddy	51.20	64.9	56.0	38.6
Maize	0.40	0.5	0.4	0.3
Kodon	0.40	0.5	0.4	0.3
Kodon mixed with arhar	0.80	1.0	0.9	0.6
b. *Pulses*	*1.20*	*1.5*	*1.3*	*0.9*
Arhar	0.80	1.0	0.9	0.6
Urad	0.40	0.5	0.4	0.3
Food-cum-Commercial Crops	*24.00*	*30.6*	*26.3*	*18.2*
Sugarcane	24.00	30.6	26.3	18.2
Non-food Crops	*0.80*	*1.0*	*0.9*	*0.6*
Fodder crops	0.80	1.0	0.9	0.6
Rabi	*53.60*	*100.0*	*58.6*	*40.5*
Foodgrains	*45.20*	*84.4*	*49.5*	*34.1*
a. *Cereals*	*36.00*	*67.2*	*39.4*	*27.2*
Wheat	32.80	61.2	35.9	24.8
Wheat mixed with barley	0.40	0.80	0.4	0.3
Barley	2.80	5.2	3.1	2.1
b. *Pulses*	*9.20*	*17.2*	*10.1*	*6.9*
Gram	0.80	1.5	0.9	0.6
Pea	0.40	0.8	0.4	0.3
Masoor	1.60	2.9	1.8	1.2
Other pulses	6.40	12.0	7.0	4.8
Food-cum-Commercial Crops	*8.40*	*15.6*	*9.1*	*6.4*
Vegetables	0.80	1.5	0.8	0.6
Oil-seeds	7.20	13.3	7.9	5.5
Others	0.40	0.8	0.4	0.3

TABLE 6.6: Land-use and Population

[*Total population of the village: 220 (1987) Area in hectares*]

	Total area of the village	Net cropped area of the village	Net cropped area in kharif	Net cropped area in rabi	Total cropped area of the village	Double cropped area of the village
	106.13	91.37	78.80	53.60	132.40	41.03
Land per head of population	0.48	0.41	0.36	0.24	0.60	0.19

Carrying Capacity of Land

At present, one hectare of the total cropped area can support 5.0 persons (5SNU), while the existing pressure of population is that of 1.7 persons only. This shows that the village land can support still much larger the number.

Land-use Planning

In view of the existing land-use and the potentials, the following scheme is proposed for planning the land-use of the village:

Land Capability Class I

This land is capable of highly intensive cultivation with paddy, sugarcane, wheat, *arhar* and some vegetables as the main crops. The conservation measures recommended are:

1. Addition of bulky organic manures to maintain soil fertility,
2. Leguminous crops to be given due place in the crop rotation,
3. Provision of proper irrigation,
4. Adoption of suitable crop rotation, viz.

   (i) Early paddy-Pea-Sugarcane,
   (ii) Medium paddy-wheat-*Lobia*, and
   (iii) Medium paddy-Berseem.

Land Capability Sub-class IIe

This type of land is capable of intensive cultivation with paddy, sugarcane, wheat, maize, *kodon, arhar* etc. as the main crops. The following steps should be taken for its conservation:

1. Addition of bulky organic manures to improve the soil properties,
2. Adequate and timely fertiliser use,
3. Adequate supply of water,
4. Adoption of scientific crop rotation with due importance of leguminous crops:

   (i) Early paddy-Pea/Mustard-Sugarcane
   (ii) Medium paddy-Wheat-*Moong*.

Land Capability Sub-class IIw

This category of land is also capable of intensive cultivation with paddy, sugarcane, wheat, *masoor*, oil-seeds etc. as the main crops. The following steps should be taken as conservation practices:

1. Addition of bulky organic manures,
2. Provision of proper surface drainage,
3. Adequate and timely fertiliser use,
4. Supply of water for irrigation, and
5. Adoption of scientific crop rotation, viz.,

   (i) Late paddy-Linseed/Lentil,
   (ii) Late paddy-zaid paddy-*Dhaincha*
   (iii) Medium paddy-Wheat-*Moong*.

Land Capability Sub-class IIIe

This category of land is capable of moderate cultivation with maize, *arhar, urad, sweet-potato,* gram and other leguminous crops. Contour bunding and cultivation of cover crops should also be given due importance together with the application of bulky organic manures and adoption of scientific crop rotation.

Land Capability Sub-class IIIw

This is also capable of moderate cultivation with paddy, *masoor*, linseed etc. as the main crops. The scientific crop rotation, viz. (i) Late paddy - zaid paddy - *dhaincha* and (ii) Late paddy - linseed/Lentil should be adopted. Other measures to be adopted are the same as given in the case of IIw.

VILLAGE MARWATIA DEEGAR

Location

Lying near the Indo-Nepal border in the gently rolling to nearly level lowlying lands, the village Marwatia Deegar is situated at 27°25′ N latitude and 83°3′ E longitude in the tahsil Naugarh of district Bansi. The nearest bus station is 7 km and the nearest railway station is 19 km. The village is thus not favourably located with respect to means of transport and communication.

Climate

No climatic data are recorded in the village. The rainfall data recorded at the tahsil headquarters of Naugarh may be taken as close approximation of the climatic conditions prevalent in the village. The average annual rainfall is 1458. 7mm, of which 2.2, 4.4 and 93.4 per cent respectively is received during the cold, hot and rainy season. The co-efficient of annual variability is 23.7 per cent, but that of monthly respectively is 53.1, 47.5, 46.9, 57.1 and 100.9 per cent for the months of June, July, August, September and October.

Soils

The soils of the village generally are silty clay loam, silty loam and sandy loam in texture (Fig. 6.3).

Land Classification

The land of the village may be grouped into the following land capability classes (Fig. 6.3):

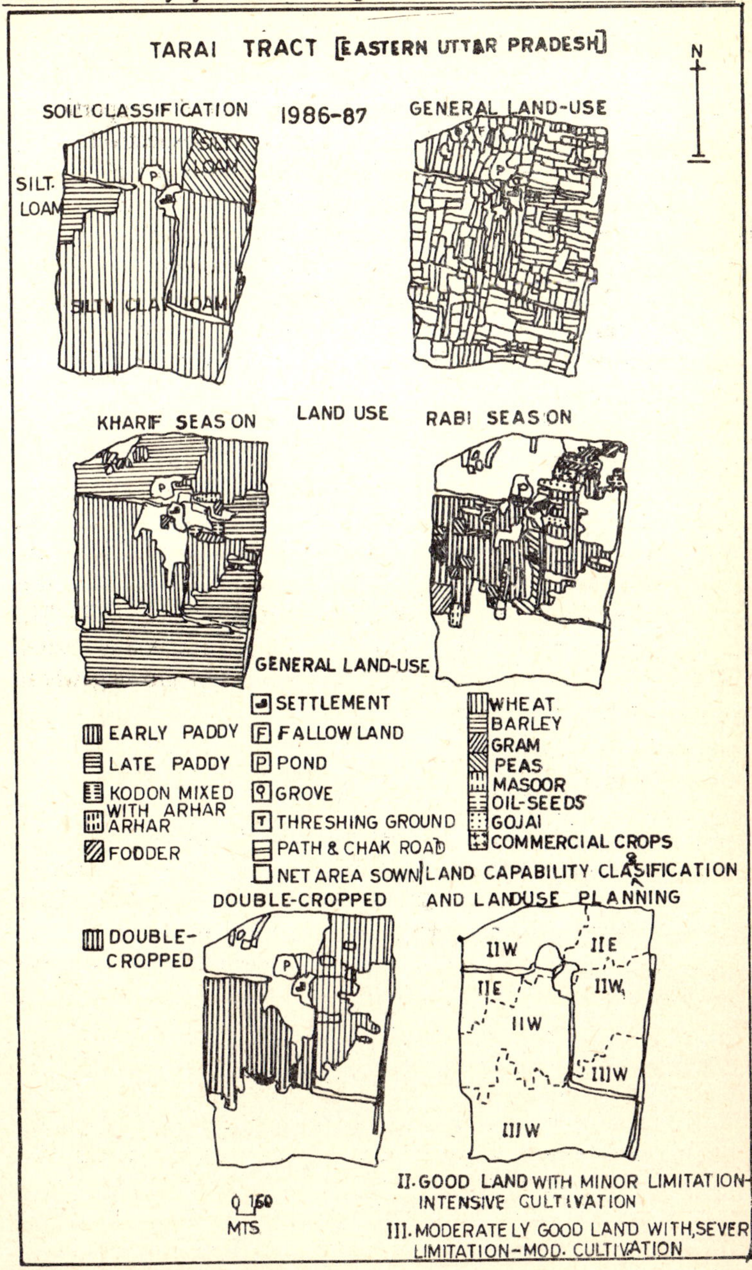

FIG. 6.3: Village Marwatia Deegar

Land Capability Sub-class IIw

As already mentioned, the soils are silty clay loam, the proportion of clay, silt and sand being 25.6, 31.4 and 43.0 per cent respectively. They are poorly to fairly drained. The slope ranges from 0 to 1 per cent in the south, and 1 to 3 per cent in the north. The permeability of soil is slow and water-holding capacity ranges between 40 and 45 per cent. The soil reaction is mildly alkaline (pH 7.5). The fertility status of the soils is low. The percentage of organic carbon is 0.3 and the amount of phosphorous and potash respectively is 10.6 and 67.2 kg/ha. The main crops which may be grown successfully are paddy, sugarcane, wheat etc.

Land Capability Sub-class IIe

There are some small patches of this land, one towards west and one towards north-east. The soils are sandy loam to silty loam. They are well drained and are subjected to minor erosion. The slope ranges between 1 and 3 per cent and permeability is moderately rapid. The water-holding capacity is moderately low. The soil reaction is neutral. The fertility status is low. The soils are suitable for growing wheat, sugarcane, arhar, barley, paddy etc.

Land Capability Sub-class IIIw

This category of land occupies the southern part of the village and consists of silty clay loam soils which are mildly alkaline in reaction. The fertility status is low and permeability is slow. The water-holding capacity is high. Other conditions are the same as in the case of IIw. Paddy, sugarcane, wheat, *masoor* etc. are the main crops suited to this type of land.

Irrigation

A little less than 40 per cent of the net cropped area is irrigated, the main source of which are ponds. The irrigated area has increased by 13.5 per cent since 1955-57.

Manures and Fertilisers

Cow-dung is used as a manure, but its quantity is very unsatisfactory.

The consumption of chemical fertilisers is hardly 15 kg per hectare.

General Land-use Pattern

Table 6.7 and Fig. 6.3 show the general land-use pattern of the village. It may be seen from them that 5.4 per cent of the total area is not available for cultivation, as it is either under water or under settlements, paths etc. Other uncultivated land excluding fallow land claims only 0.8 per cent of the total area. It was almost the same about three decades back. Fallowland is also quite insignificant, but the net cropped area claims 93.5 per cent of the total area. This indicates that there is little scope for further horizontal expansion of the cultivated land.

TABLE 6.7 General Land-use (1987)

Land-use	Area in hectares	% of the total area
A. *Area no available for cultivation*	3.06	5.4
(i) Area under water	0.81	1.4
(ii) Area under settlement etc.	2.25	4.0
B. *Other Uncultivated land excluding fallowland*	0.43	0.8
(i) Groves	0.05	0.1
(ii) Culturable wasteland	0.38	0.7
C. *Arable land*	53.17	93.8
(i) Fallowland	0.15	0.3
(ii) Net sown area	53.02	93.5
Total	56.66	100.0
Double cropped area (as % of net sown area)	18.60	35.1
Net irrigated area (as % of net sown area)	20.23	38.2
Total area of the village	56.66 hectares	

Cropland Use

Table 6.8 and Fig. 6.3 show the cropland use of the village. It is clear from them that kharif crops occupy 92.1 per cent of the net and 68.1 per cent of the total cropped area of the village. Previously, these percentages were 77.3 and 95.1 in 1955. The decreases seem due to the increasing importance of the rabi crops. Paddy is the dominant crop, claiming 96.6 per cent of the kharif, 88.9 per cent of the net and 65.8 per cent of the total cropped area. This crop was even more important in the past. Kodon and pulses are other minor crops.

Rabi crops occupy only 31.8 per cent of the total and 43.0 per cent of the net cropped area of the village. Their share was only 22.7 per cent in the total cropped area in 1955. More demand for rabi crops such as wheat and better irrigation facilities are some of the factors causing an increase in the area under rabi crops. Wheat is the main crop, occupying 15.8 per cent of the total, 21.4 per cent of the net and 49.7 per cent of the rabi cropped area of the village. It is interesting to note that in the year 1955, wheat did not appear at all on the land-use map of the village. Wheat mixed with barley, pulses and oil-seeds are other minor crops.

Double Cropping

The area cropped more than once claims 35.0 per cent of the net cropped area of the village. It was about 24.0 per cent in 1955.

Salient Features of the Cropping Pattern

An analysis of the cropland use deciphers the following salient features:

1. The village has mono-culture, as paddy is the predominant crop.
2. Kharif crops are still predominant, claiming about two-thirds of the total cropped area.
3. The subsistence nature of agriculture is explicit from the predominance of the foodgrains and comparative insignificance of the food-cum-commercial crops. About 97.0 per cent (68.1% in kharif and 28.9% in rabi) of the total cropped area is devoted to the cultivation of foodgrains. Food-cum-commercial crops occupy only 3.0 per cent of the total cropped area. The predominance was almost the same in the past.
4. Fodder crops do not appear on the cropland use map.

Land-use and Population

Table 6.9 depicts that the per capita share in the total area and net cropped area of the village respectively is 0.17 and 0.16 hectare. The share in the kharif and rabi cropped area amounts to 0.15 and 0.17 hectare respectively. Thus the per head total cropped area in the village comes to 3.22 hectare. This is the actual amount of land supporting one person. As regards the occupational structure, about 32 per cent persons are workers, all of whom depend on land for their livelihood.

TABLE 6.8: Cropland Use (1987)

(Net cropped area of the village : 53.02 hectares)
(Total cropped area of the village : 71.62 hectares)

Crops	Area in hectares	% of kharif/rabi cropped	% of net cropped area of the village	% of total cropped area of the village
1	2	3	4	5
Kharif	*48.80*	*100.0*	*91.9*	*68.0*
Foodgrains	*48.80*	*00.0*	*91.9*	*68.0*
a. Cereals	47.97	98.3	90.4	66.9
Paddy	47.14	96.6	88.9	65.8
Kodon	0.83	1.7	1.5	1.1
b. *Pulses (arhar)*	*0.83*	*1.7*	*1.5*	*1.1*
Rabi 22.82	*100.0*	*43.0*	*32.0*	
Foodgrains	*20.72*	*90.7*	*39.0*	*29.1*
a. Cereals	15.89	69.6	29.9	22.3
Wheat	11.34	49.7	21.4	16.0
Wheat mixed with barley	3.52	15.4	6.6	4.9
Barley	1.03	4.5	1.9	1.4
b. *Pulses*	*4.83*	*21.1*	*9.1*	*6.8*
Pea	2.48	10.8	4.7	3.5
Gram	2.35	10.3	4.4	3.3
Food-cum-commercial crops	*2.10*	*9.3*	*4.0*	*2.9*
Oil-seeds	1.74	8.0	3.4	2.5
Vegetables	0.36	1.3	0.6	0.4

TABLE 6.9: Land-use and Population*

[Total population of the village : 325 (1987)]

	Total area of the village	Net cropped area of the village	Net cropped area in kharif	Net cropped area in rabi	Total cropped area of the village	Double cropped area of the village
	56.66	53.02	48.80	22.82	71.62	18.60
Land per head of population	0.17	0.16	0.15	0.07	0.22	0.06

* This includes the population of some other villages also, which depends on the land of this village.

Carrying Capacity of Land

With the present management practices, the carrying capacity of land is 4.3 persons (4.3SNU) per hectare, while the existing pressure of population on per hectare of the total cropped area is almost the same. This warrants the proper utilisation of the land resources for meeting the demand of increasing population.

Land-use Planning

Keeping in view the land potentials and the conservation of the land resources, the following lines of action are proposed for planning the land-use of the village:

Land Capability Sub-class IIe

This land is capable of intensive cultivation with paddy, sugarcane, wheat, groundnut, *arhar*, barley etc. as the main crops which should be grown here. Conservation measures recommended are:

1. Addition of bulky organic manures to improve soil structure,
2. Growing of cover crops such as groundnut and leguminous crops such as *urad*, moong etc.
3. Adoption of scientific crop rotation, viz.

   (i) Early paddy/Maize-Wheat/Barley-*Moong*/*Lobia*
   (ii) Early paddy-Mustard-*Dhaincha*
   (iii) Maize-Sugarcane-x
   (iv) Maize/*Lobia*-Wheat-*Moong*.

Land Capability Sub-Class IIw

Intensive cultivation with paddy, sugarcane, wheat, peas, *masoor etc.* is recommended. The conservation measures proposed are:

1. Addition of bulky organic manures to improve soil structure,
2. Provision of surface drainage wherever feasible,
3. Sinking of tube-wells,
4. Adequate and timely fertilisers use based on soil testing recommendation.

5. Adoption of scientific crop rotation, viz.

   (i) Medium paddy-Wheat-*Moong*
   (ii) Medium paddy-Peas-Sugarcane
   (iii) *Dhaincha*/late paddy-*Lentil*-Linseed.

Land Capability Sub-Class IIIw

Moderate cultivation with paddy, pea, sugarcane, wheat, masoor etc. is recommended on this type of land. The following conservation practices are recommended:

1. Addition of bulky organic manures to improve the physical properties of the soils,
2. Provision of surface and sub-surface drainage,
3. Regulated irrigation based on crops needs and intake rates of soils,
4. Sinking of tube-wells at suitable intervals,
5. Adoption of scientific crop rotation as already recommended in IIw,
6. Adequate and timely fertiliser use, but minimum use of calcareous fertilisers.

VILLAGE MUNDILA

Location

The village Mundila of tahsil Domariaganj, district Basti, belongs to lowlying area of the tarai tract of eastern Uttar Pradesh. It is situated at 27°2′ N latitude and 82°32′ E longitude. Its location with respect to accessibility is unfavourable, particularly so in the rainy season when an extensive area surrounding it becomes waterlogged isolating it from other parts. It lies at a distance of about 35 km from the tahsil headquarters. The nearest rurban centre of Itwa is 18 km.

Climate

No climatic data are recorded in the village. The rainfall data recorded at the tahsil headquarters of Domariaganj, may be taken as close approximation of the climatic conditions of the village. The total average annual rainfall is 1310.10mm, of which 90.8 per cent is received during the rainy, 5.9 per cent during the hot and remaining 3.3 per cent during the cold season. The co-efficient of annual variability of rainfall is 44.8 per cent, which shows the highly erratic nature of rainfall. The co-efficient of monthly variability of rainfall is the highest in the month of October (147.7%). It is 68.7 for June, 52.5 for July, 62.2 for August and 92.4 per cent for the month of September. The high variabilities of the months of June and September are again very important from agricultural points of view.

Soils

The soils of the village belong to that group of soils which occur in imperfectly to moderately well drained areas. Texturally, they are silty loam to silty clay loam (Fig. 6.4).

Land Classification

The land of this village may be grouped into the following land capability classes (Fig.6.4):

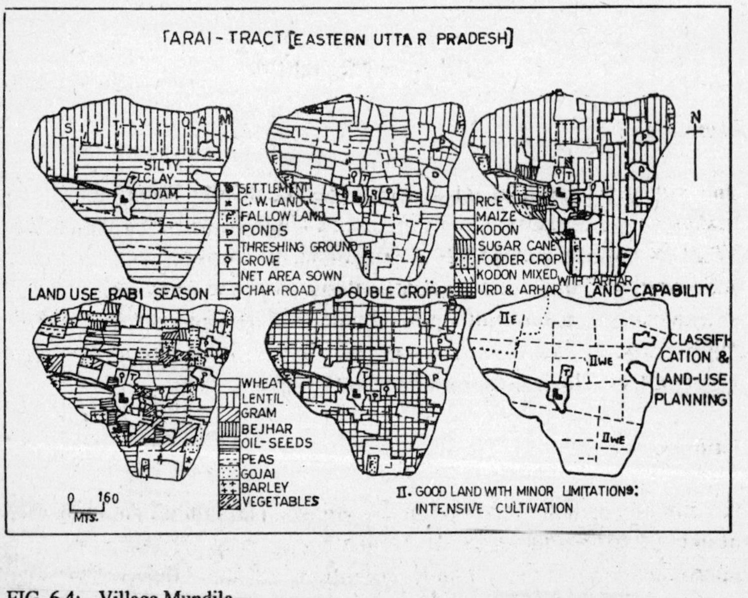

FIG. 6.4: Village Mundila

Land Capability Sub-class IIe

This land which occupies the north-western part of the village consists of silty loam fairly drained soils, slope ranging between 1 and 3 per cent, the permeability is moderately slow and water-holding capacity is moderately high. The soil reaction is mildy alkaline (pH 7.8). The proportions of clay, silt and sand respectively are 20.0, 26.0 and 54.0 per cent. The natural fertility of the soils is low in respect of phosphorous (13.4 kg/ha) and medium in that of potash (168.0 kg/ha). It is low in respect of organic carbon (0.251%). The suitable crops are paddy, sugarcane, wheat, *masoor*, maize etc. The main limitation is that of erosion.

Land Capability Sub-class IIwe

This land consists of the silty clay loam fairly to poorly drained soils which cover the rest of the village. The slope ranges between 1 and 3 per cent. The proportions of clay, silt and sand respectively are from 25.8 to 26.5, from 25.5 to 26.5 and from 47.0 to 48.7 per cent. The permeability is slow and water-holding capacity is high. The soil reaction is moderately alkaline (pH 8.0 to 8.1). The fertility status of the soils is medium in respect of organic carbon (0.458%) and potash (149.3 kg/ha). It is low in respect of phosphorous (17.9 kg/ha). The suitable crops are early and late paddy, sugarcane, wheat, oil-seeds etc. The main limitations are those of minor erosion and water-logging.

Irrigation

The facilities of irrigation are quite inadequate, as only 11.9 per cent of the net cropped area is under irrigation. Ponds are the main source of water for irrigation.

Manure and Fertilisers

The use of manure and fertilisers is quite limited, for which the economic condition and lack of due incentive, their inadequate supply together with the lack of irrigation facilities appear to be responsible.

General Land-use Pattern

Table 6.10 and Fig 6.4 give the general land-use pattern of the village, which reveal that 7.6 per cent of the total area of the village is not available for cultivation, as it is occupied by the settlements, water bodies etc. Groves and culturable wasteland claim 4.6 per cent of the total area and fallowland also covers the same extent of area. The net cropped area accounts for 83.2 per cent of the total area. An increase has been experienced in the case of this category and a corresponding decrease in the area under fallowland, culturable wasteland, groves etc. The area under settlements, path etc. has increased.

Cropland Use

Table 6.11 and Fig 6.4 indicate that 62.2 per cent of the net and 51.5 per cent of the total cropped area is devoted to the cultivation of kharif crops, among which paddy claiming 78.2 per cent of the kharif, 48.5 per cent of the net and 40.2 per cent of the total cropped area, is the main crop. Its distribution is very extensive. *Kodon* alone and mixed, maize, sugarcane etc. are other crops grown in this season, the relative importance of which such as paddy has remained almost the same.

Rabi claims 58.4 per cent of the net and 48.5 per cent of the total cropped area of the village. Wheat claiming 54.7 per cent of the rabi, 32.1 per cent of the net and 26.5 per cent of the total cropped area is the main crop, the cultivation of which also spreads over whole of the village. Wheat mixed with barley, barley mixed with gram, gram, peas, *masoor* and oil-seeds are other crops. The area under wheat has increased immensely.

Double Cropping

Double-cropped area claims 21.1 per cent of the net cropped area of the village. There has been insignificant increase in this area, which is due to inadequate means of irrigation and extensive water-logging in some years.

Salient Features of the Cropping Pattern

An analysis of the cropland use of the village shows the following salient features of the cropping pattern:

1. Kharif and rabi crops are almost of similar importance.
2. Foodgrains dominate the cropping pattern by claiming 91.0 per

cent (47.8% in kharif and 43.2% in rabi) of the total cropped area.

3. Food-cum-commercial crops occupy only 8.3 per cent of the total cropped area and thus are not very significant.
4. Non-food crops are very insignificant.

Land-use and Population

Table 6.12 gives the totals of various categories of land in the village and and also the per capita share of the villagers in these lands.

It is clear that the per head share of the total area in the village is 0.33 hectare and that in the net cropped area is 0.27 hectare. The per head share in the net cropped area in the kharif and rabi crops respectively is 0.17 and 0.16 hectare. Thus the total cropped area per head comes to 0.33 hectare. This is the actual amount of land supporting one person in the village.

Of the total population of the village, 58.1 per cent are workers, of whom 45.6 per cent are cultivators and 54.4 per cent are agricultural labourers. Thus total population depends on land resources for its livelihood.

TABLE 6.10: General Land-use (1987)

Total area of the village : 52.40 hectares

Land-use	Area in hectares	% of the total area
A. *Area not available for cultivation*	4.00	7.6
(i) Area under water	1.60	3.0
(ii) Area under settlement etc.	2.40	4.5
B. *Other Uncultivated land excluding fallowland*	2.40	4.6
(i) Groves	1.20	2.3
(ii) Culturable wasteland	1.20	2.3
C. *Arable land*	46.00	87.8
(i) Fallowland	2.40	4.6
(ii) Net sown area	43.60	83.2
Total	52.40	100.0
Double-cropped area (as % of net sown area)	9.00	21.1
Net irrigated area (as % of net sown area)	5.20	11.9

TABLE 6.11: Cropland Use (1987)

(Net cropped area of the village : 43.60 hectares)
(Total cropped area of the village : 52.60 hectares)

Crops	Area in hectares	% of Kharif/rabi cropped area	% of net cropped area of the village	% of total cropped area of the village
1	2	3	4	5
Kharif	*27.00*	*100.0*	*62.2*	*51.5*
Foodgrains	*25.00*	*92.8*	*57.6*	*47.8*
a. *Creals*	*24.60*	*91.3*	*56.7*	*47.1*
Paddy	21.00	78.2	48.5	40.3
Maize	0.80	2.9	1.8	1.5
Kodon	1.20	4.4	2.7	2.3
Kodon mixed with arhar	1.60	5.8	3.7	3.0
b. *Pulses* (Urad and *Moong*)	*0.40*	*1.5*	*0.9*	*0.7*
Food-cum-commercial crops (Sugarcane)	*1.60*	*5.8*	*3.7*	*3.0*
Non-food crops (fodder)	*0.40*	*1.40*	*0.9*	*0.7*
Rabi	*25.60*	*100.0*	*58.4*	*48.5*
Foodgrains	*22.80*	*89.1*	*52.1*	*43.2*
a. *Cereals*	*18.00*	*70.3*	*41.1*	*34.0*
Wheat	14.00	54.7	32.1	26.5
Barley	0.80	3.1	1.8	1.5
Barley mixed with gram	0.40	1.6	0.9	0.7
Wheat mixed with gram	2.00	7.8	4.5	3.8
Others	0.80	3.1	1.8	1.5
b. *Pulses*	*4.80*	*18.8*	*11.0*	*9.2*
Gram	1.60	6.3	3.7	3.1
Peas	0.80	3.1	1.8	1.5
Masoor	2.40	9.4	5.5	4.6
Food-cum-commercial crops	*2.80*	*10.9*	*6.3*	*5.3*
Oil-seeds	2.00	7.8	4.5	3.8
Vegetables	0.80	3.1	1.8	1.5

TABLE 6.12: Land use and Population

[Total population of the village : 160 (1987) Area in hectares]

	Total area of the village	Net cropped area of the village	Net cropped area in kharif	Net cropped area in rabi	Total cropped area	Double cropped area
	52.40	43.60	27.00	25.60	52.60	9.00
Land per head of population	0.33	0.27	0.17	0.16	0.33	0.0

Carrying Capacity of Land

With the present land management practices, the carrying capacity of land is 3.5 persons (3.5SNU) per hectare of the total cropped area, while existing pressure of population is 3.0 persons. If all the food production is available for consumption, there appears no problem of mal-nutrition at present, but for meeting the future demands, agricultural production has to be increased.

Land-use Planning

Keeping in view the existing land-use, land potentials and requirements of the population, the planning of the land-use of the village is envisaged as follows (Fig. 6.4):

Land Capability Sub-class IIe:

This land is capable of intensive cultivation which is recommended with due place to be given to the leguminous crops. The conservation measures recommended are :

1. Addition of bulky organic manures to improve soil structure,
2. Sinking of tubewells,
3. Adequate and timely fertilisers use based on soil testing recommendations,
4. Use of insecticides and pesticides,
5. Adoption of scientific crop rotation, viz.

   (i) Early paddy-peas-sugarcane,
   (ii) Medium paddy-Wheat-Moong/*Lobia*,
   (iii) Medium paddy-Berseem-Berseem.

Land Capability Sub-class IIw:

These are the lands which suffer from the minor hazards of submergence, water-logging, high water-table and erosion. They are thus suitable for intensive cultivation with paddy, sugarcane, wheat, oil-seeds etc. as the main crops, the cultivation of which is recommended. The following conservation measures should be adopted:

1. Addition of bulky organic manures,
2. Sinking of tubewells for assured irrigation and lowering down the water-table,
3. Adequate and timely fertiliser use,
4. Proper provision of surface drainage,
5. Leguminous crops should be grown and given due importance,
6. Adoption of scientific crop rotation, viz.

   (i) Medium paddy-wheat-moong,
   (ii) Medium paddy-wheat-sugarcane,
   (iii) Dhaincha/Late paddy-Lentil/Linseed.

Use of HYV of seeds suiting the soils of the village together with the modern technological innovations is also recommended.

VILLAGE MANGALPUR

Location

The village Mangalpur of tahsil Pharenda, district Gorakhpur, which belongs to gently rolling to nearly level lowlying lands of the tarai tract of eastern Uttar Pradesh, lies at 27°19′ N latitude and 83°18′ E longitude near the Indo-Nepal border. It is at a distance of about 25 km from the tahsil headquarters and is connected by an unmetalled road. Its accessibility becomes very poor during the rains when the streams coming out of the Nepal Tarai inundate the extensive part of this region.

Climate

No climatic data are recorded here. The rainfall data recorded at the tahsil headquarters may be taken as close approximation of the climatic conditions prevailing in the village. The average annual rainfall is

1585.8mm, of which 2.1, 6.3 and 91.6 per cent respectively is received during the cold, hot and rainy season. The co-efficient of annual variability is 26.2 per cent and that of the monthly for the months of June, July, August, September and October respectively is 59.9, 37.6, 48.0, 58.0 and 107.6 per cent. The high co-efficient of variability for the months of June and October is of great significance from agricultural point of view.

Soils

The soils of the village belong to the lowlying fairly to imperfectly drained northern belt, liable to inundation during the rains. As regard the texture, they vary from loam or silty loam to silty clay loam (Fig. 6.5).

Land Classification

The land of the village may be grouped into the following land capability classes (Fig 6.5):

Land Capability Class I

This is very good land with no limitation. It surrounds the settlement sites and consists of loam to silty loam fairly drained soils. The proportion of clay, silt and sand respectively is 17.5, 22.0 and 60.5 per cent in loam and 20.3, 27.0 and 52.7 per cent in the case of silty loam. The slope ranges up to 1 per cent or somewhat more. The permeability is moderately rapid to moderately slow and water-holding capacity is moderately low to moderately high. Soil reaction is mildly alkaline (pH 7.4). The natural fertility is low to very low in respect of organic carbon (0.125 to 0.325%) and phosphorous (8.9-17.9 kg/ha). It is medium (156.8-168.0 kg/ha) in respect of potash. The suitable crops are sugarcane, paddy, wheat, arhar etc.

Land Capability Sub-class IIe

This is good land with minor limitations of soil erosion. This is found towards west where silty loam soils predominate and slope ranges between 1 and 3 per cent. The permeability of soils is moderately slow and water-holding capacity is moderately high. The soil reaction is mildly alkaline (pH 7.4). The natural fertility is low in respect of organic carbon (0.325%), very low in respect of phosphorous (8.9 kg/ha) and medium in that of potash (156.8 kg/ha). The suitable crops are paddy, wheat, *masoor* etc.

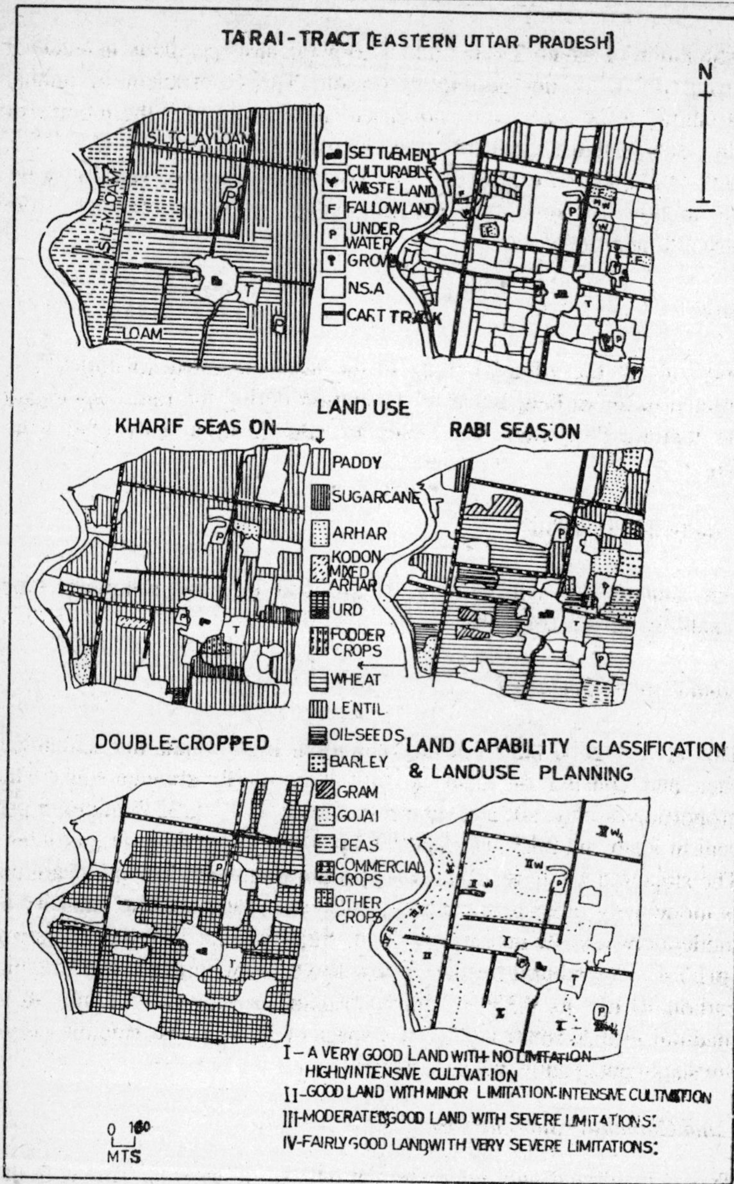

FIG. 6.5: Village Mangalpur

Land Capability Sub-class IIw

This land occupies the eastern and some portion of land in the north of the class I land. This suffers from minor hazards of water-logging and consists of silty clay loam soils with proportions of clay, silt and sand varying from 26.3 to 30.3, 26.4 to 30.0 and 43.2 to 47.3 per cent. The slope ranges between 1 and 3 per cent. The permeability of soils is slow and water-holding capacity is high. The soil reaction is mildy alkaline (pH 7.4 to 7.5). The natural fertility in respect of organic carbon is very low to low (0.125.0 - 0.350%). very low to low (8.9-17.9 kg/ha) in respect of phosphorous and medium (134.4-168.0 kg/ha) in respect of potash. Slow permeability, high water-table and submergence for 1 to 2 months are some of the problems. The suitable crops are paddy, sugarcane, *masoor*, wheat, oil-seeds, peas etc.

Land Capability Sub-class IIIw:

This land occupies the northern part of the village and has almost the same characteristic as IIw, the difference is that of degree. Here the problems of sub-mergence of land under water, slow permeability and high water-table are severe.

Land Capability Sub-class IIIe

In so many characteristics, it resembles IIe category of land, but the problem of erosion due to greater slope (which ranges between 3 and 5%) is severe. The suitable crops are arhar, sweet potato, mustard etc.

Land Capability Sub-class IVe

This sub-class is similar in characteristics to sub-classes IIe and IIIe, except slope and magnitude of erosion. The topography of soils is undulating and slopy. These soils occur on the side of the river in the west. The slope at some places is more than 5 per cent even. Consequently, erosional hazards are very severe. The suitable crops are sweet potato, arhar, oil-seeds etc.

Irrigation

About 20.0 per cent of the net sown area is irrigated. Ponds are the main sources of irrigation.

Manures and Fertilisers

The use of manure and fertilisers is quite limited and the compost is not at all in use.

General Land-use Pattern

Table 6.13 and Fig. 6.5 reveal that 11.5 per cent of the total area of the village is not available for cultivation, the major proportion of which is claimed by the area under water. The area under settlements etc. has increased in the past. Groves and culturable wastelands together claim 3.4 per cent of the total area. The area under culturable wasteland has decreased in the past. Arable land claims 85.1 per cent of the total area which includes fallow land (3.4%) and net sown area (81.7%). There is no remarkable change in the area under this category.

TABLE 6.13: General Land-use (1987)

(Total area of the village : 93.60 hectares)

Land-use	Area in hectares	% of the total area
A. *Area not available for cultivation*	10.80	11.5
(i) Area under water	7.60	8.1
(ii) Area under settlement etc.	3.20	3.4
B. *Other uncultivated land excluding fallowland*	3.20	3.4
(i) Groves	0.80	0.8
(ii) Culturable Wasteland	2.40	2.6
C. *Arable land*	79.60	85.1
(i) Fallowland	3.20	3.4
(ii) Net sown area	76.40	81.7
Total	93.60	100.0
Double-cropped area (as % of net sown area)	26.40	34.6
Net irrigated area (as % of net sown area)	18.40	20.0

Cropland Use

Table 6.14 and Fig. 6.5 show that kharif crops occupy 53.4 per cent of the total and 71.9 per cent of the net cropped area of the village. Paddy is the dominant crop, occupying 92.7 per cent of the kharif, 63.4 per cent of the net and 47.1 per cent of the total cropped area. This is extensively grown in the village. *Kodon* mixed with *arhar*, *urad*, *moong* and sugarcane are other crops. Sugarcane cultivation is mainly confined to loamy soils.

Rabi claims 46.6 per cent of the total and 65.8 per cent of the net cropped area. Wheat is the main cereal, occupying 43.3 per cent of the rabi, 27.2 per cent of the net and 20.2 per cent of the total cropped area. Its larger concentration is on fairly to well drained loamy and silty loam soils of the village. Barley and wheat mixed with barley are other cereals. The pulses occupy 30.0 per cent of the rabi, 18.8 per cent of the net and 14.0 per cent of the total cropped area. *Masoor* is the main pulse, occupying the silty clay loam soils of the village. Oil-seeds occupy only 2.7 per cent of the total cropped area of the village. There is slight increase in the rabi cropped area and in the area under wheat and *masoor*.

TABLE 6.14: Cropland Use (1987)

(Net cropped area of the village : 76.40 hectare)
(Total cropped area of the village : 102.80 hectares)

Crops	Area in hectares	% of kharif/rabi cropped area	% of net cropped area of the village	% of total cropped area of the village
1	2	3	4	5
Kharif	*54.80*	*100.0*	*71.9*	*53.4*
Foodgrains	*52.40*	*95.6*	*68.7*	*51.1*
a. *Cereals*	*50.80*	*91.7*	*66.5*	*49.4*
Paddy	48.40	88.3	63.4	47.1
Kodon mixed with *arhar*	2.40	4.4	3.1	2.3
b. *Pulses*	*1.60*	*2.9*	*2.2*	*1.7*
Urad and *moong*	1.20	2.2	1.6	1.2
Arhar	0.40	0.7	0.6	0.5
Food-cum-commercial crop (Sugarcane)	*2.0*	*4.4*	*2.6*	*1.8*
Non-food crops	*0.40*	*0.7*	*0.6*	*0.5*
Rabi	*48.00*	*100.0*	*62.8*	*46.6*
Foodgrains	*44.80*	*93.3*	*58.6*	*43.6*
a. *Cereals*	*30.40*	*63.3*	*39.8*	*29.6*
Wheat	20.80	43.4	27.2	20.2
Barley	2.80	5.8	3.7	2.7
Wheat mixed with barley	5.20	10.8	6.8	5.1
Others	1.60	3.3	2.1	1.6
b. *Pulses*	*14.40*	*30.0*	*18.8*	*14.0*
Masoor	10.40	21.7	13.6	10.1
Pea	1.60	3.3	2.1	1.6
Gram	2.40	5.0	3.1	2.3
Food-cum-commercial crops	3.20	6.7	4.2	3.0
Oil-seeds	2.80	5.9	3.7	2.7
Vegetables	0.40	0.8	0.5	0.3

Double Cropping

Double cropped area occupies 34.6 per cent of the net cropped area and it concentrates mainly on the very good and good lands. A moderate increase in this area has been observed in the past, which is due to the increasing pressure of population on land resources.

Salient Features of the Cropping Pattern

A perusal of the cropland use of the village indicates the following salient features of the cropping pattern:

1. Kharif crops are relatively a bit more important than that of the rabi crops.
2. Foodgrains occupy 94.7 per cent (51.1% in kharif and 43.6% in rabi) of the total cropped area, which shows their predominance in the cropping pattern.
3. Food-cum-commercial crops occupy only 4.8 per cent of the total cropped area and thus are quite insignificant. This substantiates the subsistence nature of farming.

Land-use and Population

Table 6.15 shows the totals of various categories of land in the village and also the per capita share of the villagers in these lands. The per head share of the total area in the village is 0.16 hectare and that of the net cropped area is 0.13 hectare. The per head share in the net cropped area in the kharif and rabi crops respectively is 0.09 and 0.08 hectare and thus the per head total cropped area actually supporting one person in the village comes to 0.17 hectare.

As regards the occupational structure, of the total population 32.1 per cent are workers. Of the total working force, 63.8 per cent are cultivators and the rest are agricultural labourers. Thus the total population depends upon land for its livelihood.

Carrying Capacity of Land

The carrying capacity of land is 3.9 persons (3.9SNU) per hectare of the total cropped area under the present management practices, i.e. one hectare of the total cropped area can support 3.9 persons, while the existing burden is that of 5.9 persons. This shows that the food resources and population are imbalanced even at present and in future, the burden of population will be still larger. This warrants a more intensive use of the land and great rise in the agricultural production to feed the present as well as the increasing population.

TABLE 6.15: Land-use and Population

[Total population of the village : 580 (1987) Area in Hectares]

	Total area of the village	Net cropped area of the village	Net cropped area in kharif	Net cropped area in rabi	Total cropped area	Double cropped area
	93.60	76.40	54.80	48.00	102.80	26.40
Land per head of population	0.16	0.13	0.09	0.08	0.17	0.04

Land-use Planning

Keeping in view the existing land-use and land potentials, the planning of the land-use of the village is envisaged as follows:

Land Capability Class I

A highly intensive cultivation with paddy, sugarcane, wheat, arhar and vegetables as the main crops is recommended for this category of land. The conservation practices recommended are:

1. Addition of bulky organic manures to maintain soil fertility,
2. Leguminous crops to be given due place in the crop rotation,
3. Provision of proper irrigation,
4. Adoption of scientific crop rotation, viz.

   (i) Early paddy-Pea-Sugarcane
   (ii) Medium paddy-Wheat-*Moong/Lobia*
   (iii) Medium paddy-Berseem.

Land Capability Sub-class IIe

On this category of land, intensive cultivation with paddy, wheat, peas, gram, *masoor*, *arhar* etc. as the main crops are recommended. The conservation measures recommended are:

1. Addition of bulky organic manures to improve the soil properties,
2. Adequately and timely use of fertilisers,
3. Adequate supply of water for irrigation
4. Adoption of scientific crop rotation, viz.

   (i) Early paddy-Pea/mustard-Sugarcane
   (ii) Medium paddy-Wheat-*Moong*.

Thus due importance should be given to the leguminous crops in the cropping pattern.

Land Capability Sub-class IIw

This land is also capable of intensive cultivation which is recommended with paddy, sugarcane, *masoor*, wheat, oil-seeds, peas etc. as the main crops. The conservation measures recommended are:

1. Addition of bulky organic manures,
2. Sinking of tube-wells to lower down the water-table,
3. Provision of surface drainage in a proper way,
4. Adequate and timely fertiliser use on the basis of soil testing recommendation,
5. Adoption of scientific crop rotation, viz.

   (i) Late paddy-Linseed/Lentil
   (ii) Late paddy-Zaid paddy-*Dhaincha*
   (iii) Medium paddy-Wheat-*Moong*.

Land Capability Sub-class IIIw

This land is capable of moderate cultivation which is recommended with special preference to crops such as paddy, *masoor* etc. The following conservation measures should be taken up earnestly:

1. The provision of proper surface and sub-surface drainage,
2. Adoption of scientific crop rotation, such as,

(i) Late paddy-Linseed/Lentil
(ii) Late paddy-Zaid paddy-*Dhaincha*
(iii) Medium paddy-Wheat-*Moong*.

Other measures to be adopted are the same as given in the case of IIw.

Land Capability Sub-class IIIe

This land is also capable of moderate cultivation with maize, *arhar*, *urad*, sweet-potato, gram and other leguminous crops as the preferable crops. Among the conservation practices, counter bunding, cultivation of cover crops, application of bulky organic manures, adoption of scientific crop rotation etc. which have already being discussed in the case of IIe, should be adopted.

Land Capability Sub-class IVe

This is the land which is capable of limited cultivation only with inter-cultivation of leguminous crops which should be grown. The conservation measures to be adopted are:

1. Contour bunding and levelling,
2. Land shaping and terraced cultivation to control soil erosion,
3. Growing of cover crops,
4. Incentive for growing orchards and afforestation,
5. Adoption of proper crop rotation,
6. Addition of bulky organic manures to improve the physical properties of the soils..

Use of HYV of seeds and adoption of modern technological innovations are also recommended.

VILLAGE DHAURAHRA

Location

The village Dhaurahra which belong to the gently rolling to nearly level lowlying lands, lies at 27°29′ N latitude and 83°28′ E longitude in the tahsil Pharenda of district Gorakhpur. It touches the Indo-Nepal border. It is connected with the rurban centre, Nautanwa by a metalled road and is only 6 km from it. Its distance from the tahsil headquarters is 48 km and that from district headquarters is 91 km. Thus its location from the accessibility point of view is not very satisfactory.

Climate

Climatic conditions are almost similar to those already discussed in the case of the village Mangalpur of the same tahsil.

Soils

The soils of the village (Fig.6.6) are silty loam, clayey loam and silty clay loam. Their genetic conditions are the same as those of other villages of this region already discussed.

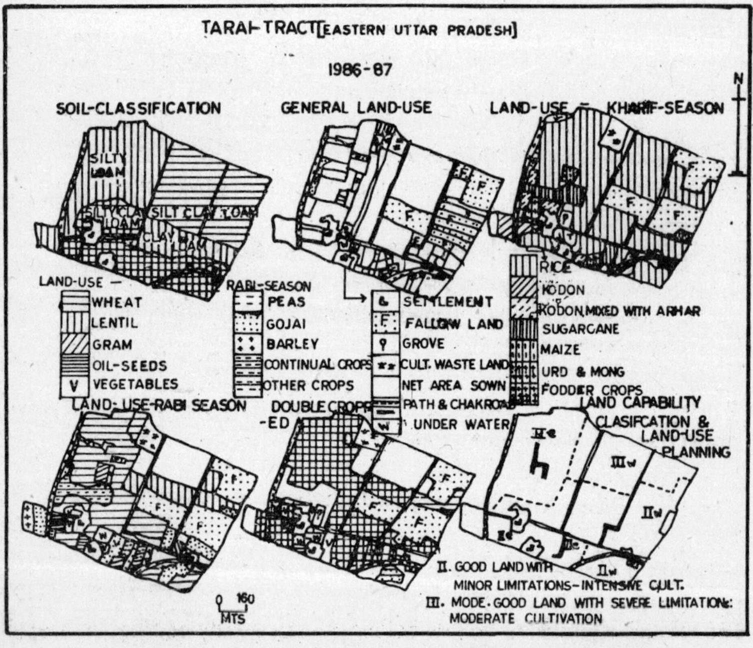

FIG. 6.6: Village Dhaurahra

Land Classification

The land of this village may be grouped into the following groups on the basis of land capability (Fig. 6.6):

Land Capability Sub-class IIe

It consists of silty loam fairly drained to well drained soils found in the western parts of the village. The slope ranges between 0 and 1 per cent and at places up to even 3 per cent. The permeability of soils is moderately slow and the water-holding capacity is moderately high. The soil reaction is mildly alkaline (pH 7.8). The natural fertility is very low in respect of organic carbon (0.125%), low in respect of phosphorous (13.4-17.9 kg/ha) and medium in respect of potash (112.0-156.0 kg/ha). Paddy, sugarcane, maize, *arhar*, wheat, mustard etc. are the suitable crops. The land suffers from minor limitations of erosions.

Land Capability Sub-class IIw

It consists of clayey loam and silty loam fairly to poorly drained soils found towards south-east. The slope ranges between 0.5 and 1 per cent. The permeability of soils is slow and water-holding capacity is high. The soil reaction is moderately alkaline (pH 7.9). The natural fertility of soil is low in respect of organic carbon (0.250%) and phosphorous (13.4 kg/ha). It is medium in respect of potash (168.0 kg/ha). The suitable crops are early and medium paddy, wheat, *masoor*, peas and barley. The minor limitation is that of water-logging.

Land Capacity Sub-class IIIw

It consists of silty clay loam soils found in the north-eastern part of the village. It is poorly drained with slope varying up to 1 per cent. The permeability is slow and the water-holding capacity is high. Submergence, water-logging and high water-table are other problems. The soil reaction is mildy alkaline (pH – 7.8). The natural fertility is low in respect of organic carbon (0.250%), very low in respect of phosphorous (8.9 kg/ha) and medium in respect of potash (156.8 kg/ha). The suitable crops are paddy, sugarcane, wheat, *masoor* etc.

Irrigation

Net irrigated area accounts for 12.9 per cent of the net sown area. The low proportion is due to dearth of suitable means of irrigation. There is enough sub-soil moisture, lessening the need of irrigation.

Use of Manure and Fertilisers

The use of manure and fertilisers is very limited, which is, to some extent, due to the apathy of farmers towards their use, but greatly due to poor economic conditions and inadequate means of irrigation.

General Land-use Pattern

Table 6.16 and Fig. 6.6 depict the general land-use pattern of the village.

TABLE 6.16: General Land-use (1987)

(Total area of the village : 94.05 hectares)

Land-use	Area in hectares	% of the total area
A. *Area not available for cultivation*	7.28	7.8
(i) Area under water	1.25	1.3
(ii) Area under settlement	6.03	6.5
B. *Other uncultivated land excluding fallowland*	3.01	3.2
(i) Groves	0.94	1.0
(ii) Culturable Wasteland	2.07	2.2
C. *Arable land*	83.76	89.0
(i) Fallowland	6.12	6.4
(ii) Net sown area	77.64	82.6
Total	94.05	100.0
Net irrigated area (as % of net sown area)	100.00	12.9
Double cropped area (as % of the net sown area)	25.60	32.9

It may be seen that about 8 per cent of the total area of the village is not available for cultivation, the greater part of which is under settlements and paths. The area under this category has increased in the past. Groves and culturable wasteland together claim 3.2 per cent of the total area, three-fourths of which is under the latter. The net sown area claims a little more than four-fifths of the total area and the fallowland claims 6.4 per cent of the total area. The area under plough has increased slightly in the past and that under fallowland has decreased.

Cropland Use

Table 6.7 and Fig 6.6 show that 55.1 per cent of the total and 73.2 per cent of the net cropped area is devoted to the cultivation of kharif crops. Paddy which is extensively grown in the village is the dominant crop, claiming 47.0 per cent of the total, 62.5 per cent of the net and 85.4 per cent of the kharif cropped area. Maize, *kodon* alone and mixed, *urad* and *moong* and sugarcane are other crops.

Rabi crops claim 44.9 per cent of the total and 59.7 per cent of the net cropped area. These has been some absolute increase in the area under rabi crops, although its proportion has decreased slightly, which seems due to the corresponding changes in the proportion of the kharif crops. Wheat is the dominant crop which shares 20.2 per cent of the total, 26.8 per cent of the net and 44.9 per cent of the rabi cropped area. Its cultivation concentrates towards west and south on the silty loam and clayey loam soils. Wheat mixed with barley and barley are other cereals, while *masoor* followed by pea and gram are the main pulses. Oil-seeds claim only 1.3 per cent of the total cropped area.

Double Cropping

Double cropped area occupies 32.9 per cent of the net cropped area, which was about 23.0 per cent some twenty years back. Double cropping concentrates mainly on class I and II lands.

Salient Features of the Cropping Pattern

A perusal of the cropland use of the village shows the following salient features of the cropping pattern:

1. Kharif crops are a bit more important than the rabi crops, which is due to the lack of irrigation facilities in the rabi season.
2. Foodgrains occupy 94.3 per cent (51.7% in kharif and 42.6% in the rabi season) of the total cropped area.
3. Food-cum-commercial crops occupy 5.1 per cent of the total cropped area and are thus quite unimportant.

4. Fodder crops are quite negligible.

TABLE 6.17: Cropland Use (1987)

(Net cropped area of the village : 77.64 hectares)
(Total cropped area of the village : 103.24 hectares)

Crops	Area in hectares	% of kharif/rabi cropped area	% of net cropped area of the village	% of total cropped area of the village area
1	2	3	4	5
Kharif	*56.86*	*100.0*	*73.2*	*55.1*
Foodgrains	*53.50*	*94.2*	*68.9*	*51.7*
a. *Cereals*	*58.70*	*92.8*	*67.9*	*51.0*
Paddy	48.56	85.4	62.5	47.0
Maize	0.32	0.6	0.5	0.3
Kodon	1.67	3.0	2.2	1.7
Kodon mixed with *arhar*	1.35	1.4	1.7	1.3
Others	0.80	1.4	1.0	0.7
b. *Pulses*	*0.80*	*1.4*	*1.0*	*0.7*
Urad and *Moong*	0.80	1.4	1.4	0.7
Food-cum-commercial crops (Sugarcane)	*2.82*	*4.9*	*3.6*	*2.8*
Non-food-crops (Fodder)	*0.54*	*0.9*	*0.7*	*0.6*
Rabi	*46.38*	*100.0*	*59.7*	*44.9*
Foodgrains	*44.09*	*95.1*	*56.7*	*42.6*
a. *Cereals*	*34.84*	*75.2*	*44.8*	*33.7*
Wheat	20.85	44.9	26.8	20.2
Barley	2.49	5.4	3.2	2.4
Wheat mixed with barley	9.46	20.4	12.2	9.2
Others	2.04	4.5	2.6	1.2
b. *Pulses*	*9.25*	*19.9*	*11.9*	*8.9*
Gram	1.10	2.4	1.4	1.1
Peas	1.24	2.7	1.6	1.2
Masoor	6.89	14.8	8.9	6.6
Food-cum-commercial crops	*2.20*	*4.9*	*3.0*	*2.3*
Oil-seeds	1.31	2.8	1.7	1.3
Vegetables	0.98	2.1	1.3	1.0

Land-use and Population

Table 6.18 shows that the per head share of the total area in the village is 0.25 hectare and that of the net cropped area is 0.20 hectare. The per head share in the net cropped area in the kharif and rabi crops respectively is 0.15 and 0.12 hectare and thus the per head total cropped area which actually supports one person in the village comes to 0.27 hectare.

As regards the occupational structure, of the total population of 382 persons, 53.2 per cent are workers. The cultivators and agricultural labourers respectively claim 67.2 and 32.8 per cent of the total working force. This indicates the total dependence of population on land.

Carrying Capacity of Land

With the present land-use practices, the carrying capacity of the land is 3.9 SNU or 1 SNU being produced from 0.25 hectare of the total cropped area, while this area per capita is 0.27 hectare. In other words, the present burden is that of 3.7 persons per hectare. This shows that the food resources and the population are fairly balanced at present, although the increasing population warrants an increase in the food production in future.

TABLE 6.18: Land-use and Population

[*Total population of the village : 382 (1987) Area in hectares*]

	Total area of the village	Net cropped area of the village	Net cropped area in kharif	Net cropped area in rabi	Total cropped area	Double cropped area
	94.95	77.64	56.86	46.38	103.24	25.60
Land per head of population	0.25	0.20	0.15	0.12	0.27	0.07

Land-use Planning

Keeping in view the existing land-use and land potential, the planning of the land use of the village is envisaged as follows:

Land Capability Sub-class IIe

This land should be devoted to intensive cultivation with paddy, sugarcane, pea, wheat, *moong* etc. as the main crops. The conservation practices suggested are:

1. Addition of bulky organic manures to improve the physical and biological properties of the soils,
2. Growing of leguminous and cover crops,
3. Proper irrigation,
4. Adequate and timely fertiliser use,
5. Adoption of scientific crop rotation, viz.,

   (i) Medium paddy-Wheat-*Moong/Lobia*
   (ii) Early paddy-Peas-Sugarcane
   (iii) Medium paddy-Berseem-Berseem.

Land Capability Sub-class IIw

This land should also be devoted to intensive cultivation with medium paddy, wheat, *masoor*, pea, barley etc. as the main crops. The conservation measures to be adopted are :

1. Addition of bulky organic manures to improve soil properties,
2. Sinking of tube-wells to lower down the water-table and provides adequate water for irrigation,
3. Provision of surface drainage where feasible,
4. Adequate and timely fertiliser use,
5. Scientific crop rotation to be adopted are:

   (i) Medium/paddy-Wheat-*Moong*/Lobia
   (ii) Early paddy-Pea-Sugarcane.

Land Capability Sub-class IIIw

This is capable of moderate cultivation with the following crop rotation:

(i) Late paddy-Linseed/Lentil
(ii) Late paddy-Zaid paddy-*Dhaincha*
(iii) Medium paddy-Wheat-*Moong*.

Other conservation measures are the same as in the case of IIw. There should be the minimum use of the calcareous fertilisers in order to avoid the increase of calcium. General recommendations are the same as already made in the case of other similar villages.

VILLAGE PURE SURAJMAN

Location

The village Surajman which belongs to the Bahraich upland, is situated at 27°50′ N latitude and 81°38′ E longitude in tahsil Nanpara of district Bahraich in the west. It lies at a distance of about 13 km from the tahsil headquarters.

Climate

No climatic data are recorded in the village. The nearest rainfall recording station is Nanpara, the rainfall data of which may be taken as close approximation of the rainfall conditions of the village. The average annual rainfall is 1185.1mm, of which 3.9, 5.6 and 90.5 per cent respectively is received during the cold, hot and rainy season. The co-efficient of annual variability is 25.2 per cent and that of monthly being 65.3 (June), 44.9 (July), 50.2 (August), 64.0 (September) and 126.3 (October). The high monthly variability in the months of September and October when the total rainfall is very insufficient, is very significant from agricultural point of view.

Soils

The loam to clayey loam soils of the village have developed on older alluvium deposits, which may be put in the category 'Degraded or leached calcium soils'.

Land Classification

The land of the village may be grouped into the following land capability mapping units (Fig. 6.7):

Land Capability Class I

This category of land consists of loamy (with clay 19.4-20.4, silt 21.2-24.2 and sand 56.4-58.4 per cent) well drained soils which are extensively distributed in the eastern part of the village. The soil reaction is mildly alkaline. The permeability is moderately rapid and the water-holding capacity is moderate. The slope ranges between zero and one per cent. The organic carbon and phosphate nutrients respectively are low and very low (0.250 to 0.312% and 8.96 kg/ha), while potash is medium (112.0 to 156.8 kg/ha). Paddy, maize, wheat, lentil, oil-seeds, sugarcane etc. are the best suited crops. This is very good land which suffers from no limitations.

Land Capability Sub-class IIe

This is good land with minor limitations of erosion. The clayey loam or loamy soils of this category are well drained with slope varying between 1 and 3 per cent. The soil reaction is mildly alkaline (pH 7.4). The permeability is moderately rapid and water-holding capacity moderately low. The fertility status is low in respect of organic carbon, very low in respect of phosphorous and medium in that of potash. The suitable crops are paddy, wheat, maize, potato, sugarcane, oil-seeds etc. which may profitably be grown.

There are three patches of this land (two towards north-east and one towards south-east).

Land Capability Sub-class IIw

The land of this category possesses clayey loam fairly to poorly drained soils. The land is mostly nearly level. The percentage of clay, silt and sand respectively is 28.4, 20.2 and 51.4. The soil reaction is mildly alkaline. The permeability is moderate and water-holding capacity is also moderate. As regards the fertility status, it is very low (0.100%) in respect of organic carbon, low (13.4 kg/ha) in respect of phosphorous and medium high (201.6 kg/ha) in respect of potash. Paddy, sugarcane wheat, lentil etc. can profitably be grown.

This land which has higher concentration towards west is subject to the limitations of water-logging.

Irrigation

Irrigation claims 23.9 per cent of the net sown area of the village. Ponds are the main sources of irrigation. There will be a hike in the hectareage of irrigated land after the Saryu canal system starts functioning. About three decades back, irrigation was not practised at all.

Manuring

Very few chemical fertilisers are used and the use of cow-dung is also very much limited.

General Land-use Pattern

Table 6.19 and Fig. 6.7 show that the area not available for cultivation claims 8.2 per cent and other uncultivated land excluding fallowland only 2.7 per cent of the total area of the village. Arable land claims 89.1 per cent of the total area. The area actually put under plough accounts for 88.5 per cent of the total area. This area has registered a slight decrease which appears due to the encroachment of the settlement sites and paths on the cultivated land.

Cropland

Table 6.20 and Fig. 6.7 reveal that the kharif crops claim 60.4 per cent of the total and 92.0 per cent of the net cropped area of the village. Paddy is the main crop, claiming 88.9 per cent of the kharif, 81.9 per cent of the net and 53.7 per cent of the total cropped area. Early paddy is widely grown due to the predominance of loamy well drained soils. Maize, *kodon* etc., are other minor crops.

Rabi crops are relatively of lesser importance, as they claim only 39.6 per cent of the total and 59.8 per cent of the net cropped area of the village. They are widely distributed on the loamy well drained soils. Wheat, occupying 53.1 per cent of the rabi, 32.1 per cent of the net and 21.0 per cent of the total cropped area, is the main crop which is followed by gram. The area under wheat has increased immensely in the past.

Double Cropping

Area cropped more than once accounts for 52.6 per cent of the net cropped area, which was only 26.4 per cent about three decades back. This increase is due to the development of irrigation in the village.

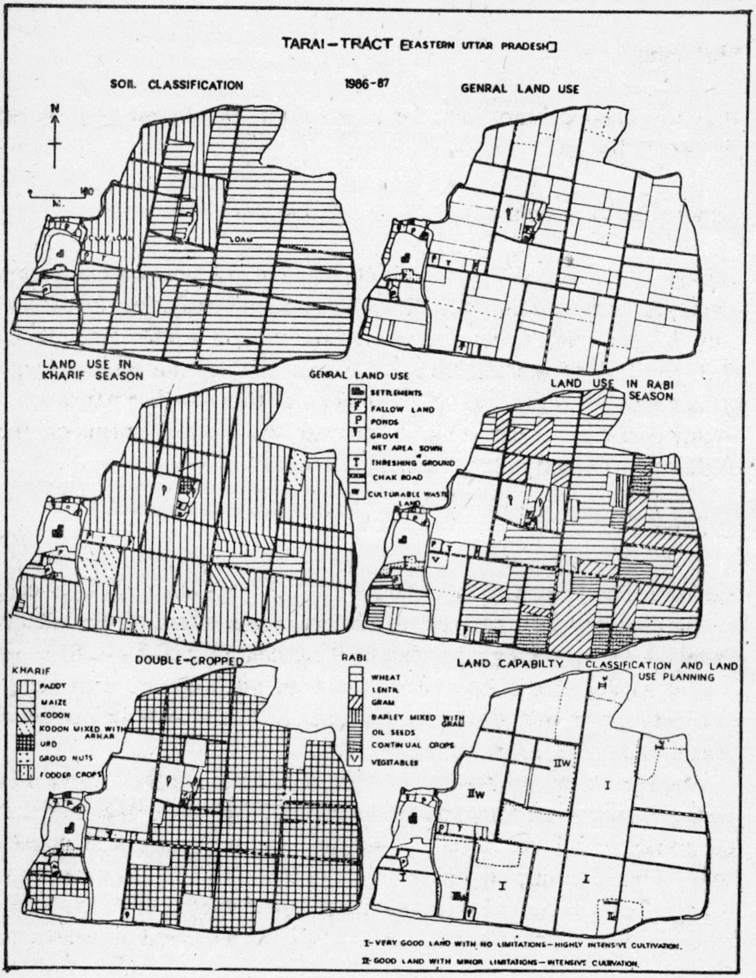

FIG. 6.7: Village Pure Surajman

Salient Features of the Cropping Pattern

A perusal of the cropland use of the village shows the following salient features of the cropping pattern:

1. Kharif crops are relatively more important than the rabi crops.
2. Foodgrains are predominant, as they occupy 97.8 per cent of the total cropped area.
3. Food-cum-commercial crops are quite insignificant.
4. Non-food crops do not exist on the cropland use map of the village.

TABLE 6.19: General Land-use (1987)

(Total area of the village : 51.49 hectares)

Land-use	Area in hectares	% of the total area
A. *Area not available for cultivation*	4.26	8.2
(i) Area under water	0.90	1.7
(ii) Area under settlement etc.	3.30	6.4
(iii) Barren land	0.06	0.1
B. *Other uncultivated land excluding fallowland*	1.33	2.7
(i) Groves	1.30	2.6
(ii) Culturable wasteland	0.03	0.1
C. *Arable land*	45.90	89.1
(i) Fallowland	0.30	0.6
(ii) Net sown area	45.60	88.5
Total	51.49	100.0
Net irrigated area (as % of the net sown area)	10.91	23.9
Double-cropped area (as % of the net sown area)	24.05	52.6

Land-use and Population

Table 6.21 shows that the totals of the various categories of land in the village and also per capita share of the villagers in these lands.

It is clear that per capita share of the total area is 0.24 and that of the net cropped area is 0.21 hectare. The per capita share in the net cropped area in the kharif and rabi crops respectively is 0.19 and 0.13 hectare. Thus the per capita share in the total cropped area comes to 0.32 hectare, which is the actual amount of land supporting one person in the village.

As regards the occupational structure, about one-third of the total population comes under workers' group, 78.7 per cent of which are cultivators and 19.1 per cent are classed as agricultural labourers. Thus 97.8 per cent belong to primary rural group.

TABLE 6.20: Cropland Use (1987)

(Net cropped area of the village : 45.60 hectares)
(Total cropped area of the village : 69.65 hectares)

Crops	Area in hectares	% of kharif/rabi cropped area	% of net cropped area of the village	% of total cropped area of the village
1	2	3	4	5
Kharif	*42.04*	*100.0*	*92.0*	*60.4*
Foodgrains	*41.55*	*98.8*	*90.9*	*59.7*
a. *Cereals*	*41.31*	*98.2*	*90.4*	*59.4*
Paddy	37.40	88.9	81.9	53.7
Maize	1.00	2.4	2.2	1.5
Kodon	0.50	1.2	1.0	0.7
Kodon mixed with *arhar*	2.4	5.7	5.3	3.5
b. *Pulses*	*0.24*	*0.6*	*0.5*	*0.3*
Food-cum-commercial crops	*0.08*	*0.2*	*0.2*	*0.1*
Non-food crops (Fodder)	*0.41*	*1.0*	*0.9*	*0.6*
Rabi	*27.60*	*100.0*	*59.8*	*39.6*
Foodgrains	*26.58*	*96.2*	*57.5*	*38.1*
a. *Cereals*	*15.37*	*55.6*	*33.6*	*22.0*
Wheat	14.66	53.1	32.1	21.0
Barley mixed with gram	0.71	2.5	1.5	1.0
b. *Pulses*	*11.21*	*40.6*	*23.9*	*16.1*
Gram	10.30	37.3	22.6	14.8
Masoor	0.81	3.3	1.3	1.3
Food-cum-commercial crops	*1.03*	*3.8*	*2.3*	*1.5*
Oil-seeds	0.90	3.3	2.0	1.3
Vegetables	0.13	0.5	0.3	0.2

Carrying Capacity of Land

The carrying capacity of land is 4.4 persons (4.4 SNU) with the existing management practices, while at present only 3.1 persons depend on one hectare of the total cropped area. Thus there is some surplus production which can meet the requirement of other adjacent villages and yield income to the inhabitants of this village. There is still scope for increasing the production in the village, provided land resources are judiciously utilised.

TABLE 6.21: Land-use and Population

[*Total population of the village : 216 (1987) Area in hectares*]

	Total area of the village	Net cropped area of the village	Net cropped area in kharif	Net cropped area in rabi	Total cropped area	Double cropped area
	51.49	45.60	42.06	27.61	69.65	24.05
Land per head of population	0.24	0.21	0.19	0.13	0.32	0.11

Land-use Planning

The following scheme of land-use planning of the village is recommended for the optimum utilisation of the land:

Land Capacity Class I

This land which is capable of highly intensive cultivation should be devoted to the cultivation of paddy, maize, wheat, sugarcane, leguminous crops, potato and vegetables. The following conservation measures should be adopted for the optimal utilisation of the land resources:

1. Addition of bulky organic manures to maintain soil productivity,
2. Leguminous crops should be given due place in the crop rotation,
3. Adoption of the scientific crop rotation, viz.

   (i) Early paddy-Pea/Mustard-Sugarcane
   (ii) Early paddy-Wheat-*Moong*
   (iii) Early paddy-Berseem

Land Capability Sub-class IIe

This is capable of intensive cultivation with almost the same crops as recommended in Class I. These crops should be grown with the adoption of the following conservation measures:

1. Addition of bulky organic manures to improve soil structure,
2. Growing of cover crops such as groundnut and soyabeen in the kharif season,
3. Use of insecticides and pesticides in the soil to control termites and grubs,
4. Provision of proper irrigation facilities,
5. Application of chemical fertilisers,
6. Adoption of scientific crop rotation, viz.

   (i) Maize/Lobia-Pea/Potato-Sugarcane
   (ii) Early paddy-Pea/Mustard-Sugarcane
   (iii) Early paddy-Wheat-*Moong*
   (iv) Early paddy-Berseem.

Land Capability Sub-class IIw

This land is also capable of intensive cultivation with paddy, sugarcane, wheat, lentil etc. as the important crops which should be grown with the adoption of the following conservation measures:

1. Addition of bulky organic manures to improve physico-biological properties of the soil,
2. Provision of drainage channels wherever needed,
3. Adequate and timely application of fertilisers,
4. Sinking of tube-wells to lower down the water-table,
5. Introduction of leguminous crops such as *moong*, *lobia* and *berseem* as a permanent feature in the cropping pattern,
6. Adoption of the scientific crop rotations, viz.

   (i) Medium paddy-Wheat-*Moong*/Lobia
   (ii) Early paddy-Pea-Sugarcane
   (iii) Late paddy-Linseed-Lentil

VILLAGE DIGAHI

Location

The village Digahi which belongs to the Gorakhpur upland region, lies at 27°19′ N latitude and 83° 40′ E longitude in tahsil Maharajganj of the district Gorakhpur. It is at a distance of about 40 km from the tahsil and 86 km from the district headquarters. It is connected by a cart-track with the metalled road and has thus moderate accessibility.

Climate

No climatic data are recorded at the village. The rainfall data recorded at the tahsil headquarters may be taken as close approximation of the climatic conditions of the village. The average annual rainfall is 1295.8mm, of which 2.8, 6.1 and 91.1 per cent respectively is received during the cold, hot and rainy season. The co-efficient of annual variability is 20.99 per cent. This is 49.8, 46.4, 49.3, 54.4 and 115.9 respectively for the months of June, July, August, September and October. The higher variability in the months of September and October may be viewed with great concern as regards the agricultural operations are concerned.

Soils

The soils of the village belong to that group which has developed in the areas of older alluvial deposits. Due to the washing out of the free calcium carbonate originally present in the soil, these soils are generally called 'Degraded or leached calcium soils'. Texturally speaking, they vary from fine sandy loam to silty loam (Fig. 6.8).

Land Classification

The land of the village may be grouped into the following land capability classes (Fig. 6.8):

Land Capability Class I

This is very good land which suffers from no limitation. This consists of very deep, friable, fine sandy loam soils which are found on level to very gently sloping lands (slope varying from zero to one per cent) with good drainage and covering an extensive area in the east. The permeability is moderately rapid and water-holding capacity is moderately low. The soil reaction is neutral (pH-7.4). The soils are highly responsive to fertilisers input and improved management practices. The natural fertility is low in respect of organic carbon (0.250-0.375%), medium high in respect of phosphorous (31.4 kg/ha) and medium in respect of potash (112 kg/ha). The best suited crops are paddy, maize, wheat, potato, sugarcane, mustard, pea etc. This land is capable of highly intensive cultivation.

Land Capability Sub-class IIw

This is good land with moderate limitation of water interfering with plant growth. It consists of very deep, somewhat poorly drained, friable silty loam soils which occupy the northern part of the village. The proportions of clay, silt and sand respectively are 16.2, 27.8 and 56.0 per cent. The land is nearly level to very gently sloping and it remains submerged with water during monsoon. The permeability is slow, but the water-holding capacity is high. The soil reaction is neutral (pH 7.4) to mildly alkaline (pH 7.5). The natural fertility of soils is low in respect of organic carbon (0.250%), medium to medium high (13.4-31.4 kg/ha) in respect of phosphorous and medium (112-179 kg/ha) in respect of potash. Submergence and water-logging, high water-table and slow permeability are the factors causing low productivity. The best suited crops are paddy, sugarcane, wheat, peas, mustard, lentil etc. This is capable of intensive cultivation.

Irrigation

Irrigation covers 67.6 per cent of the net cropped area of the village, which was about 10 per cent about three decades back. Canals are the main source of irrigation.

Manuring

The use of chemical fertilisers, green manures and cow-dung is very limited.

General Land-use Pattern

Table 6.22 and Fig. 6.8 reveal that 10.6 per cent of the total area of the village is not available for cultivation, in which settlements and paths etc. are predominant. The area under settlements has increased. Groves and culturable wasteland claim respectively 1.5 and 2.4 per cent of the total area. The net sown area claims 83.1 and fallowland 2.4 per cent of the total area. The fallowland has decreased, but the net sown area has remained stationary.

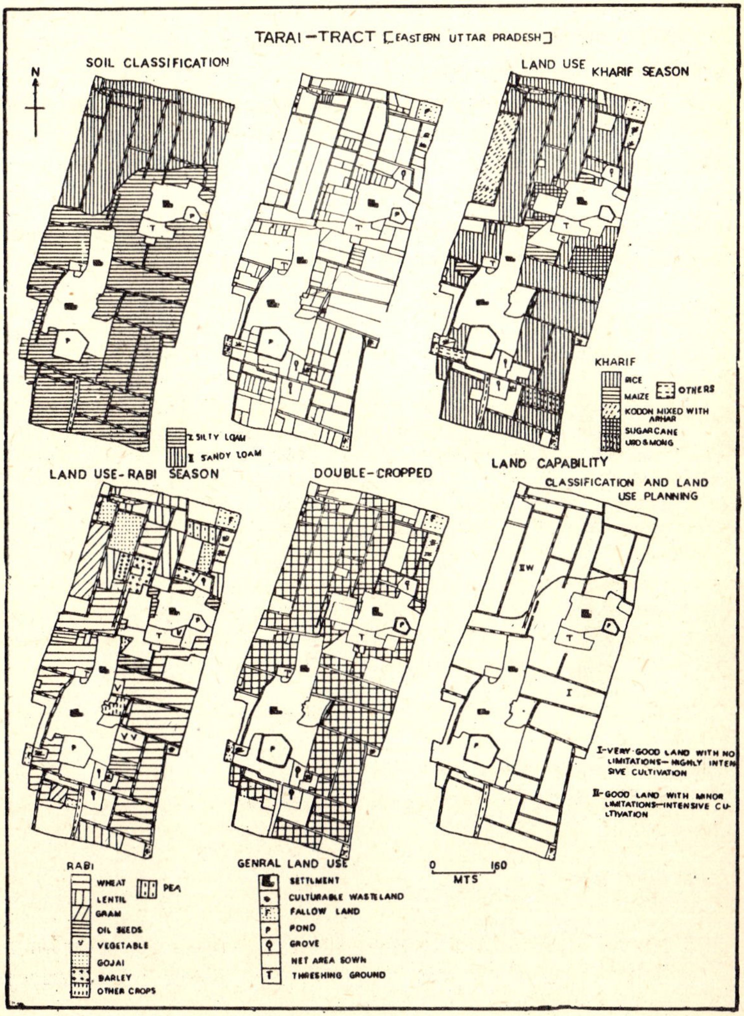

FIG. 6.8: Village Digahi

Cropland Use

Table 6.23 and Fig. 6.8 show that 54.9 per cent of the total and 91.3 per cent of the net cropped area of the village is devoted to the cultivation of kharif crops, the lion's share of which is claimed by paddy. It is extensively grown in the village. Maize, *kodon* alone and mixed, sugarcane etc. are other crops.

Rabi crops occupy only 45.1 per cent of the total and 74.9 per cent of the net cropped area of the village. Wheat is the main crop, occupying 48.2 per cent of the rabi, 36.2 per cent of the net and 21.8 per cent of the total cropped area of the village. This crop concentrates on the class I land. Wheat mixed with barley, barley, *masoor*, pea, gram, oil-seeds etc. are other crops grown in the village.

TABLE 6.22: General Land-use (1987)

(Total area of the village : 36.09 hectares)

Land-use	Area in hectares	% of the total area
A. *Area not available for cultivation*	3.85	10.6
(i) Area under water	1.05	2.9
(ii) Area under settlement etc.	2.80	7.7
B. *Other uncultivated land excluding fallowland*	1.38	3.9
(i) Groves	0.54	1.5
(ii) Culturable wasteland	0.84	2.4
C. *Arable land*	30.86	85.5
(i) Fallowland	0.86	2.4
(ii) Net sown area	30.00	83.1
Total	36.09	100.0
Net irrigated area (as % of the net sown area)	20.28	67.6
Double-cropped area (as % of the net sown area)	21.00	70.0

Double Cropping

About two-thirds of the net cropped area is cropped more than once and spreads over both the class I and II lands. About three decades back, this percentage was only a bit less.

Salient Features of the Cropping Pattern

An analysis of the cropland use of the village shows the following salient features of the cropping pattern:

1. Kharif crops are comparatively more important in the cropping pattern.
2. Foodgrains dominate the cropping pattern by claiming 90.7 per cent (53.1% in kharif and 37.6% in rabi) of the total cropped area.
3. Food-cum-commercial crops claim 8.0 per cent of the total cropped area and are thus next to foodgrains.
4. Non-food crops are quite negligible.

TABLE 6.23: Cropland Use (1987)

(Net cropped area of the village : 30.00 hecrates)
(Total cropped area of the village : 51.00 hectares)

Crops	Area in hectares	% of kharif/rabi cropped area	% of net cropped area of the village	% of total cropped area of the village
1	2	3	4	5
Kharif	*27.33*	*100.0*	*91.3*	*53.6*
Foodgrains	*26.49*	*96.9*	*88.5*	*51.9*
a. *Cereals*	*26.09*	*95.4*	*87.1*	*51.2*
Paddy	21.93	80.4	73.1	43.0
Maize	2.09	7.5	7.0	4.0
Kodon	0.45	1.6	1.5	0.9
Kodon mixed with *arhar*	1.20	4.4	4.1	2.3
Others	0.42	1.5	1.4	0.8
b. *Pulses*	*0.40*	*1.5*	*1.4*	*0.7*
Food-cum-commercial crops (Sugarcane)	*0.84*	*3.1*	*2.8*	*1.7*
Rabi	*23.67*	*100.0*	*78.9*	*46.4*
Foodgrains	*19.91*		*64.4*	*39.0*
a. *Cereals*	*14.98*		*49.9*	*29.4*
Wheat	12.03		40.1	23.6
Barley	0.91		3.0	1.8
Wheat mixed with barley	2.04		6.8	4.0
b. *Pulses*	*4.93*		*16.5*	*9.6*
Gram	*1.71*		*5.7*	*3.3*
Peas	0.49		1.6	0.9
Masoor	2.73		9.2	5.4
Food-cum-commercial crops	*3.21*		*10.7*	*6.3*
Oil-seeds	2.11		7.0	4.1
Vegetables	1.10		3.7	2.2
Non-food crops	*0.55*		*1.8*	*1.1*

Land-use and Population

Table 6.24 shows the totals of various categories of land in the village and also the per capita share of villagers in these lands.

TABLE 6.24: Land-use and Population

[Total population of the village : 600 (1987)]*

	Total area of the village	Net cropped area of the village	Net cropped area in kharif	Net cropped area in rabi	Total cropped area	Double cropped area
	36.09	30.00	27.33	22.49	49.82	19.82
Land per head of population	0.06	0.05	0.04	0.04	0.08	0.03

* The actual population of the village at the time of field study was 1599, but 999 persons were found depending on the land belonging to other adjoining villages. This number has been excluded from the population of this village.

It is clear from Table 6.24 that the per capita share in the total area and net cropped area respectively is 0.06 and 0.05 hectare and that in the kharif and rabi 0.04 and 0.04 hectare. Thus the per capita total cropped area is 0.08 hectare. This is the amount of land actually supporting one person in the village.

Of the total population of 600 persons, about 44.0 per cent belong to the working class. Very interesting feature is that agricultural labourers account for 51.8 per cent of this, while cultivators claim only 48.0 per cent.

Carrying Capacity of Land

With the existing management practices, one hectare of the total cropped area is capable of supporting 4.5 persons (4.5 SNU), but the pressure of population on this at present is 12.5 persons. This shows the great imbalance that exists between the food resources and population. It necessitates the maximum utilisation of land resources for solving the obnoxious problem of over population to some extent.

Land-use Planning

The following scheme for the optimum utilisation of the land resources is envisaged (Fig. 6.8):

Land Capability Class I

This land should be devoted to the highly intensive cultivation with paddy, wheat, sugarcane, maize, potato etc. as the main crops. The conservation measures recommended are :

1. Addition of bulky organic manures and green manuring to maintain soil productivity,
2. Leguminous crops to be given due place in the crop rotation,
3. Adoption of scientific crop rotation, viz.

   (i) Early paddy-Wheat-*Moong*,
   (ii) Early paddy-Pea/Mustard-Sugarcane
   (iii) Early paddy-Berseem

Land-use Capability Sub-class IIw

This should be devoted to intensive cultivation with paddy, wheat, sugarcane, pea, mustard etc. as the main crops. The following management practices should be adopted for the optimum utilisation and conservation of the land resources:

1. Addition of bulky organic manures to improve physico-biological properties of the soil,
2. Provision of the surface drainage wherever feasible,
3. Sinking of tube-wells to lower down the water-table and make adequate provision for irrigation,
4. Adequate and timely fertiliser use based on soil testing recommendation,
5. Introduction of leguminous crops *moong*, *lobia* and *berseem*,
6. Adoption of scientific crop rotations, viz.

   (i) Early paddy-Pea-Sugarcane
   (ii) Medium paddy-Berseem-Berseem
   (iii) Late paddy-Linseed/Lentil

Use of HYV of seeds and adoption of modern technological innovations are also recommended.

VILLAGE BISHUNPURA

Location

The village Bishunpura of tahsil Maharajganj, district Gorakhpur belongs to the Gorakhpur upland region and is situated at 27°8′ N latitude and 83°38′ E longitude. It is at a distance of about 6 km from the tahsil headquarters and 64 km from the district headquarters with which it is connected by metalled and unmetalled roads. It is thus favourably located as regards the accessibility.

Climate

The climatic conditions have already been discussed in the land-use analysis of the village Digahi.

Soils

The soils of the village also belong to the group of those soils which have developed in areas of older alluvial deposits. They vary from silty loam to silty clay loam in texture (Fig. 6.9).

Land Classification

On the basis of land capability, the land of this village may be grouped into the following capability mapping units or classes (Fig. 6.9):

Land Capability Class I

This is very good land with no limitation. This land which covers almost whole of the village except the northernmost part, consists of silty loam soils with slope varying up to 1 per cent, the permeability is moderately slow and water-holding capacity is moderately high. The soil reaction is mildly alkaline (pH 7.7 to 7.8). As regards the fertility status, organic carbon is low (0.250%), phosphorous is medium (19.4 kg/ha) and potash is high (298.7 kg/ha). The soils are well suited to the cultivation of paddy, wheat, sugarcane, vegetables etc. This land is capable of highly intensive cultivation.

Land Capability Sub-class IIw

This is good land with moderate limitation of water. It consists of silty clay loam soils of the northernmost part of the village. The soils are fairly poorly drained with slope up to 0.5 per cent. The permeability is slow and the water-holding capacity is high. The soils are moderately alkaline. The organic carbon, phosphorous and potash are low and thus the fertility status is poor. Paddy, sugarcane, wheat, barley, mustard, *masoor* etc. are the suitable crops and the land is capable of intensive cultivation.

Irrigation

Irrigation covers 66.0 per cent of the net cropped area and the canals are the main source of irrigation. About three decades back, the percentage of irrigated area was 7.7 only. This shows that there has been great increase in the irrigated area.

Manuring

The use of manures and chemical fertilisers is quite limited, for which weak economic conditions of the farmers are responsible.

General Land-use Pattern

Table 6.25 and Fig. 6.9 portray the general land-use pattern of the village.

It is clear that area not available for cultivation and other uncultivated land excluding fallowland claim respectively only 5.3 and 5.1 per cent of the total area, while arable land accounts for 89.6 per cent of it. About 84.0 per cent of the total area is under plough. This was almost the same even in the past, which shows that there is no further scope for the extension of cultivated land.

Cropland Use

Table 6.26 and Fig. 6.9 depict the cropland use of the village.

It may be seen that the kharif and the rabi crops are almost of the equal importance, claiming respectively 50.3 and 49.7 per cent of the total cropped area of the village. Paddy is the dominant crop of the kharif season, claiming 71.3 per cent of the kharif, 61.3 per cent of the net and 36.1 per cent of the total cropped area of the village. Sugarcane claims 11.5 per cent of the kharif, 9.6 per cent of the net and 5.7 per cent of the total cropped area. Maize, *kodon* alone and mixed and pulses are other crops.

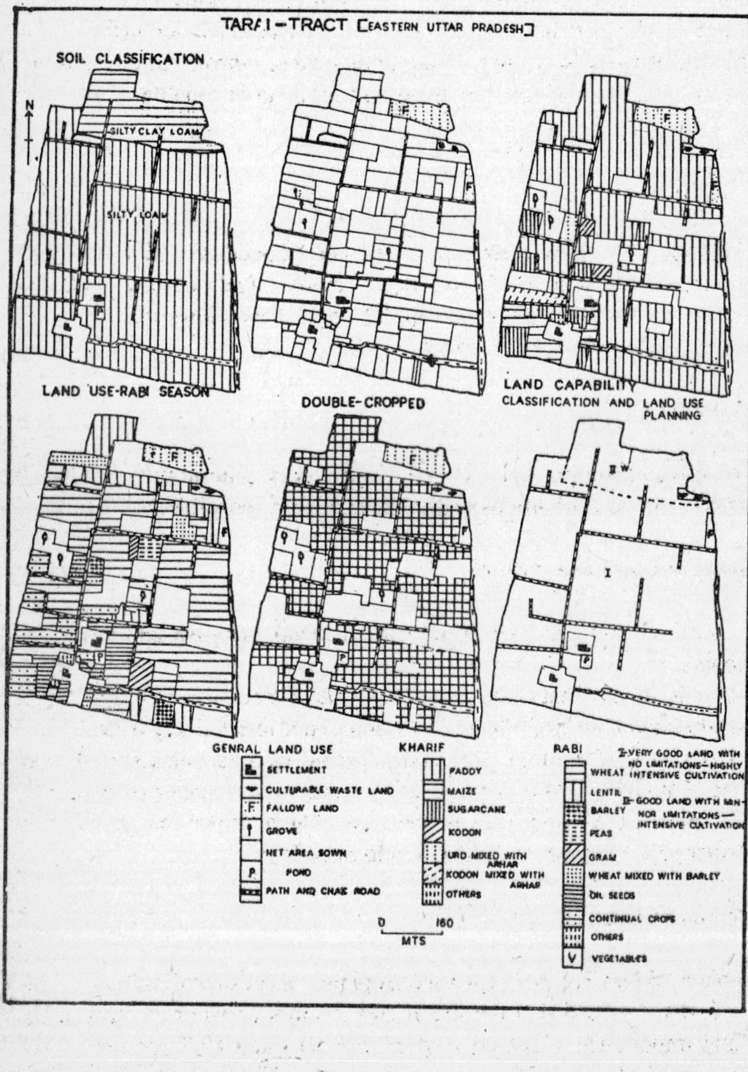

FIG. 6.9: Village Bishunpura

TABLE 6.25: General Land-use (1987)

(Total area of the village : 36.64 hectres)

Land-use	Area in hectares	% of the total area
A. *Area not available for cultivation*	1.94	5.3
(i) Area under water	0.27	0.7
(ii) Area under settlements etc.	1.67	4.6
B. *Other uncultivated land excluding fallowland*	1.87	5.1
(i) Groves	1.71	4.7
(ii) Culturable wasteland	0.16	0.4
C. *Arable land*	32.83	89.6
(i) Fallowland	2.11	5.8
(ii) Net sown area	30.72	83.8
Total	36.64	100.0
Net irrigated area (as % of net sown area)	20.29	66.0
Double cropped area (as % of net sown area)	21.43	69.7

Among the rabi crops, wheat is the most important. It claims 51.4 per cent of the rabi, 43.3 per cent of the net and 25.6 per cent of the total cropped area. Other minor crops are barley, wheat mixed with barley, gram, pea, *masoor* etc.

Double Cropping

Double-cropped area accounts for 64.9 per cent of the net cropped area of the village. This concentrates on good quality land.

Salient Features of the Cropping Pattern

An analysis of the cropland use reveals the following salient features of the cropping pattern of the village:

1. Kharif and rabi crops are equally important in the cropping pattern.
2. Foodgrains dominate the cropping pattern by claiming 85.9 per cent of the total cropped area.
3. Food-cum-commercial crops account for 11.8 per cent of the total cropped area of the village.
4. The area under non-food crops is negligible.

TABLE 6.26: Cropland Use (1987)

(Net cropped area of the village : 30.72 hecrares)
(Total cropped area of the village : 55.15 hectares)

Crops	Area in hectares	% of kharif/rabi cropped area	% of net cropped area of the village	% of total cropped area of the village
1	2	3	4	5
Kharif	*26.25*	*100.0*	*85.5*	*50.3*
Foodgrains	*22.81*	*86.9*	*74.6*	*43.8*
a. *Cereals*	*22.31*	*84.9*	*72.8*	*42.8*
Paddy	18.83	71.3	61.3	36.1
Maize	0.94	3.6	3.1	1.8
Kodon	1.00	4.0	3.2	2.0
Kodon mixed with *arhar*	1.27	4.9	4.2	2.4
Others	0.27	1.1	1.0	0.5
b. *Pulses* (*Urad* and *moong*)	*0.56*	*2.0*	*1.8*	*1.0*
Food-cum-commercial crops (Sugarcane)	*2.96*	*11.5*	*9.6*	*5.7*
Non-food crops (Fodder)	*0.42*	*1.6*	*1.3*	*0.8*
Rabi	*25.90*	*100.0*	*84.3*	*49.7*
Foodgrains	*21.90*	*84.6*	*71.3*	*41.1*
a. *Cereals*	*16.40*	*63.4*	*53.4*	*31.6*
Wheat	13.30	51.4	43.3	25.6
Barley	0.80	3.1	2.6	1.5
Wheat mixed with barley	2.30	8.9	7.5	4.5
b. *Pulses*	*5.50*	*21.2*	*17.9*	*10.5*
Gram	2.0	7.8	6.5	3.8
Peas	0.80	3.0	2.6	1.5
Masoor	2.70	10.4	8.8	5.2
Food-cum-commercial crops	*3.20*	*12.4*	*10.4*	*6.1*
Oil-seeds	2.40	9.2	7.8	4.6
Vegetables	0.80	3.0	2.6	1.5
Non-food crops	*0.80*	*3.0*	*2.6*	*1.5*

Land-use and Population

Table 6.27 shows that the per head share in the total and the net cropped area of the village respectively is 0.56 and 0.47 hectare and that in the net cropped area in the kharif and rabi crops is 0.40 hectare each. Thus the total cropped area which is the actual amount of land supporting one person in the village comes to 0.80 hectare per head.

TABLE 6.27: Land-use and Population

[*Total population of the village : 65 (1987) Area in Hectares*]

	Total area of the village	Net cropped area of the village	Net cropped area in kharif	Net cropped area in rabi	Total cropped area	Double cropped area
	36.64	30.72	26.25	25.90	52.15	21.43
Land per head of population	0.56	0.47	0.40	0.40	0.80	0.33

As regards occupational structure, about 30.0 per cent are workers who belong to primary rural group, of whom 82.0 per cent are cultivators and the remaining 18.0 per cent are agricultural labourers.

Carrying Capacity of Land

With the present management practices, one hectare of the total cropped area of the village can support 5 persons (5 SNU), while it has to support only 1.25 persons. This indicates that there is enough surplus which can still be increased if land is properly utilised.

Land-use Planning

The following suggestions are given for planning the land-use of the village (Fig. 6.9):

Land Capability Class I

This land should be devoted to the highly intensive cultivation with paddy, sugarcane, wheat, vegetables, *masoor* etc. The conservation measures to be adopted are:

1. Addition of bulky organic manures to maintain soil productivity,
2. Proper methods of application of fertilisers, specially nitrogenous fertilisers, to be adopted,

3. Some leguminous crops should also be grown and green manuring may be adopted, if feasible,
4. Adoption of scientific crop rotations, viz.

   (i) Paddy-Wheat-*Moong*
   (ii) Paddy-Pea-Sugarcane
   (iii) Paddy-Berseem-Berseem

Land Capability Sub-class IIw

Intensive cultivation with paddy, sugarcane, wheat, *masoor* etc. is recommended together with the following conservation measures:

1. Addition of bulky organic manures and green manuring to improve the physical properties of soils,
2. Provision of surface and sub-surface drainage,
3. Adequate and timely application of fertilisers based on the recommendations of soil testing results,
4. Sinking of tube-wells for lowering down the water-table,
5. Adoption of suitable crop rotations, viz.

   (i) Late paddy-Zaid paddy-*Dhaincha*
   (ii) Late Paddy-Linseed/Lentil
   (iii) Medium Paddy-Wheat-*Moong*.

In addition to the above measures, use of HYV of seeds and modern technological innovations are also recommended.

VILLAGE PATIAPAR

Location

The village Patiapar of the block Khuniaon of tahsil Domariaganj, district Basti, belongs to the slightly elevated land of Rapti-Burhi Rapti lower basin. It is situated at 27°18′ N latitude and 82°51′ E longitude at a distance of about 35 km from tahsil headquarters and 21 km from the town of Bansi. It is connected by an unmetalled road. The accessibility is adversely affected during the rainy season.

Climate

The climatic conditions of the village have already been discussed with the land-use analysis of the village Mundila.

Soils

The soils of the village belong to the group of soils which occur at slightly elevated levels that are generally outside the fluvial action of the rivers. They are loam to silty loam in texture.

Land Classification

The land of the village may be classified into the following land capability classes or mapping units (Fig. 6.10):

Land Capability Class I

This is very good land which suffers from no limitation. It consists of loamy well drained soils spreading around the settlement sites. The slope ranges up to 1 per cent or even more. The proportions of clay, silt and sand respectively are 17.25-20.50, 14.5-24.0 and 55.8-63.3 per cent. The permeability is moderately rapid and the water-holding capacity is moderately low. The soil reaction is mildly alkaline (pH 7.6 to 7.8). Organic carbon is very low to medium (0.100 to 0.700%), phosphorous is low to medium (13.4 to 22.4 kg/ha) and potash is medium (156.8 to 246.4 kg/ha). The suitable crops are paddy, maize, *arhar*, sugarcane, wheat, lentil, gram, pea etc. This land is capable of highly intensive cultivation.

Land Capability Sub-class IIe

This is good land with moderate limitations of erosion. It consists of loamy well drained soils which occupy small patches in the eastern, northern and western parts of the village. It shares most of the characteristics of the class I land, except that it is subjected to minor erosional hazards and that the slope is between 1 and 3 per cent.

Land Capability Sub-class IIw

It occupies the north-western part of the village and comprises of silty loam, somewhat fairly to poorly drained soils, the proportions of clay, silt and sand respectively being 18.0, 25.8 and 56.2 per cent. The permeability is slow and the water-holding capacity is moderately high. The soil reaction is mildly alkaline. As regards the fertility status, it is very low in respect of organic carbon, low in respect of phosphorous (13.4 kg/ha), and medium in respect of potash (246.4 kg/ha). The main crops which can be grown profitably are paddy, sugarcane, wheat, *masoor* etc. It is also capable of intensive cultivation.

Irrigation

Irrigation accounts for 50.0 per cent of the net sown area and the ponds and tube-wells are the main sources of water utilised for this purpose.

Manuring

Such as other villages, the use of manures and fertilisers is very limited, which may be ascribed to the poor economic condition of the farmers and an inadequate supply of these materials.

General Land-use Pattern

Table 6.28 and Fig. 6.10 depict the general land-use pattern of the village.

TABLE 6.28: General Land-use (1987)

(Total area of the village : 39.20 hectres)

Land-use	Area in hectares	% of the total area
A. *Area not available for cultivation*	3.60	9.1
(i) Area under water	0.80	2.0
(ii) Area under settlement etc.	2.80	7.1
B. *Other uncultivated land excluding fallowland*	2.00	5.1
(i) Groves	1.20	3.1
(ii) Culturable wasteland	0.80	2.0
C. *Arable land*	33.60	85.8
(i) Fallow land	2.00	5.1
(ii) Net sown area	31.60	80.7
Total	39.20	100.0
Net irrigated area (as % of net sown area)	15.80	50.0
Double cropped area (as % of net sown area)	17.60	55.7

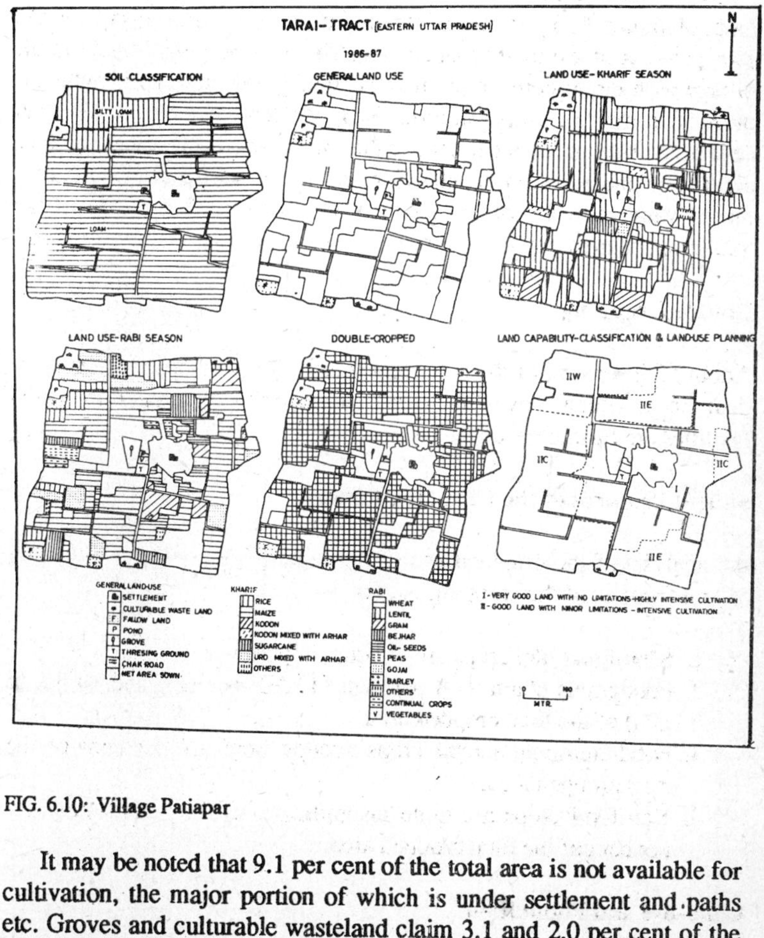

FIG. 6.10: Village Patiapar

It may be noted that 9.1 per cent of the total area is not available for cultivation, the major portion of which is under settlement and paths etc. Groves and culturable wasteland claim 3.1 and 2.0 per cent of the total area respectively. Fallowland occupies 5.1 per cent of the total area and 80.7 per cent is actually ploughed. This shows that there is still scope for the extension of cultivated land by reclaiming culturable wasteland and fallowland to cultivation.

Cropland Use

Table 6.29 and Fig. 6.10 portray the cropland use of the village. It is clear that kharif crops occupy 50.4 per cent of the total and 78.5 per cent of the net cropped area of the village, showing that kharif and rabi crops are almost of equal importance. Paddy is the most important crop, claiming 72.6 per cent of the kharif, 56.9 per cent of the net and 36.6 per cent of the total cropped area. This is widely distributed in the village with larger concentration of late paddy towards north. Wheat is the next important crop, claiming 54.1 per cent of the rabi, 41.8 per cent of the net and 26.8 per cent of the total cropped area. Kodon alone and mixed and maize in the kharif, wheat mixed with barley and barley mixed with gram are other cereals. Pulses cover 8.2 per cent of the total cropped area in rabi season.

Double Cropping

About 56.0 per cent of the net cropped area of the village is sown more than once, which may still be increased, provided proper irrigation facilities are further enhanced.

Salient Features of the Cropping Pattern

An analysis of the cropland use of the village bring out the following salient features of the cropping pattern:

1. Kharif and rabi crops are of equal importance.
2. Foodgrains claim 91.8 per cent (47.2% in kharif and 44.6% in rabi) of the total cropped area.
3. Food-cum-commercial crops occupy only 5.7 per cent of the total cropped area.
4. Non-food crops are quite insignificant, as they claim only 1.7 per cent of the total cropped area.

Land-use and Population

Table 6.30 gives the totals of various categories of land in the village and also the per capita share in these lands. It is clear that per capita share in the total area and net cropped area respectively is 0.06 and 0.05 hectare and that in the net cropped area in kharif and rabi 0.04 hectare each.

Thus the per capita of share in the total cropped area comes to 0.08 hectare. This is the actual amount of land supporting one person in the village.

TABLE 6.29: Cropland Use (1987)

(Net cropped area of the village : 31.60 hecrares)
(Total cropped area of the village : 49.20 hectares)

Crops	Area in hectares	% of kharif/rabi cropped area	% of net cropped area of the village	% of total cropped area of the village
1	2	3	4	5
Kharif	*24.80*	*100.0*	*78.5*	*50.4*
Foodgrains	*23.20*	*93.5*	*73.4*	*47.2*
a. *Cereals*	*22.80*	*91.9*	*72.1*	*46.4*
Paddy	18.00	71.6	56.9	36.6
Maize	0.40	1.6	1.3	0.8
Kodon	2.00	8.0	6.3	4.1
Kodon mixed with *arhar*	1.60	6.5	5.1	3.2
Others	0.80	3.2	2.5	1.7
b. *Pulses*	*0.40*	*1.6*	*1.3*	*0.8*
Food-cum-commercial crops (Sugarcane)	*1.20*	*4.9*	*3.8*	*2.4*
Non-food crops	*0.40*	*1.6*	*1.3*	*0.8*
Rabi	*24.40*	*100.0*	*77.2*	*49.6*
Foodgrains	*22.00*	*90.1*	*69.6*	*44.6*
a. *Cereals*	*18.00*	*73.7*	*57.0*	*36.4*
Wheat	13.20	54.1	41.8	26.8
Barley	1.20	4.9	3.8	2.4
Wheat mixed with barley	2.40	9.8	7.6	4.8
Barley mixed with gram	1.20	4.9	3.8	2.4
b. *Pulses*	*4.00*	*16.4*	*12.6*	*8.2*
Gram	1.20	4.9	3.8	2.4
Peas	0.80	3.3	2.5	1.6
Masoor	2.00	8.2	6.3	4.2
Food-cum-commercial crops	*1.60*	*6.5*	*5.1*	*3.3*
Oil-seeds	1.20	4.9	3.8	2.4
Vegetables	0.40	1.6	1.3	0.9
Non-food crops	*0.80*	*3.4*	*2.5*	*1.7*

As regards the occupational structure, about 32.0 per cent of the total population is classed as workers, 88.6 per cent of which are cultivators and 2.1 per cent are agricultural labourers. Thus 90.0 per cent of the total workers belong to the primary rural group and directly depend on land for their livelihood.

Carrying Capacity of Land

The carrying capacity of one hectare of the total cropped area is 4.1 persons (4.1SNU), i.e., one hectare can support 4.1 persons, while it has to support 12.5 persons. This shows the excessive pressure of population on land resources and necessitates their optimal utilisation for doing away with this imbalance.

TABLE 6.30: Land-use and Population

[Total population of the village : 600 (1987) Area in Hectares]

	Total area of the village	Net cropped area of the village	Net cropped area in kharif	Net cropped area in rabi	Total cropped area	Double cropped area
	39.20	31.60	24.80	24.40	49.20	17.60
Land per head of population	0.06	0.05	0.04	0.04	0.08	0.03

Land-use Planning

In view of the land potentials and their misuses, the following measures are suggested for the optimal utilisation of the land:

Land Capability Class I

Highly intensive cultivation with paddy, maize, *arhar*, sugarcane, wheat, lentil, pea etc. as the main crops is recommended. Conservation measures to be adopted are:

1. Addition of bulky organic manures to maintain soil productivity,
2. Proper provision of irrigation by sinking tube-wells,
3. Due place to be given to the leguminous crops in the cropping pattern,
4. Green manuring to be practised, if feasible,
5. Adoption of the following crop rotations:

   (i) Early paddy-Pea-Sugarcane

carbon is 0.200%, phosphorous is 8.9 kg/ha and potash 44.8 kg/ha). The soils are suitable for growing paddy, maize, *kodon*, wheat, gram, vegetables etc.

Land Capability Sub-class IIe

The soils of this group possess the same characteristics as those of the former group, except that the slope ranges between 1 and 3 per cent, the permeability is moderately rapid to rapid and water-holding capacity is moderately low. They are subjects to minor erosional hazards. This category of land is capable of intensive cultivation.

Land Capability Sub-class IIw

This land which comprises of clayey loam fairly to well drained soils is also capable of intensive cultivation. The proportions of clay, silt and sand respectively are 25.4, 18.2 and 56.4 per cent. These soils are found in the eastern part of the village. The slope varies up to 1 per cent. The soil reaction is mildly alkaline (pH 7.6). As regards the fertility status, organic carbon is low (0.400%), phosphorous is also low (17.9 kg/ha) and potash is medium (134.4 kg/ha). The permeability is moderately slow and water-holding capacity is moderately high. The suitable crops are paddy, wheat, maize, *arhar* etc.

Irrigation

About 26 per cent of the net cropped area is irrigated, the main source being river and ponds. There has been some increase in this area in the past.

Manuring

The use of fertilisers and manures is quite limited, which is due to the inadequate supply of them and lack of irrigation facilities.

General Land-use Pattern

Table 6.31 and Fig. 6.11 reveal that 5.6 per cent of the total area of the village is not available for cultivation and other uncultivated land excluding fallowland accounts for 10.4 per cent of the total area (groves 7.1 and culturable wasteland 3.3%). Arable land which includes fallowland and net area sown claims 84.0 per cent of the total area. Strangely enough, the percentage of net sown area has decreased by more than 10 per cent within two decades, which is due to an increase in the proportion of the area under groves, culturable wasteland and fallowland. The increase in the latter two heads is due to the river action.

Cropland Use

Table 6.32 and Fig. 6.11 show that the kharif crops occupy 56.8 per cent of the total and 82.1 per cent of the net cropped area of the village. This area has registered some decrease in the past, which is due to an increase in the area under rabi crops. Paddy is the main crop which claims 33.3 per cent of the kharif, 27.3 per cent of the net and 18.9 per cent of the total cropped area. This is followed by maize, *kodon* alone and mixed, pulses and oil-seeds in the kharif season, which respectively claim 10.8, 8.1 and 8.2 per cent of the total cropped area.

Rabi crops claim 43.2 per cent of the total and 62.4 per cent of the net cropped area. It has increased by 10.0 per cent in the past. Wheat is the first ranking crop which claims 62.5 per cent of rabi, 39.0 per cent of the net and 27.0 per cent of the total cropped area; Pulses, i.e., gram and *masoor* claim 10.8 per cent of the total cropped area and oil-seeds occupy 5.4 per cent. The area under wheat and pulses has increased.

Double Cropping

Area cropped more than once claims 44.5 per cent of the net cropped area, which was only 16.2 per cent two decades back. This can still be extended.

Salient Features of the Cropping Pattern

An examination of the cropland use of the village indicates the following salient features of the cropping pattern:

1. Kharif crops are comparatively more important than rabi crops, although their importance is decreasing.
2. Foodgrains occupy 83.7 per cent of the total cropped area, while food-cum-commercial crops claim only 13.6 per cent of it.
3. Non-food crops, particularly fodder crops have no place on the cropland use map of the village.

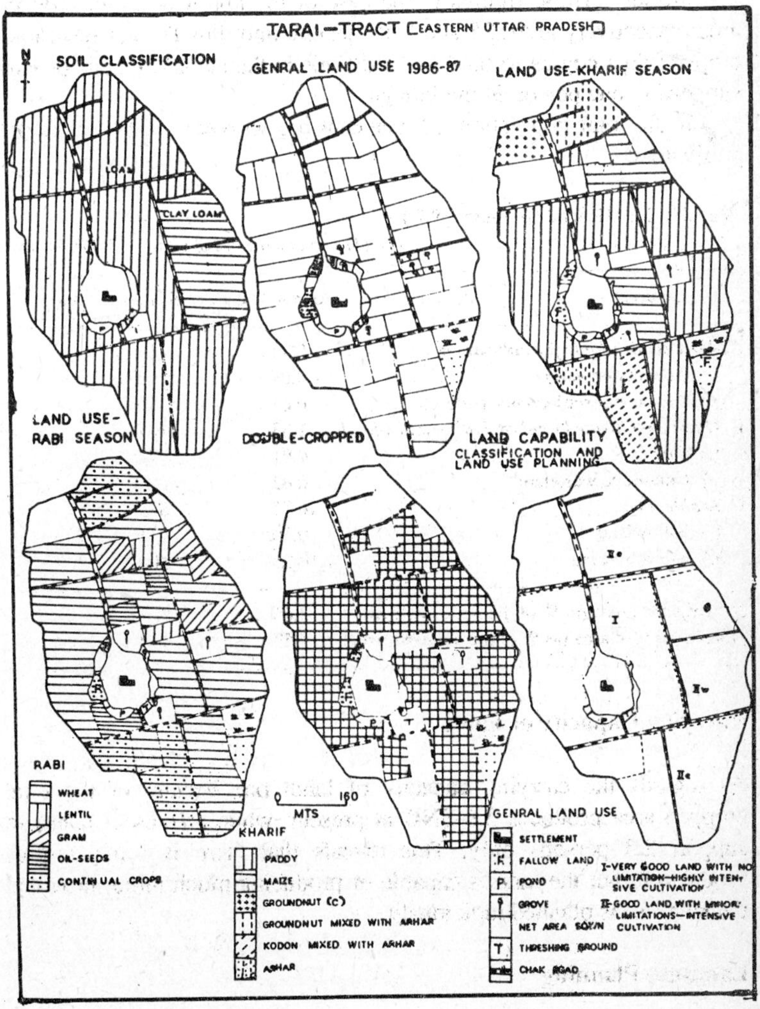

FIG. 6.11: Village Pure Shahbaz

Land-use and Population

Table 6.33 gives the total of various categories of land in the village and the per capita share of the villagers in these lands. The per capita share in the total area and net cropped area respectively is 0.39 and 0.31 hectare. These figures in the case of the kharif and rabi cropped area respectively are 0.25 and 0.19 hectare and thus the per head total cropped area comes to 0.44 hectare. This is the actual amount of land supporting one person in the village.

Of the total population, 35 per cent are workers, all of whom are cultivators.

TABLE 6.31: General Land-use (1987)

(The total area of the village : 12.79 hectres)

Land-use	Area in hectares	% of the total area
A. *Area not available for cultivation*	0.71	5.6
(i) Area under water	0.04	0.3
(ii) Area under settlements, paths etc.	0.67	5.3
B. *Other uncultivated land excluding fallowland*	1.33	10.4
(i) Groves	0.91	7.1
(ii) Culturable wasteland	0.42	3.3
C. *Arable land*	10.75	84.0
(i) Fallowland	0.50	3.9
(ii) Net sown area	10.25	80.1
Net irrigated areas (as % of the net sown area)	2.66	25.9
Double cropped area (as % of the net sown area)	4.57	44.5

Carrying Capacity of Land

As regards the carrying capacity of land, one hectare of the total cropped area produces 3.6 SNU at present, while at present it has to support 2.3 persons only. This reveals that there is some surplus production, but the land is capable of producing much more, provided the land-use is planned judiciously.

Land-use Planning

Keeping in view the land potentials and the present land utilisation, the land-use of the village may be planned on the following lines:

Land Capability Class I

Highly intensive cultivation with paddy, maize, wheat etc. is recommended. The conservation measures to be undertaken are:

1. Addition of bulky organic manures to maintain soil productivity,
2. Legumes to be given due place in the crop rotations, viz.

   (i) Maize/Lobia-Wheat-*Moong*
   (ii) Maize/Lobia-Pea/Potato-Sugarcane
   (iii) Early paddy-Pea-Sugarcane
   (iv) Early paddy-Wheat-*Moong*.

TABLE 6.32: Cropland Use (1987)

(Net cropped area of the village : 10.25 hecrares)
(Total cropped area of the village : 14.82 hectares)

Crops	Area in hectares	% of Kharif/rabi cropped area	% of net cropped area of the village	% of total cropped area of the village
1	2	3	4	5
Kharif	*8.42*	*100.0*	*82.1*	*56.8*
Foodgrains	*6.80*	*80.9*	*66.3*	*45.9*
a. *Cereals*	*5.60*	*66.6*	*54.6*	*37.8*
Paddy	2.80	33.3	27.3	18.9
Maize	1.60	19.0	15.6	10.8
Kodon mixed with *arhar*	1.20	14.3	11.7	8.1
B. *Pulses* (*arhar*)	*1.20*	*14.3*	*11.7*	*8.1*
Food-cum-commercial crops (Groundnuts alone and mixed)	*1.22*	*14.4*	*11.9*	*8.2*
Non-food crops	*0.40*	*4.7*	*3.9*	*2.7*
Rabi	*6.40*	*100.0*	*62.4*	*43.2*
Foodgrains	*5.60*	*87.5*	*54.6*	*37.8*
a. *Cereals* (Wheat)	*4.00*	*62.5*	*39.0*	*27.0*
b. Pulses	*1.60*	*25.0*	*15.6*	*10.3*
Gram	0.80	12.5	7.8	5.4
Masoor	0.80	12.5	7.8	5.4
Food-cum-commercial crops (Oil-seeds)	0.80	12.5	7.8	5.4

Land Capability Sub-class IIe

Intensive cultivation is recommended on this land. In addition to the measures suggested above, growing of cover crops such as groundnut and soyabean in kharif may be initiated.

TABLE 6.33: Land-use and Population (1987)

[*Total Population of the Village : 33 (1987) Area in hectares*]

	Total area of the village	Net cropped area of the village	Net cropped area in kharif	Net cropped area in rabi	Total cropped area	Double cropped area
	12.79	10.25	8.42	6.40	14.82	4.57
Land per head of population	0.39	0.31	0.25	0.19	0.44	0.13

Land Capability Sub-class IIw

This should also be devoted to intensive cultivation with paddy, wheat etc. as the main crops. Conservation measures to be undertaken are:

1. Addition of bulky organic manures,
2. Provision of drainage facilities,
3. Adequate and timely fertiliser use,
4. Suitable crop rotations to be adopted, viz.

   (i) Medium paddy-Wheat-*Moong*/Lobia
   (ii) Medium paddy-Berseem-Berseem
   (iii) Early paddy-Pea-Sugarcane
   (iv) Maize-Wheat-Arhar.

VILLAGE-AKBARPUR KALAN

Location

The village Akbarpur Kalan which belongs to the low lying riverine tract (Rapti-Burhi Rapti Basin) lies at 27°34′ N latitude and 82°6′ E longitude in tahsil Balrampur of district Gonda. It is at a distance of about 20 km from the tahsil headquarters, but its accessibility is very poor during the rains when the river Rapti is in full swing and it is cut from the tahsil headquarters.

Climate

No climatic data are recorded here, but the rainfall data recorded at the tahsil headquarters may be taken as close approximation of the climatic conditions prevailing in the village. They have already been discussed with the land-use analysis of the village Dhampur.

Soils

The soils of the village are predominantly fine sandy loam to loam which have developed on recent alluvial deposits. They are subject to recurring annual floods of the river Rapti. The eastern part possesses silty clay loam soils (Fig. 6.12).

Land Classification

The land of the village may be grouped into the following land capability classes (Fig. 6.12):

Land Capability Class I

This is a very good land which has few limitations and is capable of highly intensive cultivation. It comprises of fine sandy loam to loamy well drained soils which cover a considerable part of the village. The slope ranges up to 1 per cent. The permeability of soil is moderately rapid and water-holding capacity is moderately low. The soil reaction is mildly alkaline (pH 7.6). The fertility status is low in respect of organic carbon (0.225%) and phosphorous (13.4 kg\ha). It is medium in respect of potash (112 kg/ha). The best suited crops are maize, wheat, barley, oil-seeds, potato etc.

Land Capability Sub-class IIe

This is good land with moderate limitations of soil erosion. It is found in the middle part of the village and possesses almost the same ingredients as in the case of the former, except that it suffers from moderate limitations of erosion. The slope ranges between 1 and 3 per cent. The permeability is moderately rapid to rapid and water-holding capacity is moderately low to low. The suitable crops are wheat, potato, barley, legumes, oil-seeds etc.

Land Capability Sub-class IIe

This is also capable of intensive cultivation and comprises of predominantly silty clay loam fairly to poorly drained soils with some patches of silty loam fairly drained soils. The slope ranges up to 1 per cent or somewhat more. The permeability is slow and water-holding capacity is high. The soil reaction is mildly alkaline (pH 7.6). The fertility status is low (0.250%) in respect of organic carbon and phosphorous (13.4 kg/ha). It is medium in respect of potash (179.2 kg/ha). The best suited crops are late paddy, medium paddy, *masoor*, maize and sugarcane.

Land Capability Sub-class IIIe

This land suffers from severe limitation of soil erosion and is capable of moderate cultivation. It comprises of a narrow strip of land along the stream, where slope varies from 3 to 5 per cent. The soils are well drained. The permeability is rapid and water-holding capacity is low. Other ingredients are similar to those of sub-class IIe. The best suited crops are groundnut, *arhar*, sweet-potato, barley, mustard etc.

Irrigation

Irrigation claims 18.7 per cent of the net area sown. Rivers and ponds are the sources of irrigation.

Manuring

The use of cow-dung is very limited and that of the chemical fertilisers is inadequate.

General Land-use Pattern

Table 6.34 and Fig. 6.12 show that area not available for cultivation claims about 5.0 per cent of the total area of the village, the major portion of which is under settlement etc. The area under this category has increased in the past. Other uncultivated land excluding fallowland occupies 2.2 per cent of the total area. Arable land claims 92.9 per cent of the total area (fallowland - 0.6 and net sown area - 92.3%). The area under fallowland has decreased and that under net cropped has increased.

Cropland Use

Table 6.35 and Fig. 6.12 show that 56.0 per cent of the total and 84.6

per cent of the net cropped area of the village is devoted to the cultivation of kharif crops. Maize and paddy are the main crops of this season, which respectively occupy 32.2 and 23.4 per cent of the total and 57.5 and 41.7 per cent of the net cropped area of the village. Maize concentrates on the sandy loam and paddy on silty clay loam soils. The area under them has greatly decreased in the past, which is due to the increase in the importance of some rabi crops. Sugarcane, kodon mixed with *arhar* etc. are other crops.

Rabi crops occupy 44.0 per cent of the total and 66.5 per cent of the net cropped area. Wheat and *masoor* are important crops, occupying 15.3 per cent and 18.5 per cent of the total cropped area respectively. *Masoor* concentrates towards east on the silty clay loam soils and wheat on the fine sandy loam soils. Oil-seeds, barley, peas, gram alone and mixed are other crops.

Double Cropping

Double cropping covers 51.1 per cent of the net cropped area, which was about 40.0 per cent about three decades back.

Salient Features of the Cropping Pattern

A perusal of the cropland use of the village portrays the following special features of the cropping pattern:

TABLE 6.34: General Land-use (1987)

(*Total Area of the Village : 437.60 Hectares*)

Land-use	Area in hectares	% of the total area
A. *Area not available for cultivation*	21.2	4.9
(i) Area under water	4.4	1.0
(ii) Area under settlements etc.	16.8	3.9
B. *Other uncultivated land excluding fallowland*	9.9	2.2
(i) Groves	2.8	0.6
(ii) Culturable waste land	5.6	1.3
(iii) Pastures	1.5	0.3
C. *Arable land*	406.50	92.9
(i) Fallowland	2.50	0.6
(ii) Net sown area	404.00	92.3
Total	437.60	100.0
Net irrigated area (as % of net sown area)	75.6	18.7
Double cropped area (as % of net sown area)	206.40	51.1

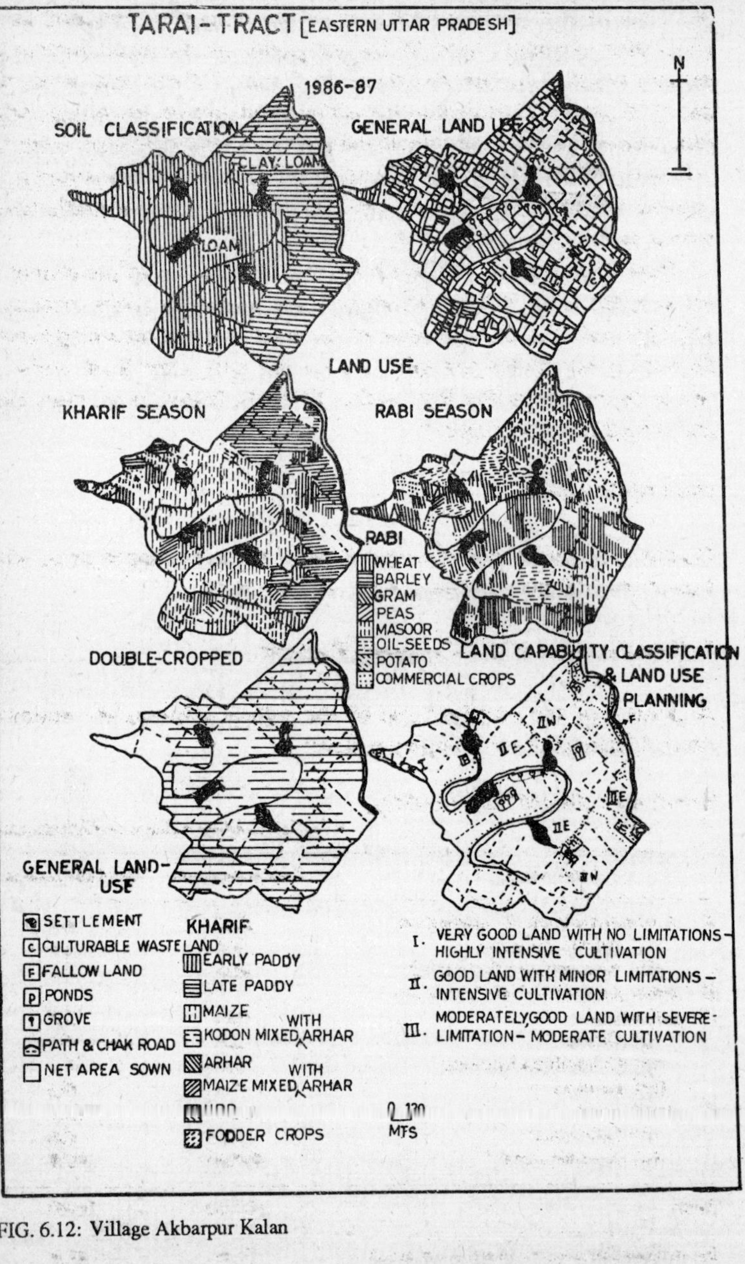

FIG. 6.12: Village Akbarpur Kalan

1. Kharif crops are comparatively more important than rabi crops; although the latter are now gradually gaining more importance.
2. Foodgrains occupy 91.9 per cent (55.9% in kharif and 36.0% in

rabi) of the total cropped area of the village.

3. Food-cum-commercial crops claim 12.3 per cent of the total cropped area, which was only 6.4 per cent three decades back. This shows that their importance is increasing.
4. Non-food crops do not appear on the cropland use map.

Land-use and Population

Table 6.36 shows the totals of various categories of land of the village and the per head share of the villagers in them. The per head share in the total area and net cropped area respectively is 0.19 and 0.17 hectare. It is 0.15 and 0.12 hectare respectively in the case of net cropped area in kharif and rabi crops. Thus the per head total cropped area which is the amount of the area actually supporting one person in the village comes to 0.27 hectare.

TABLE 6.35: Cropland Use (1987)

(Net Cropped Area of the Village : 404.00 Hectares)
(Total Cropped Area of the Village : 610.40 Hectares)

Crops	Area in hectares	% of kharif/ rabi cropped area	% of net cropped area of the village	% of total cropped area of the village
Kharif	*342.00*	*100.0*	*84.6*	*56.0*
Foodgrains	*340.80*	*99.7*	*84.3*	*55.8*
a. *Cereals*	*339.20*	*99.2*	*83.9*	*55.5*
Paddy	141.60	41.1	35.0	23.2
Maize	196.40	57.5	48.6	32.1
Kodon mixed with *arhar*	1.20	8.3	0.3	0.2
b. *Pulses (Urad* and *moong)*	*1.60*	*0.5*	*0.4*	*0.3*
Food-cum-commercial crops	—	—	—	—
Non-food crops (Fodder)	*1.20*	*0.3*	*0.3*	*0.2*
Rabi	*268.40*	*100.0*	*66.5*	*44.0*
Foodgrains	*220.00*	*82.0*	*54.5*	*36.0*
a. *Cereals*	*104.80*	*39.0*	*25.9*	*17.2*
Wheat	94.00	35.0	23.0	15.4
Barley	10.80	4.0	2.6	1.8
b. *Pulses*	*115.20*	*43.0*	*28.6*	*18.8*
Masoor	113.20	42.20	28.1	18.5
Pea	1.20	0.5	0.3	0.2
Gram	0.80	0.3	0.2	0.1
Food-cum-commercial crops	*48.40*	*18.0*	*12.0*	*8.0*
Oil-seeds	47.20	17.5	11.7	7.8
Vegetables	1.20	0.5	0.3	0.2

As regards the occupational structure of the total population 31.0 per cent are workers, 77.0 per cent of whom are cultivators and 20.0 per cent are agricultural labourers. Thus about 97.0 per cent of the total workers belong to the primary rural group and depend on land for their livelihood.

Carrying Capacity of Land

With the present management practices, one hectare of the total cropped area can support 3.6 persons (3.6SNU), while it has to support almost the same number of persons at present. The land is capable of producing more and supporting increasing number of population in future, if properly utilised.

TABLE 6.36: Land-use and Population

[Total Population of the Village : 2286 (1987) Area in hectares]

	Total area of the village	Net cropped area of the village	Net cropped area in kharif	Net cropped area in rabi	Total cropped area	Double cropped area
	437.60	404.00	342.00	268.40	610.40	206.40
Land per head of population	0.19	0.17	0.15	0.12	0.27	0.10

Land-use Planning

Keeping in view the uses and misuses of the land and its potentials, the following scheme for optimum utilisation of land resources is suggested:

Land Capability Class I

Highly intensive cultivation with maize, wheat, sugarcane, paddy, *moong* etc. is recommended. The conservation measures to be taken are:

1. Addition of bulky organic manures to maintain soil fertility,
2. Green manuring, if feasible economically,

3. Legumes to be given due place in the cropping pattern,
4. Adoption of the following crop rotations:

   (i) Maize-Lobia-Wheat-*Moong*
   (ii) Maize-Lobia-Pea/Potato/Sugarcane
   (iii) Early paddy-Wheat-*Moong*
   (iv) Early paddy-*Berseem*.

Land Capability Sub-class IIe

Intensive cultivation with maize, *arhar*, wheat, *moong*, sugarcane etc. is recommended together with the following conservation practices:

1. Adequate and timely fertiliser use based on soil testing recommendations,
2. Sinking of tube-wells for making proper provision for irrigation,
3. Growing of cover crops such as groundnut and soyabean to check soil erosion,
4. Adoption of scientific crop rotations and other measures as already recommended in the case of Class I land.

Land Capability Sub-class IIw

This should also be devoted to intensive cultivation with the following crop rotations:

(i) Late paddy-Berseem/*Masoor*
(ii) Late paddy-Zaid paddy-*Dhaincha*
(iii) Medium paddy-Wheat-*Moong*.

Other conservation practices which should be adopted are:

1. Addition of bulky organic manures to improve physico-biological properties of soil,
2. Proper provision of surface drainage,
3. Sinking of tube-wells to lower down the water-table and make provision for irrigation,
4. Adequate and timely fertiliser use on the basis of soil testing recommendations.

Land Capability Sub-class IIIe

This should be devoted to moderate cultivation with groundnut, *moong*, sweet-potato and soyabean as the main crops to minimise surface flow of water. Other measures are more or less the same as recommended in the case of sub-class IIe.

In addition to these measures, special attention should be paid towards the use of HYV of seeds, adoption of modern technological innovations and educating the farmers in agricultural science.

VILLAGE MAJHWAN

Location

The village Majhwan which belongs to the lowlying riverine tract of Burhi Rapti-Banganga and Rapti basin, lies at 27°28′ N latitude and 82°50′ E longitude in the Naugarh tahsil of district Basti. Its distance from the tahsil headquarters is 16 km It is not favourably located with respect to the means of transport and communication. The nearest railway station is at a distance of 10 km and the nearest road is at a distance of 8 km.

Climate

No climatic data are recorded at the village. The rainfall data recorded at the tahsil headquarters may be taken as close approximation of the climatic conditions of the village. They have already been discussed with the land-use analysis of the village Marwatia Deegar.

Soils

The soils of the village which have developed on the recent alluvial deposits, are subject to inundations during the rains. They are known as 'bhat'. Sometimes, they receive abundant sand deposits during the floods. Texturally speaking, the soils generally are clayey loam, silty loam and sandy loam (Fig. 6.13).

Land Classification

Based on the land capability, the land of the village has been classified into the following land capability classes or capability units having the same inherent characteristics, crop adoption, similar soil need and similar water management practices (Fig. 6.13):

Land Capability Sub-class IIw

This is good land with moderate limitations of water. It is capable of intensive cultivation and comprises of silty loam to silty clay loam soils of the southern part, which are somewhat poorly drained. The slope ranges up to 1 per cent. The permeability is moderate to moderately slow and the soil reaction is neutral to mildly alkaline. As regards the fertility status, it is very low to high in respect of organic carbon, low in respect of phosphorous and potash. These soils are suitable for growing early paddy, sugarcane, wheat, pea, gram etc.

Land Capability Sub-class IIIw

This is moderately good land with severe limitation of water-logging and high water-table. It is capable of moderate cultivation. The soils of this category of land which is found in the northern and south-eastern part of the village are silty loam to silty clay loam. They are poorly drained and occur on nearly level lands (slope varying from 1 to 3%). They remain waterlogged during the rains. The permeability is very slow and water-holding capacity is very high. As regards the fertility status, it is low in respect of organic carbon (0.196 to 0.296%), phosphorous (13.4 kg/ha) and potash (67.2 kg/ha). The soil reaction is mildly to moderately alkaline. These soils are best suited to the cultivation of paddy, wheat, barley etc.

Irrigation

About 39.4 per cent of the net sown area is irrigated, while about three decades back, it was only 1 per cent. The low percentage of irrigated land appears due to the lesser demand for irrigation as a result of higher sub-soil moisture.

Manures and Fertilisers

Their use is limited due to the poor economic condition of the farmers and the inadequate supply of these materials.

General Land-use Pattern

Table 6.37 and Fig. 6.13 reveal that 7.0 per cent of the total area is not available for cultivation. Other uncultivated land excluding fallowland accounts for 2.6 per cent, while arable land claims 90.4 per cent of the total area, almost all of which is under plough.

TABLE 6.37: General Land-use (1987)

(Total Area of the Village : 143.96 Hectares)

Land-use	Area in hectares	% of the total area
A. *Area not available for cultivation*	10.00	7.0
(i) Area under water	8.21	5.7
(ii) Area under settlements etc.	1.79	1.3
B. *Other uncultivated land excluding fallowland*	3.80	2.6
(i) Groves	1.30	0.9
(ii) Culturable waste land	2.50	1.7
C. *Arable land*	130.16	90.4
(i) Fallowland	0.27	0.2
(ii) Net sown area	129.89	90.2
Net irrigated area (as % of net sown area)	51.17	39.4
Double cropped area (as % of net sown area)	80.26	61.8

Cropland Use

Table 6.38 and Fig. 6.13 reveal that 49.1 per cent of the total and 78.2 per cent of the net cropped area of the village is devoted to the cultivation of kharif crops. A decrease has been registered in the area of kharif crops in the past, which appears due to the uncertainty of kharif crops caused by floods and water-logging. The main kharif crop is paddy which claims 80.6 per cent of the kharif, 63.8 per cent of the net and 39.5 per cent of the total cropped area of the village. Formerly, it occupied larger area. Other kharif crops are *kodon*, maize and pulses etc.

Rabi crops claim 50.9 per cent of the total and 82.2 per cent of the net cropped area of the village. Slightly higher percentage under these crops seems due to floods and water-logging of the rainy season, reducing the area under kharif crops. Barley mixed with gram, pulses, wheat alone and mixed are important crops.

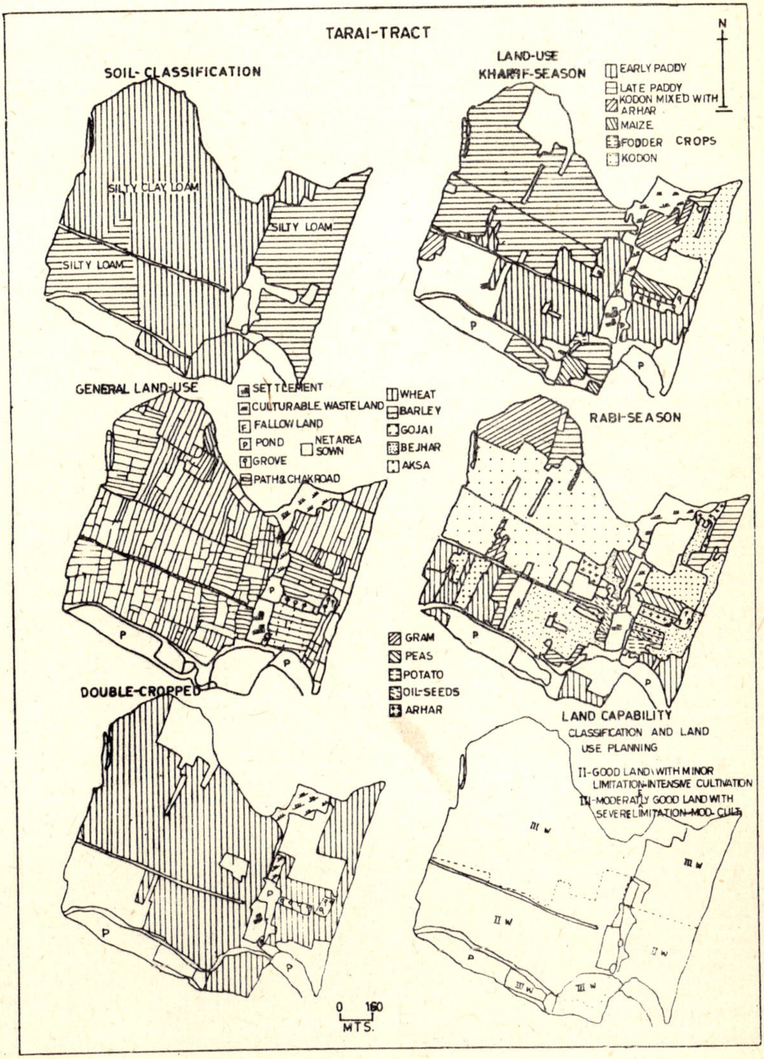

FIG. 6.13: Village Majhwan

Double Cropping

Double-cropped area accounted for 61.8 per cent of the net sown area. This was 22.0 per cent in 1955. The increase appears due to the increasing pressure of population and farmers' initiative to use the land to the maximum.

TABLE 6.38: Cropland Use (1987)

(Net Cropped Area of the Village : 129.89 Hectares)
(Total Cropped Area of the Village : 210.15 Hectares)

Crops	Area in hectares	% of kharif/ rabi cropped area	% of net cropped area of the village	% of total cropped area of the village
1	2	3	4	5
Kharif	*103.26*	*100.0*	*78.9*	*49.2*
Foodgrains	*102.92*	*97.7*	*78.7*	*49.1*
a. *Cereals*	*94.81*	*92.0*	*72.7*	*45.2*
Paddy	82.86	80.6	63.8	39.5
Maize	3.84	3.7	2.9	1.8
Kodon	8.11	7.7	6.8	3.9
Kodon mixed with *arhar*	8.11	7.7	6.8	3.9
b. *Pulses* (*Arhar* is sown mixed with *kodon* and as such there is no separate area under this head)	—	—	—	—
Food-cum-commercial crops	—	—	—	—
Non-food crops	*0.34*	*0.3*	*0.2*	*0.1*
Rabi	*106.89*	*100.0*	*82.3*	*50.8*
Foodgrains	*104.33*	*97.5*	*80.2*	*49.6*
a. *Cereals*	*56.50*	*52.9*	*43.4*	*26.9*
Wheat	14.18	13.3	10.9	6.8
Wheat mixed	8.51	7.9	6.6	4.4
Barley	8.42	7.9	6.5	4.0
Barley mixed with gram	25.45	23.8	19.4	12.2
b. *Pulses*	*47.77*	*44.6*	*3.8*	*22.7*
Pea	2.37	2.2	1.8	1.2
Gram	3.18	3.0	2.4	1.5
Masoor	0.88	0.8	0.6	0.4
Others	41.34	38.6	32.0	19.7
Food-cum-commercial crops (vegetables)	*2.56*	*2.5*	*2.0*	*1.2*

Salient Feature of Cropping Pattern

A perusal of cropland use depicts the following special features of the cropping pattern:

1. Kharif and rabi crops are equally important in the cropping pattern.
2. Foodgrains occupy dominant position in the cropping pattern, claiming 89.0 per cent of the total cropped area.
3. Food-cum-commercial crops occupy only 1.20 per cent of the total cropped area. The rest is occupied by other crops.

Land-use and Population

Table 6.39 shows the totals of various categories of land in the village and the per head share of the villagers in these lands. It is clear that the per head share in the total area and in net cropped area respectively is 0.22 and 0.20 hectare. It is 0.16 hectare each in the net cropped area in kharif and rabi crops. Thus the per head share in the total cropped area is 0.32 hectare. This is the actual amount of land supporting one person in the village.

As regards the occupational structure, about 45.0 per cent of the total population belongs to the working class, about 60.0 per cent of whom are cultivators and the rest are agricultural labourers. This shows exclusive dependence of the population on land.

Carrying Capacity of Land

With the present management practices, one hectare of the total cropped area produces 2.9 SNU, while it has to support 3.1 persons. This indicates that the population is under-nourished and it necessitates an enhancement in the agricultural production by proper utilisation of the land resources.

Land-use Planning

The following measures for the optimum planning of the land-use in accordance with land potentials and requirement of population are recommended (Fig. 6.13):

Land Capability Sub-class IIw

Intensive cultivation with paddy sugarcane, wheat and pea etc. is recommended together with the following conservation practices:

1. Addition of bulky organic manures to improve physico-biological properties of the soils,
2. Provision of surface drainage,
3. Adequate and timely fertiliser use on the basis of soil testing recommendation, especially nitrogenous fertilisers,
4. Irrigation of crops only in accordance with their needs to avoid a rise in water-table,
5. Adoption of the following crop rotations:

(i) Paddy-Wheat-*Moong/Lobia*
(ii) Paddy-Pea-Sugarcane
(iii) Early paddy-Mustard-*Dhaincha.*

TABLE 6.39: Land-use and Population

[*Total Population of the Village : 660 (1987) Area in hectares*]

	Total area of the village	Net cropped area of the village	Net cropped area in kharif	Net cropped area in rabi	Total cropped area	Double cropped area
	143.96	129.89	103.26	106.89	218.15	80.26
Land per head of population	0.22	0.20	0.16	0.16	0.32	0.12

Land Capability Sub-class IIIw

Moderate cultivation with paddy, wheat, *masoor* and barley etc. with the following crop rotations is recommended:

(i) Late paddy-Linseed/Lentil
(ii) Late paddy-Zaid paddy-*Dhaincha*
(iii) Medium paddy-Wheat-*Moong*
(iv) Medium paddy-Pea-Sugarcane.

All the management practices enumerated in sub-class IIw should be followed in this capability sub-class.

VILLAGE SONBARSA

Location

The village Sonbarsa which belongs to the Great Gandak and Little Gandak basin in tahsil Hata of district Hata, lies at 27°12′ N latitude and 83°52′ E longitude. It is about 30 km from the town of Padrauna, but its distance from its own tahsil headquarters is about 70 km. Its accessibility is greatly reduced during the rains when the larger area is inundated by the river Gandak.

Climate

No climatic data are recorded at the village. The nearest rainfall recording station is Padrauna, the rainfall data of which may be taken as close approximation of the climatic conditions of the village. The average annual rainfall is 1268.3mm here, of which 2.9 per cent is received during the cold, 6.4 per cent during the hot and 90.7 per cent during the rainy season. The co-efficient of annual variability of rainfall is 25.3 per cent. The monthly variability is the highest in the month of October, which is very significant from the agricultural point of view. In the month of June, it is 65.7, in July 46.4, in August 51.9 and in September 57.3 per cent.

Soils

The soils of the village belong to the calcium soils which are lowlying and consequently liable at times to inundation during high floods or heavy monsoon. The texture varies from silty loam to silty clay loam (Fig. 6.14). The water-table is very near the surface.

Land Classification

Based on its capability, the land of the village may be grouped into the following capability classes or mapping units (Fig. 6.14):

Land Capability Class I

This is very good land which suffers from few limitations and is capable of highly intensive cultivation. It spreads over the eastern part of the village and comprises of sandy loam to silty loam fairly to well drained soils. The slope ranges between 0.5 and 1 per cent or even up to 3 per cent The permeability is moderate to slow and water-holding capacity is moderate to high. The soil reaction is mildly alkaline. The fertility status is medium in respect of organic carbon (0.500%) and potash (112.0 kg/ha). It is very low in respect of phosphorous (8.9 kg/ha). The soils are adapted to a wide range of crops, such as early paddy, sugarcane, wheat, mustard, maize, vegetables etc.

Land Capability Sub-class IIw

This is good land with moderate limitation of water-logging and high water-table. It comprises of silty loam fairly drained soils and covers a large portion of the village in the middle part. It resembles the previous class in many respects, but submergence to water is somewhat serious. The fertility status is low in respect of potash (67.2 to 89.6 kg/ha) and organic carbon (0.250 to 0.375%) and it is very low in respect of phosphorous (8.96 kg/ha). It is capable of intensive cultivation.

Land Capability Sub-class IIIw

This is fairly good land with severe limitations of submergence and is capable of moderate cultivation. It comprises of silty clay loam poorly drained soils occupying the western and southern parts of the village. These soils occur on nearly level to very gently sloping land liable to inundation during the monsoon. The permeability is slow and water-holding capacity is high. The soil reaction is mildly alkaline. The fertility status is low in respect of all the major nutrients. Paddy, sugarcane, wheat, *masoor* etc. are suitable crops.

Irrigation

About 30 per cent of the net cropped area of the villages is irrigated by irrigation channels recently developed.

Manuring

The use of manures and fertilisers is fairly satisfactory and needs an enhancement.

General Land-use Pattern

Table 6.40 and Fig. 6.14 reveal that area not available for cultivation and other uncultivated land excluding fallowland respectively account for 3.8 and 1.9 per cent of the total area of the village. Fallowland claims only 2.0 per cent, while 92.3 per cent of the area is under plough. There has been only a slight increase in this latter category in the past.

Cropland Use

Table 6.41 and Fig. 6.14 show that kharif crops claim 54.8 per cent of the total and 83.5 per cent of the net cropped area of the village. Paddy is the main crop which claims 75.7 per cent of the kharif, 63.3 per cent of the net and 41.5 per cent of the total cropped area and spreads widely on all types of land in the village. Maize and kodon mixed with *arhar* are other important cereals and sugarcane is the only food-cum-commercial crop.

Rabi crops claim 45.2 per cent of the total and 69.0 per cent of the net cropped area. Wheat is the dominant crop, as it occupies 59.6 per cent of the rabi, 41.[illegible] per cent of the net and 26.9 per cent of the total cropped area.

TABLE 6.40: General Land-use (1987)

(Total Area of the Village : 85.50 Hectares)

Land-use	Area in hectares	% of the total area
A. *Area not available for cultivation*	3.15	3.18
(i) Area under water	1.40	1.6
(ii) Area under settlements etc.	1.75	2.2
B. *Other uncultivated land excluding fallowland*	1.65	1.9
(i) Groves	0.95	1.1
(ii) Culturable waste land	0.70	0.8
C. *Arable land*	80.70	94.3
(i) Fallowland	1.70	2.0
(ii) Net sown area	79.00	92.3
Total area	95.50	100.0
Net irrigated area (as % of net sown area)	24.00	30.3
Double cropped area (as % of net sown area)	41.50	52.5

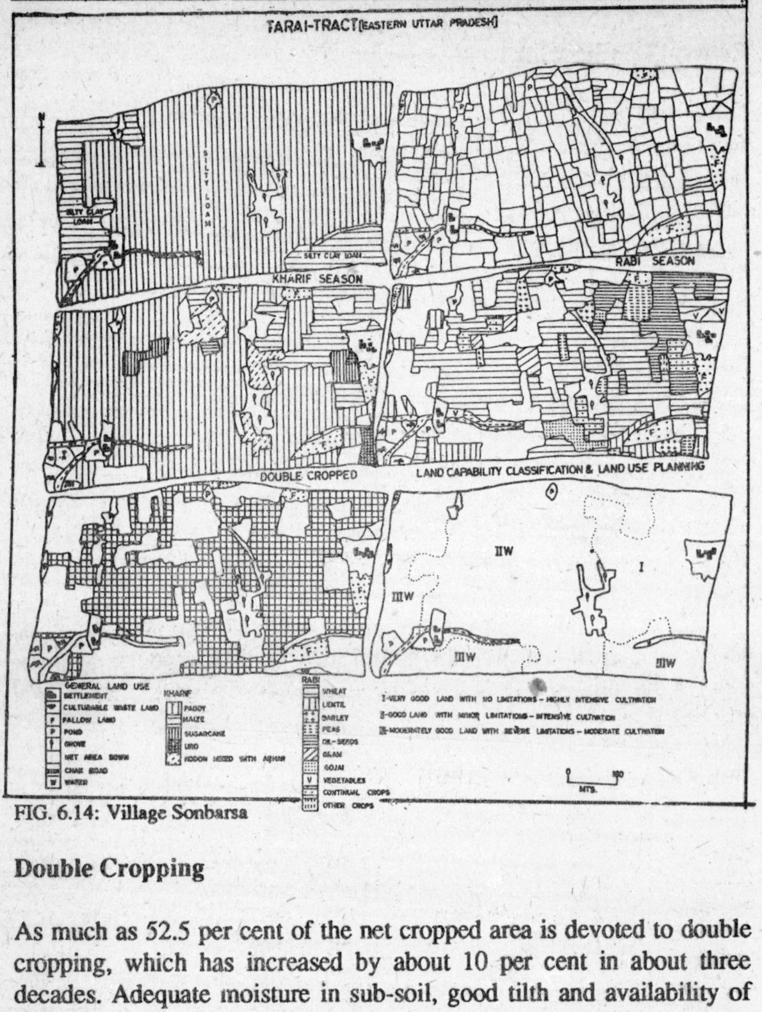

FIG. 6.14: Village Sonbarsa

Double Cropping

As much as 52.5 per cent of the net cropped area is devoted to double cropping, which has increased by about 10 per cent in about three decades. Adequate moisture in sub-soil, good tilth and availability of water for irrigation are some of the reasons for this high percentage which can further be increased.

Salient Features of the Cropping Pattern

An analysis of the cropland use of the village brings out the following special features of the cropping pattern:

1. Kharif crops are comparatively more important than rabi crops.

2. Foodgrains dominate the cropping pattern by claiming 90.9 per cent of the total cropped area (52.3% in kharif and 38.6% in rabi).
3. Food-cum-commercial crops claim 9.1 per cent of the total cropped area, of which larger portion is claimed by rabi crops.
4. Non-food crops do not appear on the cropland use map.

TABLE 6.41: Cropland Use (1987)

(Net Cropped Area of the Village : 79.00 Hectares)
(Total Cropped Area of the Village : 120.50 Hectares)

Crops	Area in hectares	% of kharif/ rabi cropped area	% of net cropped area of the village	% of total cropped area of the village
1	2	3	4	5
Kharif	*66.00*	*100.0*	*83.5*	*84.8*
Foodgrains	*63.00*	*95.4*	*79.7*	*52.3*
a. *Cereals*	*61.50*	*93.1*	*77.8*	*51.1*
Paddy	50.00	75.7	63.3	41.5
Maize	6.50	9.8	8.2	5.4
Kodon mixed with *arhar*	5.00	7.6	6.3	4.2
b. *Pulses (Urad* and *moong)*	*1.50*	*2.3*	*1.9*	*1.2*
Food-cum-commercial crops (Sugarcane)	*3.00*	*4.6*	*3.8*	*2.5*
Rabi	*54.50*	*100.0*	*68.9*	*45.2*
Foodgrains	*46.50*	*85.3*	*58.9*	*38.6*
a. *Cereals*	*37.00*	*67.9*	*46.8*	*30.6*
Wheat	32.50	59.6	41.1	26.9
Barley	2.00	3.7	2.5	1.6
Wheat mixed with barley	2.50	4.6	3.2	2.1
b. *Pulses*	*9.50*	*17.4*	*12.1*	*8.0*
Gram	2.50	4.6	3.2	2.1
Pea	2.50	4.6	3.2	2.1
Masoor	4.50	8.2	5.7	3.8
Food-cum-commercial crops	*8.00*	*14.7*	*10.0*	*6.6*
Oil-seeds	6.00	11.0	7.5	5.0
Vegetables	2.00	3.7	2.5	1.6

Land-use and Planning

Table 6.42 shows the totals of various categories of land in the village and the per head share of the villagers in these lands. The per head share of the total area and the net cropped area respectively is 0.12 and 0.11 hectare. It is 0.09 and 0.08 hectare in the case of the kharif and rabi cropped area respectively. Thus the actual amount of land supporting one person comes to 0.17 hectare.

Of the total population, about 30.0 per cent is engaged in work. The cultivators claim only 31 per cent of the total workers and about 60 per cent are agricultural labourers. This indicates the predominance of the landless agricultural labourers and heavy pressure of population on land.

Carrying Capacity of Land

One hectare of the total cropped area of the village can support 4.6 persons with the existing management practices, but the present burden is that of 5.9 persons, which warrants the enhancement of the agricultural production and land-use planning in accordance with the land capabilities.

TABLE 6.42: Land-use and Population

[*Total Population of the Village : 710 (1987) Area in hectares*]

	Total area of the village	Net cropped area of the village	Net cropped area in kharif	Net cropped area in rabi	Total cropped area	Double cropped area
	85.50	89.00	66.00	54.50	120.50	41.50
Land per head of population	0.12	0.11	0.09	0.08	0.17	0.06

Land-use Planning

The proposals for future planning are as follows, which include the optimal utilisation of land according to its capability and some conservation measures (Fig. 6.14):

Land Capability Class I

Highly intensive cultivation with paddy, sugarcane, wheat, maize, mustard and vegetables is recommended together with the following conservation practices:

1. Addition of bulky organic manures to maintain soil productivity,
2. Green manuring, if feasible,
3. Adoption of the following crop rotations:

   (i) Early paddy-Barley/Wheat-*Moong*/*Lobia*
   (ii) Early paddy-*Berseem*
   (iii) Maize-Sugarcane.

Land Capability Sub-class IIw

Intensive cultivation with almost the same crop rotations as recommended in class I is recommended. In addition to the conservation practice recommended in class I, the following are also recommended:

1. Provision of proper surface drainage,
2. Adequate and timely fertiliser use,
3. Working out the methods, time and frequency of irrigation in accordance with the intake rate of the soils, effective root-depth and evaporation needs of the crops.

Land Capability Sub-class IIIw

Moderate cultivation with the following crop rotations is recommended:

(i) Late paddy-*Berseem*/*Masoor*,
(ii) Late paddy-Zaid paddy-*Dhaincha*,
(iii) Medium paddy-Wheat-*Moong*.

The conservation measures to be recommended are:

1. Addition of bulky organic manures to improve the physical properties of soils,
2. Balanced and timely application of fertilisers which may give better response to water-logged conditions in accordance with

the soil tests,

3. Minimum use of calcium fertilisers to check calcium content increase in the soil,
4. Proper provision of surface drainage,
5. Controlled and restricted irrigation according to crops needs.

In addition to the measures suggested for different types of land, use of HYV of seeds and adoption of modern technological innovations together with training of the farmers in these practices is also recommended.

PART IV

SUMMARY

7

SUMMARY OF MAIN FINDINGS AND RECOMMENDATIONS

THE EAST-WEST narrow, elongated strip of the Tarai of the eastern Uttar Pradesh, touching the Indo-Nepal boundary, which is low with surface entirely criss-crossed by innumerable streams and *nalas*, is marked by high water-table and fertile soils with excessive water-logging and flood hazards in many parts. Its unhealthy malarial climate is characterised by excessive heat and humidity and moderately heavy rainfall. It is essentially an agricultural region. It is marked by comparatively a lower density of population, poor means of transport and communication. It has suffered for long from stagnation and underdevelopment, although it is well endowed with rich land resources which have remained untapped until recently. It promises to become one of the prosperous regions not only of the State of Uttar Pradesh, but also that of the country as a whole, provided its land-use is planned systematically, judiciously and prudently. The region is a distinct homogeneous unit for a geographical investigation and analysis.

The spatio-temporal investigation of the general land-use of the region reveals that the forests occupy 4.8 per cent of the total area as against 17.3 per cent in the State of Uttar Pradesh and 22.1 per cent in the country. In about one-third area, the proportions are above the regional average (reaching as high as 15.5%), while in the remaining two-thirds, it is below it (reaching as low as below 1%). The forest cover has declined during 1955-57 to 1985-87, the reduction being to the tune of 3.8 per cent of the total area, which is due to the destruction of forests as a result of agricultural colonisation and felling of trees in order to meet the increasing demand for timber and firewood. The area under this category needs expansion in order to maintain the ecological balance.

Area not available for cultivation claims 11.1 per cent of the total area of the region, as compared to the State's average of 11.6 per cent and 13.3 per cent that of the country. There are some micro-level variations in the extent of this category of land. There has been an overall increase of 3.9 per cent in the area of this category of land, but the micro-level study shows that some blocks have experienced decrease to some extent. The high plus changes are in those blocks where growth rate of population has generally been higher and where due to some development schemes, the area under settlements and road etc. has increased and which still needs some expansion.

Other uncultivated land excluding fallowland which includes culturable wasteland, area under miscellaneous trees and groves, pastures and grasslands, accounts for 2.3 per cent of the total area, which was 3.9, 7.7 and 9.5 per cent respectively in 1975-77, 1965-67 and 1955-57. Culturable wasteland claims 1.5 per cent of the total area. The small extent of this category of land apparently shows that the scope for future extension of cultivated land by reclaiming this land is very limited.

The area under this category has decreased greatly (from 8.4% in 1955-57 to 1.5 per cent in 1985-87). The outstanding and high reductions have been experienced in those areas where due to malarial conditions, deep-rooted grasses, extensive water-logging and sparse population, this category of land was very extensive. The magnitude of decrease appears a spontaneous reaction to the inflow of population, rapid reclamation measures and resulting stepping up of cultivation. This land has to be reclaimed either to cultivation and groves or to afforestation, settlement etc.

Land under miscellaneous trees and groves possesses negligible area (0.6%), as against the State's average of 1.9 per cent. This has only slightly increased, although greater expansion by reclaiming culturable wasteland is desirable. Permanent grasslands and pastures again claim only 0.2 per cent of the total area. Some expansion in the area of this category is needed.

Fallowland which includes the old and new fallowland accounts for 4.2 per cent of the total area, which is about two-thirds of the State's average of 6.5 per cent. There are some micro-level disparities. The percentage of the total area involved in change in the case of this category of land is only 0.4. This shows the slow process of reclamation of this land to cultivation. Leaving the ploughable land as fallow is a great misuse of the land resources. Early and effective measures are needed to reclaim this land to agriculture and horticulture.

Net area sown occupies the largest share (77.6%), as against 57.8 and 46.2 in Uttar Pradesh and India respectively. There are wide sub-regional variations (62.3 to 86.1%). There is high positive correlation between population pressure and percentage of net area sown and negative moderate correlation between this category of land and the culturable wasteland, fallowland and forested land. As regards the volume of change, there has been an increase of 6.7 per cent in the area under this category from (70.9% in 1955-57 to 77.6% in 1985-87), although some areas have registered some decrease also. The increases range between 0.5 and 24.5 per cent. It may be mentioned here that in some areas, there is very little scope for expansion of cultivated land and primary reliance ought to be placed on more intensive use of the cultivated land. But in some areas where the proportions of fallowland and culturable land are high, this can be expanded.

As regards the overall volume of change, in about 19.0 per cent of the blocks, the changes have been very slow (below 5%). These are the blocks where there were already high concentrations of net sown area and the area under culturable wasteland, fallowland and forested land was not very extensive or adequate efforts have not been made to increase the area under the former category. In about 42 per cent blocks, the land-use has been moderately dynamic (5-10%). About 19 per cent blocks belong to the dynamic (10-15%) and the rest to the high dynamic category (above 15%). These are the blocks which touch the northern border of India, which formerly possessed extensive lowlying and swampy areas infested with reeds, tall grasses, wide-spread forest, culturable waste-lands, fallowlands, unhealthy climatic conditions and sparse population. Agricultural colonisation inviting great inflow of population has resulted in great expansion of net sown area and corresponding reduction in other categories of land. As regards the decennial variations in the overall volume of change, the latter decades show more dynamism which appears due to the greater efforts to step up agricultural colonisation during this period.

As regards the intensity of cropping, the intensity index for the region is 149.1 per cent, as against 145.0 per cent of the State of Uttar Pradesh and 125.0 per cent that of India. This shows that the region enjoys a moderate position with respect to intensity of cropping and agricultural efficiency. There are micro-regional variations (from 119.9 to 177.5%) in the intensity of cropping, moderately high to high intensities being wide-spread. Although irrigation plays an important role in determining the degree of intensity, it does not enjoy the same dominance everywhere. There is low positive co-efficient of correlation between the net irrigated area and the intensity of cropping. The presence of fertile soils with high content of moisture in the sub-soil lessening the need of irrigation is a very important factor in this respect. Another important factor is the pressure of population on the agricultural land. It may be mentioned here that it does not rule out the necessity of extending irrigation facilities in this region. There are still large areas where double cropping can be extended efficiently, provided irrigation facilities are made available. Besides, irrigation is needed for the required use of the HYV of seeds, chemical fertilisers and thus for boosting up the per hectare yields from the land already put to double cropping.

During the span of three decades, an addition of 17.3 per cent of the net area sown has shifted to double cropping in the region as a whole, but there are micro-regional variations (plus 2.0 to 44.0% and minus 0.3 to 27.3%), for which the variations in the ecological conditions are responsible. The changes have been more phenomenal during the latter decades, which appears due to better irrigation facilities, better farm technology and infrastructure, an increasing rate of growth of population and comparatively greater measures to combat recurring floods and water-logging menace, which has made extensive areas unsuitable for a second crop and thus decreased the intensity of cropping.

Net irrigated area accounts for 32.6 per cent of the net sown area, as compared to 57.1 per cent in the case of State of Uttar Pradesh and 29.7 per cent in that of the country as a whole. There are considerable micro-regional variations (from 1.5 to 82.4%). The overall pattern shows that higher concentration of irrigated area is found towards east and low to very low towards west, which is mainly due to the abundance of irrigation facilities in the Gandak Command Area. The optimum utilisation of the agricultural land resources warrants sufficiently high increase in the net irrigated area, for which there is enough scope. The canal irrigation has already occupied the eastern part and is being spread over the western part, although it is not very much suited to the hydrological conditions of some areas and has very severe hazards in its womb. It is, therefore, suggested that the possibilities for extending tube-well irrigation, for harnessing the waters of ponds, depressions and other surface sources be explored and genuine efforts be made to develop them.

The cropland use analysis of the study area reveals that kharif crops are more important than rabi crops, as they occupy 84.1 per cent of the net and 56.4 per cent of the total cropped area, as against 63.8 and 42.8 per cent respectively in the case of the latter. They, thus, dominate the agricultural landscape everywhere with some spatial variations. There have been important changes in the relative strength of the kharif and rabi area. Generally, the rabi percentages have increased due to an increase in the irrigation facilities and decreases in the kharif area.

Foodgrains which include cereals and pulses play an excessively predominant role in the cropping pattern of the region by occupying 90.0 per cent of the total cropped area. This may be attributed mainly to their importance in sustaining the population and also due to their climatic and edaphic preferences. It may be noted here that kharif foodgrains are more important, as they claim 51.3 per cent of the total cropped area, while this percentage in the case of the rabi crops is only 38.7.

The area under foodgrains has registered a decrease of 3.1 per cent, which appears due to an increase in the area under food-cum-commercial crops.

Cereals form the major constituent of the foodgrains by claiming 80.1 per cent of the total cropped area. Kharif cereals claim 49.8 and the rabi 30.3 per cent of the total cropped area. Rice and wheat are the principal cereals followed by maize, *kodon* etc. in the kharif and wheat mixed, barley alone and mixed in the rabi season. As regards the volume of change, the region as a whole has registered a decrease to the tune of 2.9 per cent, although there are some micro-regional variations.

As already mentioned, rice is the principal crop (occupying 64.7% of the net and 43.4% of the total cropped area, while in Uttar Pradesh, it occupies 21.2 per cent the total cropped area and in the country as a whole 23.2 per cent) and has been the staple food of the majority of people. It ranks first in 30 out of 31 blocks. The predominance of this corp is due to the predominance of clayey, clayey loam or heavy loam soils which are well suited to the cultivation of this crop, the high water-table and abundance of moisture in the sub-soil, the lowlying lands that favour impounding of water which is beneficial for the aquatic crop, and hot and humid climate with moderately high rainfall.

It may be mentioned here that the concentration of this crop is not uniform throughout the region, as it varies from 28.2 to 58.7 per cent of the total cropped area, albeit the area of low concentration is not so extensive. The analysis of volume of change shows that the region has registered a minor decrease (1.1% of the total cropped area) during the study period, but the micro-regional analysis shows that 45.0 per cent of the blocks have registered increases also. The hectareage has risen from 3,80,700 (1955-57) to 4,53,842 (1985-87) showing a percentage increase of 19.2.

Wheat ranks second by claiming 28.7 per cent of the total cropped area, as against 33.1 per cent in the State of Uttar Pradesh and 12.9 per cent in the country as a whole and now forms quite an important item in human diet in many parts. Its second rank is taken only in 3 blocks by some other crops. In one block, it ranks first. Although the Tarai climatic conditions may not be considered very congenial for wheat cultivation, its superior adaptability to environmental conditions being a winter crop, the prevalence of clayey loam to sandy loam soils with sufficient sub-soil moisture together with the irrigation facilities being available in some parts, and an improvement in drainage conditions are some of the factors which have provided sufficient incentive for the extension of hectareage under this crop.

The spatial distribution exhibits micro-regional variations (from 21.6 to 40.8% of the total cropped area). It is inverse to that of rice. Wheat is grown both as dry as well as an irrigated crop.

The hectareage under this crop has risen from 87,366 (1955-57) to 2,99,834 (1985-87), i.e. by 343.1 per cent, which is much more than that of rice. This appears due to an increase in the intensity of irrigation, reclamation of waterlogged areas, improvement in drainage conditions and increasing importance of this crop in the diet of the people. Some blocks have experienced increases more than 400 per cent. As regards the volume of change, 18.5 per cent of the total cropped area has been added to the wheat hectareage with some micro-regional variations.

Maize is the third ranking crop occupying 4.2 per cent of the total cropped area (as against 5.1% in the State of Uttar Pradesh). Its share varies from zero to 19.9 per cent in the region. The hectareage under this crop has decreased by 24.5 per cent. As regards the volume of change, the region as a whole has registered a decrease to the tune of 2.6 per cent with some sub-regional variations.

The region grows a variety of pulses, for example, *arhar*, *urad*, *moong* etc. in the kharif and gram, pea, *masoor* etc. in the rabi season. Kharif and rabi pulses account for 1.5 and 8.4 per cent of the total cropped area respectively. There are considerable micro-regional

Land Capability Class I

This is very good land with few limitations and comprises of very deep, friable, fine sandy loam to loamy or silty loam soils underlain with loamy to clayey loam textured friable to firm sub-soils with well drained conditions (slope ranging from zero to one per cent). The permeability is moderately rapid and water-holding capacity is moderately low. The soil reaction is mildly alkaline to neutral. The fertility status is low to medium. The soils are highly responsive to fertiliser inputs and improved management practices.

Suitability

The land is well suited to the cultivation of wheat, maize, paddy, sugarcane, oil-seeds, vegetables etc.

Land-use Planning

This should be put to highly intensive cultivation with the suitable crops as mentioned above. Addition of bulky organic manures in order to maintain soil productivity and green manuring, if economically feasible, with due place to legumes is recommended. Suitable crop rotations are:

(i) Medium paddy-Wheat-*Moong*,
(ii) Early paddy-Pea/Mustard-Sugarcane,
(iii) Maize-Wheat-*Arhar*,
(iv) Maize+Lobia-Wheat-*Moong*,
(v) Maize+*Lobia*-Pea/Potato/Sugarcane
(vi) Medium paddy-Wheat-*Moong/Lobia*,
(vii) Early paddy-Pea-Sugarcane.

Land Capability Sub-class IIe

This is good land with moderate limitations and comprises of very deep, very friable well drained sandy loam to fine sandy loam and silty loam soils underlain with loam to sandy clay loam sub-soils. The slope ranges between 1 and 3 per cent. The soil reaction is neutral to mildly alkaline. The permeability is moderately rapid to rapid and water-holding capacity is moderately low to low. The fertility status is low to very low. Improvable problems are : (i) erosion susceptibility, and (ii) low fertility status, while the inherent problems are : (a) Unfavourable texture and (b) moderately rapid to rapid permeability.

Suitability

Suitable crops are wheat, maize, *arhar*, paddy, potato, barley, oil-seeds, groundnut etc.

Land-use Planning

The land should be intensively cultivated with suitable crops. The following conservation practices are recommended to minimise the limitations:

(i) Addition of bulky organic manures to improve their physico-biological properties,
(ii) Adoption of soil conservation measures, such as contour bunding and levelling,
(iii) Growing of cover crops such as groundnut, soyabean, *moong* etc. to minimise soil erosion,
(iv) Use of insecticides and pesticides to control termites and grubs,
(v) Application of nitrogenous fertilisers on the soil testing basis for adequate growth of crops,
(vi) Provision of line irrigation channels to avoid seepage,
(vii) Adoption of scientific crop rotations, viz.

(a) Medium paddy-Wheat-*Moong*,
(b) Early paddy-Pea/Mustard-Sugarcane,
(c) Early paddy/Maize-Wheat/Barley-*Moong/Lobia*,
(d) Maize-Sugarcane - x,
(e) Early paddy-Mustard-*Dhaincha*,
(f) Maize/Lobia-Wheat-*Moong*.

Land Capability Sub-class IIwe

This land suffers from moderate limitations of unfavourable water conditions and erosion. This comprises of silty clay loam fairly to poorly drained soils. The slope ranges between 1 and 3 per cent. The permeability is slow and water-holding capacity is high. The soil reaction is moderately alkaline. The fertility status is from low to moderate.

Problems

The main problems are submergence, water-logging, high water-table and erosion.

Suitability

Suitable crops are early and late paddy, sugarcane, wheat, oil-seeds etc.

Land-use Planning

Intensive cultivation with suitable crops together with the following conservation measures should be adopted:

(i) Addition of bulky organic manures,
(ii) Sinking of tube-wells for secured irrigation,
(iii) Adequate and timely fertiliser use,
(iv) Proper provision of surface drainage,
(v) Leguminous crops to be grown and given due importance,
(vi) Adoption of scientific crop rotations, viz.

(a) Early paddy-Wheat-*Moong*,
(b) Medium paddy-Wheat-Sugarcane,
(c) *Dhaincha*/Late paddy-Lentil/Linseed.

Land Capability Sub-class IIw

This is good land which suffers from moderate limitations of unfavourable water conditions and comprises of very deep somewhat fairly to poorly or imperfectly drained silty loam to silty clay loam or clayey loam soils. The slope ranges from zero to one per cent and at places even up to 3 per cent. The soils remain saturated during monsoon and water-table fluctuates from less than one metre to 3 metres throughout the year. The permeability is slow to very slow and water-holding capacity is moderately high to high. The soil reaction is mildly alkaline and fertility status is generally low to very low.

Suitability

The soils are well-suited to the cultivation of paddy, sugarcane, wheat, lentil, vegetables, pea etc.

Problems

Improvable problems are surface and sub-surface drainage and low fertility, while the inherent problem is that of moderately slow permeability.

Land-use Planning

This land should be devoted to the cultivation of suitable crops with intensive methods of cultivation. The following measures are suggested to minimise the limitations:

(i) Addition of bulky organic manures to improve physico-biological properties of the soils,
(ii) Provision of drainage channels along with irrigation channels,
(iii) Sinking of tube-wells to lower down the water-table,
(iv) Proper cleaning, deepening and straightening of natural drains to increase the water carrying capacity of them,
(v) Provision of water for irrigation according to the intake rates of the soils,
(vi) Introduction of leguminous crops such as lobia, berseem and *moong* as a permanent feature in the cropping pattern,
(vii) Adequate and timely fertiliser use based on soil-testing recommendations and split application of nitrogenous fertilisers,
(viii) Adoption of scientific crop rotations, viz.

(a) Early paddy-Pea-Sugarcane,
(b) Late paddy-Lentil/Linseed x,
(c) Late paddy-Zaid paddy-*Dhaincha*,
(d) Medium paddy-Wheat-*Moong/Lobia*.

Land Capability Sub-class IIIe

This sub-class is similar in many characteristics to sub-class IIe, except in slope and the magnitude of erosion. The slope ranges from 3 to 5 per cent and undulating topography is also common, which results in severe erosion. The water-table fluctuates from 2-3 to 3-6 metres. The water-holding capacity is low and permeability is rapid.

Suitability

The suitable crops are groundnut, *arhar*, potato and sweet potato, maize, mustard etc.

Problems

Improvable problems are: (i) erosion susceptibility, (ii) slopy and undulating land, and (iii) low fertility; while the inherent ones are (a) light texture, (b) rapid permeability and (c) low water-holding capacity.

Land-use Planning

In addition to the measures already suggested in the case of IIe, land shaping and terraced cultivation is also recommended.

Land Capability Sub-class IIIw

This is moderately good land with severe limitations. It comprises of very deep fairly to poorly drained silty loam to silty clay loam soils underlain with silty clay loam to silty clay sub-soils. It resembles IIw category in many respects. The soils remain saturated during the monsoon and water-table fluctuates from 0.5 to 2 metres. Depending on the same inherent characteristics, similar crop rotations, similar soil needs and similar water management practices, they may be placed in capability units 1 and 2.

Suitability

The soils of this category are suitable for growing paddy, sugarcane, wheat etc.

Problems

The inherent problems are slow to very slow permeability and unfavourable texture, while the improvable problems are water-logging and high water-table and low fertility.

Land-use Planning

The following steps should be taken in addition to those suggested in the case of IIw for planning the use of this category of land:

(i) Only medium paddy be grow taken to give a margin for land preparation for wheat,
(ii) Adoption of scientific crop rotations, viz.

(a) Medium paddy-Wheat-*Moong*,
(b) Medium paddy-Pea-Sugarcane
(c) *Dhaincha*-Late paddy-Lentil/Linseed,

(iii) Adoption of improved fertilisers application techniques for deep water regime and minimum use of calcareous fertilisers to avoid further increase in pH.

Land Capability Sub-class IVe

This is fairly good land with very severe limitations of soil erosion. It comprises of very deep, friable well to excessively drained sandy loam to silty loam soils. It resembles the sub-classes IIe and IIIe in many respects, except the slope and magnitude of erosion. The topography is undulating and slopy, the slope at some places is more than 5 per cent. Consequently, erosional hazards are very severe.

Suitability

The land is suitable for limited cultivation and tree plantation with sweet potato, potato, oil-seeds, *arhar* and groundnut, as the main crops.

Land-use Planning

The measures already suggested in IIe and IIIe should be adopted. Inter-cultivation of leguminous crops and growing of orchards together with afforestation are recommended specially.

Land Capability Sub-class IVw

This is the land which suffers from very severe limitations of excessive water. Although this category of land does not appear in the sample villages, it comprises of very deep and very poorly drained silty clay loam to silty clay soils. The soils mostly occur in lowlying and water-logged conditions on nearly level to very gently sloping-lands which generally remain submerged up to October or even later and are found in some parts of the region. Other conditions are almost the same as found in the case of IIIw.

Suitability

This is suitable for late or Zaid paddy and lathyrus and at some places for wheat also.

Land-use Planning

In addition to the measures already suggested in the case of IIIw, the following recommendations are also made:

(i) Construction of protective bunds on the sides of nalas to check overflow of water,
(ii) Minimum use of calcium fertilisers to check calcium content increases in the soil,
(iii) Adoption of scientific crop rotations, viz.

(a) Late paddy-Zaid paddy-*Dhaincha*,
(b) Jute-*Masoor*/Linseed,
(c) Medium paddy-Wheat-*Moong*.

Land Capability Class V

This is the land with limitations other than erosion, for example, wetness. Generally, it is difficult to remove these limitations and reclaim the land to cultivation. At places, it may be utilised for 'boro' cultivation, but generally, it is suitable for wild life or pastures.

Land Capability Class VI

This is unsuitable for cultivation due to very severe erosion. This is found immediately on the banks of *nalas*. This should be utilised for afforestation, pastures or wild-life.

Land Capability Class VII

Although this class of land does not appear on the land-use maps of the sample villages, the forest-clad lands of the northern parts of Gonda and Bahraich districts may be placed in this category. They are unsuitable for cultivation, but may be utilised for forestry and grazing with extreme care to prevent erosion.

Special mention may be made here of the flood and water-logging hazards hampering the optimum use of the land in the region. In addition to the measures already discussed in the preceding pages to combat this debacle, the following recommendations are also made:

1. Efforts should be made to augment the natural drainage by removing the existing bunds across the streams.
2. Levees should be constructed to confine the main streams to their water channels in addition to the straightening and deepening of the courses to hasten the down-stream flow.
3. Reservoirs to hold excess rain water in the main catchment area of the sub-Himalayan zone should be made at suitable places where the valleys are broad. In the upper reaches, the stream courses should be 'trained' by dams and embankments reinforced by planting trees.
4. The adaptable and effective land-use measures should be adopted for an upstream control for flood reduction.

Last but not the least, a large amount of rural infrastructures, viz. adequate system of rural roads, extension of electric power into rural areas, development of improved marketing and storage facilities, both for inputs and agricultural products laying the base for further and sustained diffusion and expanded use of the new technology, has to be created. To enable and encourage the significant number of the brightest, most imaginative, innovative and ambitious people of the country-side to take up the modern scientific techniques of efficient land utilisation, general and professional education should be imparted.

The help of the government in the form of loans and subsidies, experts, launching of beneficiary-oriented programmes and in making the people plan-conscious is very much needed. Tenure system has to be changed, so that the tillers of the soil enjoy the full ownership of the land they till.

It has always to be kept in mind that the land resources have to be used according to their capability optimally and wisely keeping in view the local and national demands and their conservation for the generations to come. It hardly needs any emphasis that mother earth is always generous and never denies to feed her children whose primary duty is not to prejudice her with their follies which make her body torn off, impoverished and sticky.

APPENDICES

APPENDIX I

Physical, Mechanical and Chemical Analytical Data of Different Soils

(*Average Figures*)

Soil Categories	Village	Depth in inches	Physical constant		Mechanical Analysis			Textural class	Chemical Analysis Available (kg/ha)				
			W.H.C	B.D.	Clay	Silt	Sand		pH	E.C	O.C	P_2O_5	K_2O
1	2	3	4	5	6	7	8	9	10	11	12	13	14
1. *Soils of the undulating topography*	Baghawa	0-9	26.72	1.393	5.58	14.83	81.92	LS	7.7	1.2	.158	4.93	100.53
		9-18	35.2	1.244	5.75	13.00	81.25	LS	7.7	1.2	.050	13.44	100.80
2. *Soil of the Gently sloping to Nearly level lowlying lands:*													
(i) *Western Region*	Koilahwa	0-9	31.57	1.49	22.00	19.83	58.17	L	7.6	1.1	.562	11.20	257.8
		9-18	41.93	1.35	19.50	20.00	60.50	L	7.6	1.1	.262	15.68	218.5
	Shankarpur	0-9	44.25	1.12	29.50	24.00	46.50	CL	7.7	1.0	.400	11.20	160.0
		9-18	44.25	1.12	27.00	23.00	50.00	CL	7.8	1.0	.420	13.20	185.00

(*Contd.*)

(Contd.)

1	2	3	4	5	6	7	8	9	10	11	12	13	14
	Dhampur	0-9	55.02	1.13	24.37	24.28	51.44	L	7.8	1.2	.201	16.80	131.40
		9-18	40.11	1.13	30.37	21.13	49.51	CL	7.8	1.1	.350	10.70	134.40
	Jumuli Kalan	0-9	35.89	1.02	17.00	20.00	63.00	L	7.6	1.0	.392	13.44	134.80
		9-18	51.90	.95	20.50	29.87	49.63	SIL	7.6	1.0	.577	11.20	126.67
	Udaipur	0-9	55.00	1.33	37.00	22.00	41.00	CL	7.9	1.2	.220	22.00	155.00
		9-18	51.00	1.48	41.00	20.00	39.00	CL	7.8	1.2	.130	18.00	140.00
	Kewalpur	0-9	49.34	1.22	28.13	18.25	53.62	CL	7.8	1.2	.212	15.68	140.00
		9-18	44.00	1.58	26.25	21.25	52.50	CL	7.8	1.2	.538	20.16	151.20
	Achakapur	0-9	52.00	1.23	34.75	23.75	41.50	CL	7.9	1.2	.200	20.16	145.60
		9-18	39.80	1.58	31.50	22.50	56.00	CL	7.9	1.2	.125	15.68	162.40
(ii) *Eastern Region*													
	Darhwal	0-9	44.60	1.56	13.60	27.40	59.00	SIL	7.6	1.0	.126	13.44	44.8
		9-18	50.33	1.26	12.00	23.60	64.00	L	7.7	1.2	.098	8.96	44.8
	Darjipur	0-9	53.70	1.14	26.00	30.00	43.00	SICL	7.8	.9	.476	13.40	67.2
		9-18	36.45	1.51	15.5	27.00	57.50	SIL	7.8	1.2	.028	17.92	44.8
	Siswa Bujurg	0-9	37.62	1.43	23.5	33.0	43.5	SIL	7.7	1.0	.532	8.96	67.2
		9-18	43.27	1.31	16.0	25.0	59.0	L	7.6	1.1	.140	4.48	44.8
	Shohratgarh	0-9	42.63	1.42	31.2	21.8	47.0	CL	7.7	1.4	.364	8.96	134.4
		9-18	45.56	1.32	28.0	26.0	46.0	SICL	7.5	1.1	.140	13.44	122.00
	Nibi	0-9	47.46	1.16	29.6	23.0	47.0	CL	7.4	0.9	.168	14.48	67.2
		9-18	40.96	1.39	19.0	29.5	51.5	SIL	7.3	1.1	.196	17.92	112.0
	Mundila	0-9	50.25	1.07	26.2	28.1	47.7	SICL	8.0	1.2	.458	17.92	149.32
		9-18	54.16	1.08	27.5	30.8	41.7	SICL	7.9	1.1	.250	8.49	179.20

1	2	3	4	5	6	7	8	9	10	11	12	13	14
	Jhuria	0-9	48.00	1.15	32.0	24.0	44.0	CL	7.9	1.1	.280	17.46	150.00
		9-18	50.00	0.96	34.0	24.0	42.0	CL	7.9	1.2	.220	15.40	130.00
	Mangalpur	0-9	38.30	1.21	26.4	28.2	45.4	SICL	7.4	1.0	.340	13.44	151.20
		9-18	33.28	1.09	22.7	25.2	52.1	SIL	7.4	1.0	.390	11.71	153.0
	Dhaurahra	0-9	45.76	1.06	23.4	26.4	50.2	SIL	7.8	1.1	.187	11.20	134.20
		9-18	45.16	1.20	24.4	26.3	49.4	SIL	7.9	1.1	.208	13.44	149.0
3. *Soil of the Upland*													
	(i) *Bahraich Upland*												
	Pure Surajman	0-9	38.73	1.36	24.4	20.7	54.9	L	7.5	1.0	.175	11.20	179.2
		9-18	46.83	1.24	25.4	22.9	51.7	L	7.5	1.0	.245	8.96	156.6
	Gaundaria	0-9	42.38	1.29	18.9	25.0	56.1	SIL	7.3	1.0	.500	13.44	148.00
		9-18	41.22	1.37	23.2	22.0	54.8	L	7.3	0.9	.567	8.96	810.6
	(ii) *Gorakhpur Upland*												
	Bishunpura	0-9	40.34	1.37	21.6	26.6	51.8	SIL	7.7	1.1	.412	24.64	156.00
		9-18	41.79	1.35	21.2	26.6	52.2	SIL	7.6	1.0	.287	20.16	672.0
	Ardauna	0-9	30.00	1.33	10.6	24.2	65.2	LS	8.3	1.1	.504	26.18	560.00
		9-18	32.00	1.42	11.6	22.2	66.2	L	8.4	1.1	.252	26.18	67.20
	Purushottampur	0-9	43.80	1.09	24.0	17.7	58.7	L	7.7	1.2	.292	14.33	168.00
		9-18	42.16	1.15	16.0	23.0	61.0	L	7.8	1.2	.125	13.44	156.80
	Kolhwa	0-9	38.48	1.18	22.0	19.8	58.2	L	7.5	1.0	.208	14.87	147.77
		9-18	34.84	0.97	19.5	20.0	60.5	L	7.7	1.1	.250	17.92	445.00
	Barnahwa	0-9	48.68	1.09	24.4	25.6	50.0	SIL	7.5	1.0	.233	17.92	130.90
		9-18	49.90	1.15	22.0	29.4	48.6	SIL	7.5	1.0	.250	11.20	123.20
	Araji Varia	0-9	30.39	1.47	15.2	21.3	65.5	L	7.7	1.4	.165	13.44	179.20
		9-18	29.98	1.42	14.9	25.3	63.1	L	7.5	1.0	.192	22.40	672.20

(Contd.)

(*Contd.*)

1	2	3	4	5	6	7	8	9	10	11	12	13	14
	Digahi	0-9	29.32	1.41	16.7	26.6	56.7	SIL	7.7	1.1	.252	22.40	145.60
		9-18	29.91	1.48	14.9	24.1	61.1	L	7.7	1.1	.250	26.88	104.00
4. *Soil of the flat but slightly elevated topography*													
	(i) *Trans-Rapti Flatland of Bahraich District*												
	Lakshmanpur	0-9	38.00	1.45	24.0	22.0	54.0	L	7.7	1.0	.390	18.00	175.00
		9-18	36.00	1.40	24.0	20.0	56.0	L	7.6	0.9	.250	15.00	165.00
	(ii) *Flatland of the lower Rapti-Burhi Rapti Doab (Basti district)*												
	Takia	0-9	42.00	1.49	22.7	20.5	56.8	L	7.7	1.1	.169	14.50	140.30
		9-18	44.00	0.97	26.5	23.5	50.0	CL	7.7	1.1	.125	17.92	112.00
	Patiapar	0-9	40.80	1.06	19.3	17.9	62.8	L	7.7	1.2	.300	16.64	134.13
		9-18	42.70	1.03	19.3	23.6	57.1	L	7.8	1.2	.200	15.68	218.40
	Narainpur	0-9	33.50	1.21	19.3	15.7	65.0	L	7.7	1.1	.075	8.96	179.20
		9-18	42.43	1.16	17.9	19.7	62.4	L	7.7	1.1	.183	14.93	182.93
	(iii) *Banganga-Januwar Flatland of Basti District*												
	Sukrauli	0-9	40.33	1.27	26.5	27.0	46.5	SICL	7.7	1.1	.196	4.48	44.80
		9-11	42.19	1.29	20.5	36.5	43.5	SIL	7.7	1.1	.028	10.75	67.20
	Jagmohani	0-9	42.31	1.39	15.5	41.0	43.5	SIL	7.7	1.1	.196	10.75	54.80
		0-18	39.81	1.40	23.3	33.7	43.0	SIL	7.7	1.2	.098	8.06	67.20

1	2	3	4	5	6	7	8	9	10	11	12	13	14
5.	*Soils of the Lowlying Riverine Topography*												
	(i) *Ghaghra Basin*												
	Chahelawa	0-9	20.26	1.59	3.0	12.0	85.0	LS	7.8	1.1	.025	17.92	134.00
		9-18	21.00	1.58	2.0	9.0	89.0	LS	7.7	0.9	.020	5.92	130.40
	(ii) *Bhakla-Rapti Basin*												
	Pure Shahbaz	0-9	32.05	1.37	19.6	21.1	59.3	L	7.7	1.1	.300	13.44	89.60
		9-18	34.06	1.58	31.2	22.1	56.1	L	7.7	1.1	.150	11.95	74.70
	Tigharwa	0-9	41.66	1.36	19.6	26.7	53.7	SIL	7.5	0.9	.175	13.44	171.10
		0-18	33.04	1.37	15.3	24.6	60.1	L	7.5	0.9	.575	8.96	112.00
	Rewalia	0-9	47.58	1.20	25.3	22.5	52.2	L	7.6	1.0	.437	8.96	168.20
		0.18	42.07	1.40	17.8	27.0	55.2	SIL	7.7	1.0	.281	24.66	145.60
	Mahrauli	0-9	43.90	1.24	18.7	17.8	63.5	L	7.7	1.0	.469	8.96	89.60
		9-18	44.58	1.64	19.4	20.0	60.6	L	7.7	1.1	.368	8.96	93.30
	(iii) *Rapti-Burhi Rapti Basin*												
	Akbarpur Kalan	0-9	44.76	1.19	19.8	22.5	57.7	L	7.6	1.0	.283	14.93	173.07
		9-18	39.14	1.23	17.5	23.1	59.4	L	7.6	1.0	.288	8.96	139.50
	Jaitapur	0-9	42.05	1.12	20.3	22.3	57.4	L	7.7	1.1	.230	17.92	150.04
		9-18	35.81	1.16	16.0	22.0	62.0	L	7.8	1.1	.750	17.92	256.60
	(iv) *Burhi Rapti-Banganga & Rapti Basin*												
	Imlia	0-9	39.82	1.24	19.6	22.4	58.0	SIL	7.8	1.1	.317	13.47	104.53
		9-18	43.72	1.14	20.1	25.5	54.4	SIL	7.8	1.0	.512	11.20	173.60

(Contd.)

(Contd.)

1	2	3	4	5	6	7	8	9	10	11	12	13	14
	Majhwan	0-9	34.62	1.26	15.5	30.5	54.0	SIL	7.7	1.1	.284	14.48	54.80
		9-18	54.98	1.13	28.0	30.5	41.5	SICL	7.7	1.1	.518	11.20	56.00
	Deoria	0-9	52.22	1.07	19.1	30.0	50.9	SIL	7.8	1.1	.437	13.44	134.40
		9-18	46.68	1.08	26.0	21.8	52.2	CL	7.6	1.0	.212	13.44	117.60
	Ucharia	0-9	48.39	1.28	21.5	24.5	54.0	L	7.6	0.9	.140	4.40	44.80
		9-18	47.00	1.27	20.0	27.7	52.3	SIL	7.6	0.9	.271	4.78	44.80
(v)	*Great Gandak & Little Gandak Basin*												
	Kalahani Khurd	0-9	31.45	1.39	14.5	24.8	60.7	SIL	7.7	1.1	.125	24.64	134.40
		9-18	31.72	1.36	15.7	25.7	58.6	SIL	7.6	1.0	.166	13.09	134.40
	Dharmpur	0-9	40.10	1.36	13.8	35.8	50.4	SIL	7.7	1.1	.475	16.13	80.54
		9-18	31.30	1.46	13.3	30.6	56.1	SIL	7.7	1.1	.312	15.68	44.80
	Sonbarsa	0-9	36.96	1.45	13.7	37.1	49.2	SIL	7.8	1.1	.375	13.44	89.80
		9-18	39.97	1.38	14.2	39.4	46.6	SIL	7.8	1.1	.312	8.96	112.00
	Jaichhapra	0-9	37.84	1.16	17.8	37.2	45.0	SIL	7.7	1.1	.350	13.44	89.60
		9-11	44.73	1.32	17.7	37.2	47.1	SIL	7.8	1.3	.183	14.42	19.47
	Bishunpura Khard	0-9	42.39	1.23	16.1	34.8	49.1	SIL	7.7	1.1	.250	19.41	298.67
		9-18	40.67	1.25	17.3	31.2	31.5	SIL	7.7	1.0	.187	20.16	48.67
	Mansachhapra	0-9	36.16	1.34	20.3	29.1	50.6	SIL	7.7	1.0	.167	25.39	59.73
		9-18	40.32	1.38	18.8	33.6	47.6	SIL	7.7	1.2	.250	17.92	44.80

Soil Analysis Results of the Soil Samples of the Villages Selected for Intensive Study

1	2	3	4	5	6	7	8	9	10	11	12	13	14
	Dhampur	0-9	57.20	1.09	25.3	25.0	49.7	L	8.0	1.2	.275	13.44	156.00
		9-18	58.50	1.15	25.7	26.8	47.5	L	7.7	1.2	.200	8.96	123.20
		0-9	29.60	1.43	11.2	12.0	76.8	SL	7.0	0.6	.420	6.4	50.00
		9-18	28.50	1.50	14.2	13.0	72.8	L	7.2	0.5	.540	8.0	70.00
		0-9	47.40	1.09	21.5	19.5	59.0	L	7.6	1.1	.100	22.40	100.80
		9-18	61.80	1.09	35.0	13.5	51.5	CL	7.8	1.1	.500	13.44	145.80
		0-9	53.68	1.09	26.3	19.2	54.5	CL	7.6	1.1	.675	13.44	156.80
		9-18	38.58	1.10	35.0	24.8	40.2	CL	7.9	0.9	.144	15.8	60.00
	Kewalpur Kalan	0-9	36.88	1.40	20.3	20.7	59.0	L	7.9	1.2	.300	17.92	156.80
		9-18	40.00	1.30	24.5	21.3	54.2	L	7.8	1.1	.250	16.00	150.00
		0-9	48.00	1.10	26.7	23.3	50.0	CL	7.9	1.2	.260	15.00	160.00
		9-18	48.80	0.98	26.0	23.3	50.7	CL	7.8	1.2	.250	22.40	168.00
		0-9	43.00	1.15	27.0	26.0	47.0	SICL	8.0	0.6	.360	16.50	100.00
		9-18	44.00	1.10	26.0	28.0	46.0	SICL	7.9	0.8	.300	16.90	100.00
	Marwatia Deegar	0-9	42.65	1.30	25.6	31.4	43.0	SICL	7.5	2.0	.308	10.65	67.20
		9-18	45.56	1.32	28.0	26.0	46.0	SICL	7.5	1.1	.140	13.44	112.00
		0-9	39.67	1.30	17.0	30.4	52.0	SIL	7.4	1.2	.308	4.48	44.80
		9-18	34.32	1.30	18.0	27.2	54.8	SIL	7.5	0.8	.345	7.20	50.00
		0-9	29.60	1.40	11.0	12.2	76.8	SL	7.0	0.6	.400	6.30	52.00
		9-18	38.00	1.30	16.0	25.0	59.0	L	7.6	1.1	.140	4.48	44.80
	Mundila	0-9	53.76	1.08	26.5	26.5	47.0	SICL	8.1	1.2	.375	17.92	123.20
		9-18	49.76	1.12	26.3	25.5	48.2	SICL	8.0	1.2	.250	22.46	156.80
		0-9	50.00	1.08	20.0	26.0	54.0	SIL	7.9	1.2	.250	8.96	179.20
		9-18	54.16	1.10	21.0	27.0	52.0	SIL	7.8	1.1	.250	8.80	170.00

(Contd.)

(Contd.)

1	2	3	4	5	6	7	8	9	10	11	12	13	14
	Mangalpur	0-9	33.76	1.19	26.2	26.4	47.4	SICL	7.4	1.0	.125	8.96	134.40
		9-18	41.76	1.08	30.2	26.5	43.3	SICL	7.4	1.0	.125	8.96	134.40
		0.9	33.28	1.09	22.6	25.2	52.2	SIL	7.4	1.1	.340	13.44	151.20
		9-18	34.00	1.10	21.5	28.2	50.3	SIL	7.4	1.1	.390	11.41	153.07
		0-9	24.00	1.19	19.5	20.0	60.5	L	7.5	1.1	.140	16.20	160.00
		9-18	24.44	1.17	17.5	22.0	60.5	L	7.5	1.0	.125	17.62	168.00
	Dhaurahra	0-9	41.60	1.18	20.3	26.5	53.2	SIL	7.7	1.1	.125	13.44	112.00
		9-18	41.60	1.15	21.5	28.3	50.2	SIL	7.8	1.1	.125	8.96	156.00
		0-9	53.60	0.95	26.5	26.2	47.3	SICL	8.0	1.2	.250	8.96	156.80
		9-18	54.00	1.10	27.0	28.0	45.0	SICL	8.0	1.1	.300	10.20	150.00
		0-9	44.00	1.12	28.0	24.0	48.0	CL	7.9	1.1	.260	13.00	165.00
		9-18	45.00	0.97	28.2	23.3	48.5	CL	7.9	1.1	.250	13.44	168.00
	Pure Surajman	0-9	34.74	1.43	20.4	20.2	48.4	L	7.6	1.1	.250	8.96	156.80
		9-18	41.04	1.42	19.4	24.2	56.4	L	7.6	0.9	.312	8.96	112.00
		0-9	42.73	1.29	28.4	20.2	51.4	CL	7.1	1.1	.100	13.44	201.00
		9-18	48.28	1.12	27.4	22.2	50.4	CL	7.4	1.0	.262	8.96	80.00
	Digahi	0-9	30.10	1.40	16.2	27.8	56.0	SIL	7.7	1.1	.250	13.44	179.20
		9-18	28.46	1.42	17.2	25.4	57.4	SIL	7.7	1.1	.250	20.44	130.00
		0-9	29.20	1.48	20.0	10.0	70.0	SIL	7.7	1.0	.250	26.88	112.00
		9-18	30.10	1.50	16.2	24.0	59.8	L	7.7	1.0	.375	31.36	112.00
	Bishunpura	0-9	45.34	1.30	24.2	25.6	49.2	SIL	7.7	1.1	.125	22.40	89.60
		9-18	42.20	1.32	25.2	25.6	50.2	SIL	7.6	1.0	.250	22.40	89.60
		0-9	50.00	1.10	26.0	26.7	47.3	SICL	7.9	1.1	.250	15.40	100.00
		9-18	52.00	1.15	26.5	27.5	46.0	SICL	7.8	1.1	.270	14.30	90.00

(Contd.)

1	2	3	4	5	6	7	8	9	10	11	12	13	14
	Patiapur	0-9	41.80	0.96	20.5	15.3	64.2	L	7.6	1.1	.125	22.40	201.00
		9-18	44.44	1.07	20.5	21.5	58.0	L	7.7	1.2	.125	17.92	190.40
		0-9	40.96	1.10	18.0	25.8	56.2	SIL	7.8	1.1	.275	13.44	240.00
		9-18	41.76	1.08	20.3	24.0	55.7	SIL	7.7	1.1	.270	13.40	220.00
	Pure Shahbaz	0-9	38.06	1.29	25.4	18.2	56.4	CL	7.6	1.0	.400	17.92	134.40
		9-18	40.28	1.51	25.4	20.2	54.4	CL	7.6	1.0	.250	8.96	67.20
		0.9	26.05	1.46	12.0	18.0	70.0	SL	7.7	1.0	.200	8.96	44.80
		9-18	30.89	1.53	17.8	22.0	60.2	L	7.8	1.0	.150	8.96	67.20
	Akbarpur	0-9	30.00	0.89	20.0	10.0	70.0	SL	7.6	1.0	.225	13.44	112.00
		9-18	35.00	1.31	15.0	16.0	69.0	L	7.5	1.0	.230	12.20	130.00
		0-9	33.80	1.39	15.7	16.3	68.0	L	7.5	1.0	.375	17.22	168.00
		9-18	35.00	1.32	14.5	20.0	65.5	L	7.5	1.0	.250	8.96	156.80
		0-9	59.68	1.28	27.0	25.0	48.0	SICL	7.6	1.1	.250	13.44	179.20
		9-18	43.28	1.14	20.5	26.3	53.2	SIL	7.8	1.1	.150	8.96	123.20
	Majhwan	0-9	34.62	1.26	15.5	30.5	54.0	SIL	7.7	1.1	.284	14.40	54.00
		9-18	38.66	1.37	19.0	35.0	46.0	SIL	7.8	1.4	.280	6.27	67.00
		0-9	64.61	1.04	30.0	28.0	42.0	SICL	7.6	1.0	.840	8.96	67.20
		9-18	45.36	1.23	26.0	33.0	41.0	SICL	7.8	1.1	.196	13.41	44.80
		0-9	29.00	1.03	20.0	10.0	70.0	SL	7.5	1.0	.200	10.20	100.00
		9-18	48.39	1.08	21.5	24.5	54.0	L	7.6	0.9	.140	4.48	44.80

Contd.

(Contd.)

1	2	3	4	5	6	7	8	9	10	11	12	13	14
	Sonbarsa	0-9	35.44	1.51	14.2	34.4	51.4	SIL	7.8	1.1	.500	8.96	112.00
		9-18	39.48	1.44	13.2	37.2	49.8	SIL	7.7	1.1	.250	8.96	130.00
		0-9	60.00	1.14	29.0	28.0	43.0	SICL	7.6	1.0	.300	9.20	80.00
		9-18	38.48	1.40	13.2	39.8	47.0	SIL	7.8	1.1	.250	17.92	67.20
		0-9	28.00	1.20	18.0	18.0	72.0	SL	7.6	1.0	.220	9.40	120.00
		9-18	45.10	1.30	21.0	25.0	54.0	L	7.6	0.9	.150	9.46	70.00

As given in Sharma, S.C., 'Land Capability Classification and Land-use, Planning — A Case Study' in Noor Mohd, (ed.) *Perspectives in Agricultural Geography* Vol. III, 1981, in classifying the land into capitability classes, the All India Standard Capability Classification Pattern has been followed, according to which the soils having same degree of limitations are grouped into a class, these bearing the same nature of limitations within a class as a sub-class, which further are sub-divided into capability units on the basis of inherent characteristics, similar crop adaptation and similar soil and water management practices etc.

W.H.C. — Water-Holding Capacity, B.D — Bulk Density, E.C. — Electrical Conductivity, O.C. — Organic Carbon, P_2O_5 — Phosphorous, K_2O — Potash, LS — Loamy Sand, L — Loamy, CL — Clay Loam, SIL-Silty Loam, SICL — Silty Clay Loam.

APPENDIX II

Population Growth

Blocks	Population Number				Growth rate			
	1951	1961	1971	1981	1951-61	1961-71	1971-81	1951-81
1	2	3	4	5	6	7	8	9
Bahraich								
1. Mihipurwa	61375	69875	97999	161103	13.8	40.2	64.4	162.5
2. Nawabganj	58806	65703	76941	92428	11.7	17.1	20.1	57.2
3. Jamunaha	77862	82711	91725	114035	6.2	10.9	24.3	46.5
4. Hariharpur Rani	76133	87245	92928	137113	14.6	6.5	47.5	80.1
5. Sirsia	97994	110623	125868	130111	12.9	13.8	3.4	32.8
Gonda								
6. Haraiya Satgharwa	93137	103815	108250	134525	11.5	4.3	24.3	44.4
7. Tulsipur	97678	110069	115114	131314	12.7	4.6	14.1	34.4
8. Gainsari	80979	89760	96753	123883	10.8	7.8	28.0	53.0
9. Pachperwa	77649	85881	96921	110229	10.6	12.8	13.7	42.0
10. Balrampur	28024	31539	33701	40718	12.5	6.8	20.8	45.3

(Contd.)

(Contd.)

1	2	3	4	5	6	7	8	9
Basti								
11. Itwa	70312	71481	79611	98701	1.7	11.4	24.0	40.4
12. Bhanwapur	79371	83154	94010	111427	4.8	13.0	18.5	40.4
13. Khuniaon	79945	83136	87326	113845	4.0	5.0	30.4	42.4
14. Barhni	69227	78856	89827	103220	13.9	13.9	14.9	49.1
15. Birdpur	69290	76323	86763	105387	11.1	13.7	21.8	52.1
16. Naugarh	74437	84923	97441	98864	14.0	14.7	1.5	32.8
17. Jogia	54317	60982	68343	79136	12.3	12.1	15.8	45.7
18. Uska	71474	78676	68072	104137	10.1	11.9	18.2	45.7
19. Bansi	38955	50594	56279	64113	29.9	11.2	13.9	64.6
Gorakhpur								
20. Nautanwa	79201	86740	93247	116177	9.8	1.2	24.6	46.7
21. Laxmipur	66579	73659	80769	107335	10.6	9.6	32.9	61.2
22. Brijmanganj	66686	73130	86642	99781	9.7	18.5	15.2	49.6
23. Pharenda	55209	66735	73136	93029	20.9	9.6	27.2	68.5
24. Nichlaul	74972	83573	87963	126122	11.5	4.9	43.4	68.2
25. Mithaura	78981	89034	103019	124178	12.7	15.7	20.5	57.2
26. Maharajganj	69817	78672	86729	119934	12.7	10.2	38.3	71.8
27. Ghughli	67393	76839	90539	110037	14.0	17.8	21.5	63.3
28. Siswa	73783	80314	89823	106396	8.8	11.8	18.5	44.2
Deoria								
29. Khadda	69968	84582	99508	123218	20.9	17.6	23.8	76.1
30. Nebua Naurangia	66713	72284	85544	105672	8.3	18.3	23.5	58.4
31. Bishunpura	69918	75173	91542	113542	7.5	21.5	24.3	62.4
Grand Total	2196185	2446081	2751863	3399709	11.4	12.5	23.5	54.8

APPENDIX III

Population Distribution

Blocks	1981		1971		1961		1951	
	% of the total popu-lation of the regi-on	% of the total area of the regi-on	% of the total popu-lation of the regi-on	% of the total area of the regi-on	% of the total popu-lation of the regi-on	% of the total area of the regi-on	% of the total popu-lation of the regi-on	% of the total area of the regi-on
Bahraich								
1. Mihipurwa	4.7	6.6	3.6	7.2	2.9	7.2	2.8	7.2
2. Nawabganj	2.7	2.7	2.8	2.8	2.7	2.8	2.7	2.8
3. Jamunaha	3.4	3.8	3.3	3.7	3.4	3.7	3.8	3.8
4. Hariharpur Rani	4.0	3.7	3.4	3.6	3.6	3.6	3.5	3.6
5. Sirsia	3.8	5.5	4.6	5.5	4.5	5.5	4.5	5.5
Gonda								
6. Haraiya Satgharwa	4.1	5.5	3.9	5.4	4.2	5.4	4.2	5.4
7. Tulsipur	3.9	4.5	4.2	4.6	4.5	4.6	4.4	4.6
8. Gainsari	3.6	5.1	3.5	5.1	3.6	5.1	3.7	5.1
9. Pachperwa	3.2	4.3	3.5	4.4	3.5	4.4	3.5	4.4
10. Balrampur	1.2	1.3	1.2	1.3	1.3	1.3	1.3	1.3
Basti								
11. Itwa	2.9	3.0	2.9	3.0	2.9	3.0	3.2	3.0
12. Bhanwapur	3.3	3.1	3.4	3.0	3.4	3.0	3.6	3.0
13. Khuniaon	3.4	2.6	3.2	2.6	3.4	2.6	3.6	2.6
14. Barhni	3.0	3.4	3.3	3.2	3.2	3.2	3.2	3.2
15. Birdpur	3.1	2.7	3.2	2.4	3.1	2.4	3.2	2.4
16. Naugarh	2.9	2.4	3.8	2.5	3.5	2.5	3.4	2.5
17. Jogia	2.3	2.0	2.5	1.9	2.5	1.9	2.5	1.9
18. Uska	3.1	2.4	3.2	2.4	3.2	2.4	3.2	2.4
19. Bansi	1.9	1.7	2.0	1.7	2.1	1.7	1.8	1.7
Gorakhpur								
20. Nautanwa	3.4	4.2	3.4	4.1	3.6	4.1	3.6	4.1
21. Laxmipur	3.1	3.4	2.9	2.9	3.0	2.9	3.0	2.9
22. Brijmanganj	2.9	2.2	3.1	2.3	3.0	2.3	3.0	2.3
23. Pharenda	2.7	2.6	2.7	2.6	2.7	2.6	2.8	2.6
24. Nichlaul	3.8	3.6	3.2	3.5	3.4	3.5	3.4	3.5
25. Mithaura	3.7	2.7	3.7	2.6	3.6	2.6	3.6	2.6
26. Maharajganj	3.5	2.4	3.2	2.4	3.2	2.4	3.2	2.4
27. Ghughli	3.2	2.0	3.3	1.9	3.1	1.9	3.1	1.9
28. Siswa	3.1	2.3	3.3	2.2	3.3	2.2	3.4	2.2
Deoria								
29. Khadda	3.7	3.6	3.6	4.4	3.5	4.4	3.2	4.4
30. Nebua Naurangia	3.1	2.3	3.1	2.3	3.0	2.3	3.0	2.3
31. Bishunpura	3.3	2.4	3.3	2.5	3.1	2.5	3.2	2.5
Total region	100.0	100.0	100.0	100.0	100.0	100.0	100.0	100.0

APPENDIX IV A

Arithmetic Density of Population

(*Persons per sq km*)

Block	Density				Variation			
	1951	1961	1971	1981	1951-61	1961-71	1971-81	1951-81
1	2	3	4	5	6	7	8	9
Bahraich								
1. Mihipurwa	93	106	149	270	13	43	121	177
2. Nawabganj	229	254	299	378	25	45	79	149
3. Jamunaha	227	242	269	333	15	27	53	106
4. Hariharpur Rani	230	264	280	414	34	16	134	184
5. Sirsia	196	221	252	260	25	31	8	64
Gonda								
6. Haraiya Satgharwa	188	210	219	268	22	9	49	80
7. Tulsipur	232	261	272	319	29	11	47	87
8. Gainsari	174	192	207	267	18	15	60	93
9. Pachperwa	191	211	239	281	20	28	42	90
10. Balrampur	244	275	294	355	31	19	61	111
Basti								
11. Itwa	256	260	290	359	4	30	69	103
12. Bhanwapur	288	302	342	402	14	40	60	114
13. Khuniaon	341	355	372	490	14	17	118	149
14. Barhni	235	268	305	239	33	37	34	104
15. Birdpur	321	354	402	437	33	48	35	116
16. Naugarh	333	380	436	445	47	56	9	112
17. Jogia	314	353	395	443	39	42	48	129
18. Uska	336	367	411	487	31	44	76	151
19. Bansi	250	325	361	404	75	36	43	154
Gorakhpur								
20. Nautanwa	213	233	251	302	20	18	51	89
21. Laxmipur	249	275	302	347	26	27	45	98
22. Brijmanganj	318	339	420	491	31	71	71	173
23. Pharenda	233	282	310	395	49	28	85	162
24. Nichlaul	236	264	275	395	28	11	120	159
25. Mithaura	374	433	433	521	43	59	88	190
26. Maharajganj	320	363	403	557	43	40	154	237
27. Ghughli	379	432	511	621	53	79	110	242
28. Siswa	359	391	440	521	32	49	81	162
Deoria								
29. Khadda	172	209	245	376	37	36	131	204
30. Nebua Naurangia	325	353	416	514	28	63	98	189
31. Bishunpura	313	336	408	522	23	72	114	209
Total region	240	268	301	376	28	33	76	136

APPENDIX IV B

Physiological Density of Population*

(Persons per sq km)

Block	Density				Variation			
	1951	1961	1971	1981	1951-61	1961-71	1971-81	1951-81
1	2	3	4	5	6	7	8	9
Bahraich								
1. Mihipurwa	226	243	224	411	17	-19	187	185
2. Nawabganj	308	325	358	473	17	33	115	165
3. Jamunaha	300	305	318	421	5	13	103	121
4. Hariharpur Rani	294	329	340	511	35	11	171	217
5. Sirsia	354	368	324	334	14	-44	10	-20
Gonda								
6. Haraiya Satgharwa	293	302	294	395	9	-8	101	102
7. Tulsipur	317	346	336	417	29	-10	81	100
8. Gainsari	259	267	261	371	8	-6	110	112
9. Pachperwa	286	296	306	380	10	10	74	94
10. Balrampur	312	350	366	491	38	16	125	179
Basti								
11. Itwa	303	313	342	431	10	29	89	128
12. Bhanwapur	347	362	427	510	15	65	83	163
13. Khuniaon	429	466	458	606	17	12	148	177
14. Barhni	289	333	375	414	44	42	39	125
15. Birdpur	390	428	488	545	38	60	57	155
16. Naugarh	411	473	538	562	62	65	24	151
17. Jogia	365	417	568	528	52	51	-40	163
18. Uska	410	456	503	608	46	47	105	198
19. Bansi	346	440	496	508	96	54	12	162
Gorakhpur								
20. Nautanwa	294	318	399	368	24	81	-31	74
21. Laxmipur	396	433	487	416	37	54	-54	20
22. Brijmanganj	375	407	479	574	32	72	95	199
23. Pharenda	388	460	480	472	72	20	-8	84
24. Nichlaul	346	378	361	542	32	-17	181	196
25. Mithaura	400	436	505	635	36	69	130	235
26. Maharajganj	411	445	494	709	24	49	215	288
27. Ghughli	467	525	593	726	58	68	133	259
28. Siswa	439	463	503	632	24	40	129	193
Deoria								
29. Khadda	288	237	354	602	-51	117	248	314
30. Nebua Naurangia	411	418	479	596	7	61	117	185
31. Bishunpura	378	407	476	636	29	69	160	258
Total region	338	369	387	485	31	18	98	147

* For working out physiological density, the figures for net sown area belong to 1955-57, 1965-67, 1975-77 and 1985-87. Hence these years have been mentioned in the text instead of 1951, 1961, 1971 and 1981.

APPENDIX V

Pressure of Population

(Per capita land in hectare)

Blocks	Total area	Net sown area	Double cropped area	Total cropped area
1	2	3	4	5
Bahraich				
1. Mihipurwa	A* 0.37	0.24	0.10	0.34
	B* 0.67	0.44	0.19	0.63
	C* 0.94	0.41	0.16	0.57
	D* 1.07	0.44	0.15	0.59
2. Nawabganj	A 0.26	0.21	0.10	0.31
	B 0.33	0.28	0.08	0.36
	C 0.39	0.31	0.05	0.36
	D 0.44	0.32	0.06	0.38
3. Jamunaha	A 0.30	0.24	0.12	0.36
	B 0.37	0.31	0.10	0.41
	C 0.41	0.33	0.11	0.44
	D 0.44	0.33	0.12	0.45
4. Hariharpur Rani	A 0.24	0.20	0.12	0.32
	B 0.36	0.29	0.15	0.44
	C 0.38	0.30	0.15	0.45
	D 0.43	0.34	0.17	0.51
5. Sirsia	A 0.38	0.31	0.14	0.45
	B 0.39	0.31	0.11	0.42
	C 0.45	0.27	0.07	0.34
	D 0.51	0.28	0.10	0.38
Gonda				
6. Haraiya Satgharwa	A 0.37	0.25	0.15	0.40
	B 0.45	0.34	0.14	0.48
	C 0.47	0.32	0.11	0.43
	D 0.53	0.34	0.13	0.47
7. Tulsipur	A 0.31	0.24	0.14	0.38
	B 0.37	0.30	0.17	0.47
	C 0.38	0.29	0.10	0.39
	D 0.43	0.31	0.14	0.45
8. Gainsari	A 0.37	0.27	0.15	0.42
	B 0.48	0.38	0.14	0.52
	C 0.52	0.37	0.11	0.48
	D 0.57	0.39	0.13	0.52

(Contd.)

(Contd.)

1	2	3	4	5
9. Pachperwa	A 0.36	0.26	0.16	0.42
	B 0.42	0.33	0.13	0.46
	C 0.47	0.34	0.11	0.45
	D 0.52	0.35	0.14	0.49
10. Balrampur	A 0.28	0.20	0.11	0.31
	B 0.34	0.27	0.09	0.36
	C 0.36	0.29	0.18	0.47
	D 0.41	0.32	0.24	0.56
Basti				
11. Itwa	A 0.28	0.23	0.11	0.34
	B 0.34	0.29	0.12	0.41
	C 0.38	0.32	0.09	0.41
	D 0.39	0.33	0.09	0.42
12. Bhanwapur	A 0.25	0.20	0.10	0.30
	B 0.29	0.23	0.10	0.33
	C 0.33	0.28	0.08	0.36
	D 0.35	0.29	0.09	0.38
13. Khuniaon	A 0.20	0.16	0.09	0.25
	B 0.27	0.22	0.08	0.30
	C 0.28	0.22	0.07	0.29
	D 0.29	0.23	0.06	0.29
14. Barhni	A 0.29	0.24	0.07	0.31
	B 0.33	0.26	0.08	0.34
	C 0.37	0.30	0.08	0.38
	D 0.42	0.35	0.10	0.45
15. Birdpur	A 0.23	0.18	0.10	0.28
	B 0.25	0.20	0.04	0.24
	C 0.28	0.23	0.06	0.29
	D 0.31	0.25	0.07	0.32
16. Naugarh	A 0.22	0.18	0.08	0.26
	B 0.23	0.19	0.06	0.25
	C 0.26	0.21	0.06	0.27
	D 0.30	0.24	0.08	0.32
17. Jogia	A 0.22	0.19	0.09	0.28
	B 0.25	0.17	0.07	0.24
	C 0.28	0.24	0.08	0.32
	D 0.32	0.27	0.12	0.39
18. Uska	A 0.21	0.16	0.09	0.25
	B 0.24	0.20	0.06	0.26
	C 0.27	0.22	0.08	0.30
	D 0.30	0.24	0.08	0.32

(Contd.)

(Contd.)

1		2	3	4	5
19. Bansi	A	0.24	0.20	0.09	0.29
	B	0.28	0.20	0.10	0.30
	C	0.30	0.23	0.13	0.36
	D	0.40	0.29	0.19	0.48
Gorakhpur					
20. Nautanwa	A	0.33	0.27	0.07	0.34
	B	0.40	0.25	0.13	0.38
	C	0.43	0.31	0.10	0.41
	D	0.47	0.34	0.11	0.45
21. Laxmipur	A	0.29	0.24	0.07	0.31
	B	0.33	0.20	0.09	0.29
	C	0.36	0.23	0.06	0.29
	D	0.40	0.25	0.07	0.32
22. Brijmanganj	A	0.20	0.17	0.08	0.25
	B	0.24	0.21	0.03	0.24
	C	0.29	0.24	0.03	0.27
	D	0.31	0.27	0.03	0.30
23. Pharenda	A	0.25	0.21	0.06	0.27
	B	0.32	0.21	0.08	0.29
	C	0.35	0.22	0.07	0.29
	D	0.43	0.26	0.07	0.33
24. Nichlaul	A	0.25	0.18	0.12	0.30
	B	0.36	0.28	0.14	0.42
	C	0.38	0.26	0.07	0.33
	D	0.42	0.29	0.07	0.36
25. Mithaura	A	0.19	0.16	0.12	0.28
	B	0.23	0.20	0.10	0.30
	C	0.27	0.23	0.07	0.30
	D	0.30	0.25	0.07	0.32
26. Maharajganj	A	0.18	0.14	0.11	0.25
	B	0.25	0.20	0.14	0.34
	C	0.27	0.22	0.06	0.28
	D	0.31	0.24	0.08	0.32
27. Ghughli	A	0.40	0.14	0.09	0.23
	B	0.19	0.17	0.10	0.27
	C	0.23	0.19	0.05	0.24
	D	0.26	0.21	0.05	0.26
28. Siswa	A	0.19	0.16	0.10	0.26
	B	0.28	0.20	0.11	0.31
	C	0.25	0.21	0.07	0.28
	D	0.28	0.23	0.06	0.29

(Contd.)

(Contd.)

1	2	3	4	5
Deoria				
29. Khadda	A 0.27	0.16	0.05	0.21
	B 0.41	0.28	0.03	0.31
	C 0.48	0.42	0.03	0.45
	D 0.58	0.35	0.04	0.39
30. Nebua Naurangia	A 0.19	0.17	0.03	0.20
	B 0.24	0.21	0.07	0.28
	C 0.28	0.24	0.03	0.27
	D 0.30	0.24	0.04	0.28
31. Bishunpura	A 0.19	0.16	0.07	0.23
	B 0.24	0.21	0.06	0.27
	C 0.29	0.24	0.05	0.29
	D 0.32	0.26	0.04	0.30
Regional average	A 0.27	0.21	0.10	0.31
	B 0.33	0.26	0.10	0.36
	C 0.37	0.27	0.08	0.35
	D 0.41	0.29	0.09	0.38

* A = 1985-87, B = 1975-77, C = 1965-67, D = 1955-57.

APPENDIX VI

Irrigation

Blocks	Net irrigated area in hectares	As % of the net area sown	Area irrigated by different sources as % of the net irrigated area				
			Surface water			Underground water	
			Canals	Ponds etc.	Others	Tube-wells	Other wells
1	2	3	4	5	6	7	8
Bahraich							
1. Mihipurwa	*A. 4188.5	10.7	2.6	1.0	1.6	94.7	0.1
	*B. 1926.0	4.4	—	27.0	—	51.1	21.9
2. Nawabganj	A. 4292.0	22.0	17.1	12.4	6.7	60.4	3.4
	B. 2886.0	13.3	—	17.3	—	72.9	9.8
3. Jamunaha	A. 5471.0	20.2	0.5	7.5	2.9	84.4	4.6
	B. 3237.0	11.2	—	11.3	—	79.3	9.4
4. Hariharpur Rani	A. 2311.0	8.6	—	4.7	7.3	81.8	6.2
	B. 2119.0	7.7	—	9.1	17.4	50.1	23.4
5. Sirsia	A. 1564.0	3.8	29.8	20.3	24.2	24.6	1.1
	B. 819.0	2.1	—	10.4	11.8	64.4	13.4
Gonda							
6. Haraiya Satgharwa	A. 1802.0	5.3	4.7	12.2	6.5	57.4	19.2
	B. 1436.0	3.9	—	7.5	6.1	76.1	10.3
7. Tulsipur	A. 2040.5	6.5	16.4	7.5	12.8	46.6	16.7
	B. 1140.0	3.3	17.6	2.6	23.1	47.7	9.0
8. Gainsari	A. 3313.0	9.9	9.9	13.7	10.4	55.1	10.9
	B. 2702.0	7.4	13.9	6.2	6.4	69.8	3.7
9. Pachperwa	A. 445.5	1.5	30.0	14.4	21.2	22.8	11.6
	B. 432.0	1.4	43.8	17.1	12.5	25.7	0.9
10. Balrampur	A. 3458.0	41.7	0.9	6.9	2.3	75.4	14.5
	B. 228.0	2.5	—	32.9	2.2	56.1	8.8
Basti							
11. Itwa	A. 10876.0	47.5	—	21.6	68.2	4.1	6.0
	B. 9449.0	40.5	—	18.0	62.6	14.0	5.4
12. Bhanwapur	A. 11464.0	52.4	—	4.3	18.8	69.9	7.0
	B. 11924.0	54.2		15.3	54.7	23.5	6.5
13. Khuniaon	A. 12562.0	66.8	—	5.6	30.2	54.2	10.0
	B. 10604.0	55.6	—	10.2	45.0	31.7	13.1

(Contd.)

(Contd.)

1	2	3	4	5	6	7	8
14. Barhni	A. 5319.0	21.3	16.8	6.0	40.7	2.2	34.7
	B. 3197.0	20.1	20.1	4.7	74.9	—	0.3
15. Birdpur	A. 8383.5	43.3	41.3	25.5	10.5	12.2	10.5
	B. 5556.0	28.8	16.0	5.0	78.8	—	0.2
16. Naugarh	A. 9508.0	54.0	49.5	4.4	12.9	20.3	12.9
	B. 7089.0	39.1	32.6	5.9	60.4	—	1.1
17. Jogia	A. 4664.0	31.1	29.2	16.6	9.6	35.2	9.5
	B. 2580.0	17.7	40.1	5.9	52.7	—	1.3
18. Uska	A. 8164.5	47.7	13.3	12.2	24.1	27.0	23.3
	B. 6003.0	34.3	16.1	10.2	73.2	—	0.5
19. Bansi	A. 4304.0	34.1	—	4.9	55.8	25.0	14.3
	B. 6394.0	56.3	—	—	19.1	70.4	10.5
Gorakhpur							
20. Nautanwa	A. 12062.5	32.3	20.6	9.2	33.1	33.9	9.2
	B. 7267.0	31.1	6.5	14.0	8.6	13.7	57.2
21. Laxmipur	A. 14267.5	55.3	19.1	16.0	12.9	48.6	3.6
	B. 10487.0	63.2	3.5	29.6	42.4	11.4	13.1
22. Brijmanganj	A. 10205.0	69.4	0.3	17.6	11.8	64.5	5.7
	B. 8274.0	45.7	—	22.3	63.9	1.3	12.5
23. Pharenda	A. 11525.5	58.5	—	8.8	6.5	75.6	9.1
	B. 9088.0	59.6	—	24.7	32.6	30.8	11.9
24. Nichlaul	A. 10067.0	43.3	41.7	6.4	4.4	41.0	6.5
	B. 10629.0	43.7	35.8	8.8	7.1	42.9	5.3
25. Mithaura	A. 13653.0	69.8	83.7	3.5	0.6	8.0	4.2
	B. 12456.0	61.0	78.3	11.6	0.5	4.8	4.8
26. Maharajganj	A. 13950.0	82.4	84.6	1.7	1.6	9.3	2.8
	B. 12484.0	71.1	68.1	5.9	2.1	15.8	8.1
27. Ghughli	A. 11646.0	76.8	70.1	2.0	1.6	24.4	1.9
	B. 9959.0	65.2	87.6	7.2	3.8	1.0	0.4
28. Siswa	A. 11435.0	67.9	74.5	2.4	0.8	21.6	0.7
	B. 10151.0	56.8	71.4	7.7	3.0	15.5	2.4
Deoria							
29. Khadda	A. 3714.0	18.2	44.4	7.0	22.8	22.2	3.5
	B. 8117.0	28.9	79.5	0.6	2.3	14.9	2.7
30. Nebua Naurangia	A. 6046.0	34.1	79.7	7.9	3.7	4.5	4.2
	B. 5361.0	30.0	89.2	0.8	1.1	6.1	2.8
31. Bishunpura	A. 6054.0	33.9	69.2	2.0	2.8	25.0	1.0
	B. 2583.0	13.5	41.2	0.9	0.9	51.9	5.1
Grand Total	A.228757.0	32.6	32.8	8.9	14.9	37.4	6.0
	B.186577.0	26.3	31.0	11.4	27.6	21.3	8.7

* A = 1985-87; B = 1975-77

APPENDIX VII

General Land-use

(As % of Total Area)

Block	Year	Total area in hactare	Forest	Area not available for cultivation	Culturable waste land	Groves & gardens	Grass lands	Fallow-land	Net sown area	Double cropped area	Net irrigated area
1	2	3	4	5	6	7	8	9	10	11	12
Bahraich											
1. Mihipurwa	*A.	59546	15.5	12.5	0.7	0.5	—	5.0	65.8	40.4	10.7
	*B.	65820	18.0	10.7	2.5	0.7	0.1	1.5	66.5	44.0	4.4
	*C.	65850	21.2	9.0	16.3	0.7	NA	9.1	43.7	41.2	0.4
	*D.	65950	22.0	6.7	21.4	0.6	NA	8.0	41.3	32.9	—
2. Nawabganj	A.	24457	6.0	9.2	0.5	1.8	—	2.7	79.8	50.2	22.0
	B.	25727	7.2	5.1	1.1	1.6	0.1	1.3	83.6	28.3	13.4
	C.	25865	7.9	6.5	3.9	1.4	NA	1.2	78.1	15.2	7.1
	D.	25714	9.7	6.2	6.0	1.0	NA	1.3	75.2	19.0	4.0
3. Jamunaha	A.	34275	3.6	11.9	0.5	0.8	—	4.2	79.0	50.2	20.2
	B.	34128	2.9	8.2	0.7	1.2	0.2	2.1	84.6	35.9	11.2
	C.	34136	5.8	7.9	3.6	1.1	NA	2.1	79.5	33.7	1.1
	D.	34262	9.1	6.2	6.0	1.0	NA	1.8	75.8	36.7	2.6

1	2	3	4	5	6	7	8	9	10	11	12
4. Hariharpur Rani	A.	33094	1.5	11.6	0.9	0.6	—	4.2	81.2	62.1	8.6
	B.	33094	0.5	11.6	1.1	0.9	0.2	3.0	82.7	49.9	7.7
	C.	33094	0.9	11.3	2.9	0.7	NA	3.9	80.3	48.4	—
	D.	33043	1.5	10.7	4.9	0.7	NA	3.7	78.5	50.9	—
5. Sirsia	A.	50028	8.8	7.1	2.6	0.6	—	3.1	77.8	47.9	3.8
	B.	50028	8.9	7.1	2.5	1.6	0.1	2.1	77.7	52.9	2.1
	C.	50028	13.3	8.2	15.9	0.6	NA	2.9	60.1	39.8	0.1
	D.	50028	16.5	7.0	15.5	0.5	NA	5.3	55.2	35.7	—
Gonda											
6. Haraiya Satgharwa	A.	50103	12.3	10.5	1.3	0.6	—	7.4	67.9	58.8	5.3
	B.	49479	12.2	9.4	1.7	0.8	0.02	1.5	74.4	40.4	3.9
	C.	49469	13.0	7.5	6.2	0.7	NA	3.0	69.5	33.3	0.8
	D.	49470	15.3	7.1	10.0	0.7	NA	2.7	64.2	39.6	0.3
7. Tulsipur	A.	41114	3.0	11.5	1.1	1.1	—	6.8	76.5	57.4	6.5
	B.	42350	2.6	10.3	1.8	0.9	0.1	3.4	80.9	56.9	3.3
	C.	42147	5.0	9.5	6.3	0.7	NA	3.0	75.5	35.0	1.3
	D.	42138	7.3	9.2	6.7	0.9	NA	2.9	73.0	44.3	0.9
8. Gainsari	A.	46456	8.7	10.5	0.8	0.9	—	7.2	71.9	55.2	9.9
	B.	46744	6.0	9.1	2.0	0.8	0.2	3.2	78.7	36.9	7.4
	C.	46650	8.2	8.1	7.1	0.8	NA	3.6	72.2	28.6	1.8
	D.	46590	10.8	7.9	8.7	1.8	NA	4.3	67.3	32.9	0.5
9. Pachperwa	A.	39248	7.6	10.5	1.0	1.2	—	7.9	73.8	61.4	1.5
	B.	40563	7.5	8.9	1.7	0.9	0.1	2.7	78.0	40.0	1.4
	C.	40700	8.8	7.9	7.3	1.1	NA	4.2	70.7	32.8	0.7
	D.	40663	10.4	7.5	10.8	0.8	NA	1.6	66.8	39.8	1.2

(Contd.)

(Contd.)

1	2	3	4	5	6	7	8	9	10	11	12
10. Balrampur	A.	11456	0.5	12.2	0.9	0.8	—	6.7	78.9	48.7	38.2
	B.	11456	0.6	12.5	0.9	1.2	0.0	4.4	80.3	33.6	2.5
	C.	11490	0.7	15.1	1.3	1.1	NA	3.3	78.5	62.6	2.4
11. Itwa	A.	27481	0.1	7.3	3.8	1.2	—	4.2	83.4	46.8	47.5
	B.	27481	0.1	7.5	3.0	1.2	0.8	2.5	84.8	48.9	40.5
	C.	27476	0.2	7.8	4.1	1.5	NA	3.1	83.2	30.1	—
	D.	27481	0.3	6.5	4.3	1.5	NA	2.5	84.4	27.3	0.1
12. Bhanwapur	A.	27711	0.01	11.8	2.9	1.6	—	4.7	89.0	49.2	52.4
	B.	27457	0.03	11.8	2.0	1.6	1.0	3.4	80.2	43.6	54.2
	C.	27532	0.1	8.7	3.4	2.0	NA	2.3	83.5	30.9	—
	D.	27547	0.1	8.4	3.3	2.0	NA	2.8	83.3	30.7	—
13. Khuniaon	A.	23231	0.9	8.8	4.4	1.6	—	3.4	80.9	57.4	66.8
	B.	23468	0.9	8.6	3.7	1.5	1.2	2.8	81.2	39.6	55.6
	C.	23452	0.3	7.2	7.8	1.6	NA	3.4	79.5	33.7	—
	D.	23450	1.9	6.2	8.1	1.6	NA	2.7	79.5	28.4	—
14. Barhni	A.	30416	0.2	10.2	2.3	1.3	—	4.0	82.0	29.8	21.3
	B.	29440	0.1	10.8	4.1	0.9	—	2.6	81.5	31.4	13.3
	C.	29443	0.1	10.8	4.5	0.7	NA	3.4	80.5	27.5	—
	D.	29443	0.1	9.0	5.3	0.7	NA	3.4	81.5	30.1	—
15. Birdpur	A.	24126	1.4	11.4	2.7	0.5	—	3.8	80.2	56.9	43.3
	B.	24592	0.8	9.2	6.6	0.4	—	0.6	82.4	20.9	31.2
	C.	21591	0.9	9.3	6.2	0.1	—	0.9	82.6	25.7	—
	D.	21593	0.9	8.0	6.8	0.1	—	1.2	82.4	27.2	—

1	2	3	4	5	6	7	8	9	10	11	12
16. Naugarh	A.	22238	2.2	11.1	3.2	1.2	—	3.2	79.1	47.4	54.0
	B.	22356	0.8	8.1	7.2	1.4	—	1.4	81.1	31.8	39.1
	C.	22356	0.8	8.1	7.6	1.4	—	1.7	80.4	31.3	—
	D.	22358	0.9	8.1	8.2	0.5	—	1.4	80.9	34.2	—
17. Jogia	A.	17863	0.3	9.8	1.4	1.0	—	3.7	83.8	50.0	31.1
	B.	17296	0.4	8.5	2.8	1.4	—	2.8	84.3	35.2	17.7
	C.	17296	0.4	8.5	3.4	1.1	—	[illegible]	84.6	35.5	—
	D.	17298	0.4	8.8	3.4	0.5	—	0.8	86.1	44.1	—
18. Uska	A.	21401	0.3	12.8	2.2	1.1	—	3.6	80.0	53.8	47.7
	B.	21426	0.2	10.1	3.7	1.9	—	2.4	81.7	33.2	34.3
	C.	21427	0.2	10.3	4.2	1.7	—	3.1	80.5	36.2	—
	D.	21428	0.2	9.9	4.7	1.3	—	2.6	81.3	32.7	—
19. Bansi	A.	15588	0.1	10.5	1.9	1.6	—	5.0	80.9	46.3	34.1
	B.	15588	0.1	11.4	2.9	2.6	—	10.2	72.8	50.3	56.3
	C.	15588	—	11.6	1.7	2.8	—	10.5	73.4	43.2	45.8
	D.	15588	—	10.8	1.8	2.8	—	12.3	72.3	48.0	40.9
Gorakhpur											
20. Brijmanganj	A.	20320	0.2	9.2	1.3	0.2	—	3.5	85.6	47.0	69.4
	B.	20933	0.1	7.6	1.7	1.5	0.03	2.7	86.4	13.9	45.8
	C.	20950	0.1	6.6	3.6	1.4	NA	2.6	85.7	13.7	0.2
	D.	20970	0.2	6.7	3.9	1.4	NA	3.1	84.7	11.0	0.03

(Contd.)

(Contd.)

1	2	3	4	5	6	7	8	9	10	11	12
21. Laxmipur	A.	30920	5.8	7.6	1.0	0.1	—	2.0	83.5	29.2	55.3
	B.	26768	23.9	8.0	2.0	1.3	0.2	2.7	62.0	45.5	63.2
	C.	26428	24.6	6.2	2.1	1.1	NA	2.6	63.4	24.9	—
	D.	26781	25.2	5.2	3.3	1.3	NA	2.4	62.6	27.0	2.1
22. Nautanwa	A.	38416	6.0	7.7	1.1	—	—	—	82.1	23.9	32.3
	B.	37100	18.5	6.1	1.7	1.9	0.1	8.6	63.1	51.8	31.1
	C	37213	18.9	4.6	0.9	0.7	NA	1.5	73.4	32.6	1.0
	D.	37241	19.4	4.2	2.0	0.7	NA	1.4	72.3	34.2	0.8
23. Pharenda	A.	23553	6.3	7.0	0.8	0.1	—	2.3	83.5	30.1	58.5
	B.	23632	21.2	7.4	2.1	1.5	0.3	3.0	64.5	41.2	59.5
	C.	23630	28.6	5.3	1.6	1.0	NA	2.1	61.4	30.9	0.6
	D.	23650	29.1	4.7	2.7	1.1	NA	2.2	60.2	28.1	0.6
24. Nichlaul	A.	31920	10.6	10.9	1.6	—	—	4.0	72.9	63.7	43.3
	B.	31920	8.7	8.3	2.3	0.9	0.2	3.4	76.2	49.1	43.7
	C.	31686	9.6	5.1	8.1	0.9	NA	6.6	69.7	26.7	0.4
	D.	31827	0.1	4.3	11.8	0.9	NA	4.7	68.2	25.8	0.3
25. Mithaura	A.	23815	3.4	11.5	1.0	1.3	—	1.9	82.1	77.5	69.8
	B.	23815	2.6	6.7	2.6	1.3	0.1	0.9	85.8	53.3	51.0
	C.	23823	2.7	4.6	2.5	1.3	NA	3.2	85.7	31.4	—
	D.	23840	2.9	5.5	4.0	1.5	NA	3.3	82.8	30.4	—
26. Maharajganj	A.	21546	6.1	11.7	1.4	—	—	2.3	78.5	77.4	82.5
	B.	21546	5.0	7.5	2.3	1.0	0.2	1.9	82.1	71.2	71.1
	C.	21703	6.3	3.9	2.3	2.3	NA	3.8	81.4	29.6	—
	D.	21827	6.6	5.8	5.2	3.0	NA	3.5	75.9	33.4	0.4
27. Ghughli	A.	17721	0.04	9.9	1.3	—	—	3.2	85.6	67.8	76.8
	B.	17721	0.3	7.0	1.0	2.4	0.2	2.1	86.2	62.0	65.2
	C.	17777	0.3	6.2	4.6	1.3	NA	5.2	82.4	26.1	—
	D.	17801	0.4	6.4	6.3	1.3	NA	4.4	81.2	23.3	0.1

1	2	3	4	5	6	7	8	9	10	11	12
28. Siswa	A.	20434	0.6	12.2	1.7	—	—	3.1	82.4	65.0	67.9
	B.	20434	0.4	6.6	1.6	2.2	0.3	1.5	87.4	53.9	86.8
	C.	20528	0.5	5.7	3.7	1.6	NA	4.0	84.5	32.5	0.2
	D.	20529	0.6	6.3	5.2	1.8	NA	4.2	81.9	27.2	0.2
Deoria											
29. Khadda	A.	32795	1.2	32.0	1.7	0.3	—	2.5	62.3	28.5	18.2
	B.	40612	1.8	15.4	1.4	1.2	1.4	9.6	69.2	11.0	28.9
	C.	40383	3.3	7.9	18.7	1.1	NA	6.9	62.1	11.9	—
	D.	40612	3.9	8.0	20.4	1.0	NA	6.8	59.7	11.3	—
30. Nebua Naurangia	A.	20570	0.1	9.5	1.3	0.2	—	2.8	86.1	19.4	34.1
	B.	20571	0.02	7.9	1.8	1.1	0.6	1.7	86.9	32.5	30.0
	C.	20490	0.1	8.9	3.5	1.1	NA	2.0	84.4	13.4	—
	D.	20528	0.2	7.8	6.5	2.1	NA	4.3	79.1	17.3	—
31. Bishunpura	A.	21733	0.3	10.7	2.8	0.6	—	3.6	82.2	45.8	33.9
	B.	22372	0.3	5.9	2.1	1.9	1.3	2.1	85.9	29.4	13.5
	C.	22353	0.7	5.1	5.5	1.6	NA	4.4	82.7	22.9	15.7
	D.	22353	1.0	5.1	6.4	1.5	NA	3.3	82.7	16.9	11.4
Total	A.	903560	4.8	11.1	1.5	0.6	0.2	4.2	77.6	49.4	32.0
	B.	912907	8.2	9.0	2.3	1.3	0.3	3.0	77.9	40.2	26.3
	C.	912954	7.5	7.8	6.6	1.1	**	3.8	73.2	29.9	—
	D.	913503	8.6	7.2	8.4	1.1	**	3.8	70.9	32.1	—

* A = 1985-87, B = 1975-77, C = 1965-67, D = 1955-57.

** Included in the forested area.

NA = Not Available

APPENDIX VIII

Overall Cropping Pattern

(1985-87 Average)
(As % of the total cropped area)

Blocks	Cereals			Pulses			Total foodgrains			Fruits & vegetables	Oil-seeds	Sugar-cane	Non-food crops
	Kharif	Rabi	Total	Kharif	Rabi	Total	Kharif	Rabi	Total				
1	2	3	4	5	6	7	8	9	10	11	12	13	14
Bahraich													
1. Mihipurwa	48.5	25.9	74.4	1.2	11.6	12.8	49.7	37.5	87.2	0.7	4.7	5.0	1.1
2. Nawabganj	54.3	25.5	79.8	2.8	15.0	17.8	57.1	40.5	97.6	0.7	1.0	—	0.7
3. Jamunaha	50.6	26.1	76.7	2.7	16.0	18.7	53.3	42.1	95.4	0.2	3.4	0.03	0.4
4. Hariharpur Rani	50.7	30.3	81.0	1.3	13.3	14.6	52.0	43.6	95.6	0.5	3.3	0.06	0.3
5. Sirsia	54.9	23.9	78.8	1.2	11.8	13.0	56.1	35.7	91.8	0.8	7.0	0.02	0.5
Gonda													
6. Haraiya Satgharwa	47.6	28.8	76.4	4.1	11.6	15.7	51.6	40.4	92.0	0.7	6.6	0.4	0.3
7. Tulsipur	44.6	25.1	69.7	4.3	11.0	15.3	48.9	36.1	85.0	0.9	7.7	6.2	0.2
8. Gainsari	43.9	25.2	69.1	2.3	11.2	13.5	46.2	36.4	82.6	0.8	6.9	9.3	0.2
9. Pachperwa	50.7	25.7	76.4	1.3	12.2	13.5	52.0	37.9	89.9	1.0	5.8	2.4	0.1
10. Balrampur	42.3	30.2	72.5	4.4	12.4	16.8	46.7	42.6	89.9	1.5	3.6	4.9	0.2

1	2	3	4	5	6	7	8	9	10	11	12	13	14
Basti													
11. Itwa	58.4	31.9	90.3	0.5	4.1	4.6	58.9	36.0	94.9	1.1	2.4	0.9	0.5
12. Bhanwapur	58.9	31.4	90.3	1.0	3.2	4.2	59.9	34.6	94.5	1.5	1.7	2.0	0.3
13. Khuniaon	51.0	37.4	88.4	0.7	5.6	6.3	51.7	43.0	94.7	1.5	1.8	1.5	0.3
14. Barhni	59.0	28.6	87.6	1.3	4.9	6.2	60.3	33.5	93.8	1.4	2.5	2.9	—
15. Birdpur	60.1	29.7	89.7	1.3	3.9	5.2	61.4	33.6	95.0	0.5	3.2	—	0.4
16. Naugarh	57.3	31.9	89.2	1.2	4.4	5.6	88.4	36.3	94.7	1.9	2.7	0.03	0.3
17. Uska	49.9	34.4	84.3	0.5	8.6	9.1	50.4	43.0	93.4	2.1	2.9	0.4	0.7
18. Jogia	48.2	36.1	84.3	0.7	7.8	8.5	48.9	43.9	92.8	1.9	2.7	1.3	0.4
19. Bansi	43.9	43.6	87.5	0.1	7.4	7.5	44.0	51.0	95.0	3.2	1.3	0.3	0.7
Gorakhpur													
20. Nautanwa	55.6	28.1	83.7	1.0	9.5	10.5	56.6	37.6	94.2	1.8	3.3	0.1	0.1
21. Laxmipur	51.8	38.9	90.7	1.0	3.2	4.2	52.8	42.1	94.9	1.8	2.5	0.2	0.2
22. Brijmanganj	49.0	38.1	87.1	0.9	3.3	4.2	49.9	41.4	91.3	2.5	3.0	2.7	0.3
23. Pharenda	47.1	40.3	87.4	1.2	3.8	5.0	48.3	44.1	92.4	2.2	2.2	2.2	0.1
24. Nichlaul	44.6	30.5	75.1	0.7	8.9	9.6	45.3	39.4	84.7	1.5	3.1	9.9	0.3
25. Mithaura	50.8	34.7	85.5	0.3	8.7	9.0	51.1	43.4	94.5	1.5	1.8	1.6	0.2
26. Maharajganj	50.2	38.4	88.6	0.2	4.9	5.1	50.4	43.3	93.7	1.8	1.5	2.5	0.2
27. Ghughli	45.5	38.2	83.7	0.2	4.4	4.6	45.7	42.6	88.3	2.5	1.5	7.1	0.1
28. Siswa	46.0	35.5	81.6	0.8	4.8	5.6	46.8	40.3	87.2	2.0	1.5	6.8	0.2
Deoria													
29. Khadda	38.9	25.0	63.9	1.5	4.0	5.5	40.4	29.0	69.4	2.4	3.5	22.4	1.7
30. Nebua Naurangia	37.5	23.7	61.2	0.5	2.5	3.0	38.0	26.2	64.2	2.9	4.1	26.4	1.3
31. Bishunpura	40.4	24.8	65.2	0.8	1.8	2.6	42.2	26.6	68.8	3.3	1.9	26.1	0.6
Total Region	49.8	30.3	80.1	1.5	8.4	9.9	51.3	38.7	90.0	1.0	3.7	4.2	0.3

APPENDIX IX

Ranking Crops

Blocks	Ranks																			
	1				2				3				4				5			
	A	B	C	D	A	B	C	D	A	B	C	D	A	B	C	D	A	B	C	D
1	2	3	4	5	6	7	8	9	10	11	12	13	14	15	16	17	18	19	20	21
Bahraich																				
1. Mihipurwa	R31.6	R29.9	M27.0	M28.3	W25.0	M25.3	R25.7	R24.6	M15.8	W18.4	W11.4	W11.4	RP11.6	RP7.2	RP9.9	RP8.7	S5.0	O6.1	WB3.7	O7.8
2. Nawabganj	R50.5	R46.6	R49.8	R46.6	W24.7	W20.0	W16.6	RP16.0	RP15.0	RP12.8	RP14.8	W12.1	KP2.8	K3.7	M5.8	WB9.8	M2.7	M2.9	WB4.3	M6.0
3. Jamunaha	R41.4	R38.2	R31.5	R31.0	W25.2	W19.3	M16.0	M14.5	RP16.0	M12.4	W13.6	W13.9	M7.9	RP10.8	RP13.5	RP12.9	O3.4	K4.3	WB9.1	K4.9
4. Hariharpur Rani	R30.4	M33.7	M34.3	M30.6	W28.6	W27.6	W19.6	W19.7	M19.9	R19.9	R18.9	W17.2	RP13.3	RP5.1	RP8.0	B13.0	O3.3	O4.5	B7.2	RP9.0
5. Sirsia	R40.4	R39.7	R35.0	R34.5	W21.6	M22.6	M20.3	M18.3	M13.6	W17.8	W15.1	W17.1	RP11.8	RP7.3	RP7.1	RP8.8	O7.0	O6.3	B6.5	B7.4
Gonda																				
6. Haraiya Satgharwa	R28.2	R37.5	R38.3	R36.2	W26.1	W21.0	M13.5	M12.2	M16.9	M14.1	W11.2	RP11.0	RP11.6	RP6.4	RP8.9	WB9.5	O6.6	O4.5	WB5.6	W8.5
7. Tulsipur	R37.6	R34.4	R38.0	R37.4	W22.8	W20.2	M11.5	RP10.8	RP11.0	K10.6	W10.8	W9.9	O7.7	M7.8	RP7.5	M9.8	S6.2	RP6.2	B5.1	WB7.4
8. Gainsari	R39.7	R44.9	R47.5	R46.4	W23.2	W18.3	W 8.0	B 8.2	RP11.2	RP8.6	K6.7	RP8.1	S9.3	O6.8	RP6.6	W7.9	O6.9	S5.6	B6.5	K6.9
9. Pachperwa	R46.5	R51.9	R49.2	R47.0	W22.7	W16.9	B 9.5	RP12.0	RP12.2	RP7.7	W6.4	B7.6	O5.8	O7.3	RP6.1	K6.6	K2.7	K4.0	K5.6	W6.2
10. Balrampur	R37.7	R28.1	R31.8	R29.6	W28.5	W19.7	RP12.4	W13.7	RP12.4	M13.1	W12.1	RP10.6	S4.9	RP11.5	M11.8	M9.7	KP4.4	K4.9	B6.8	B8.9
Basti																				
11. Itwa	R50.3	R50.3	R51.7	R49.2	W29.4	W27.8	W15.8	WB10.4	K7.7	K8.3	K8.3	W9.9	RP4.4	RP5.0	RP8.1	K9.6	O2.4	O2.0	WB7.4	B7.1
12. Bhanwapur	R54.5	R53.8	R52.5	R50.8	W29.3	W28.1	W18.1	W12.3	K3.7	RP4.8	RP6.1	WB8.5	RP3.2	K3.7	WB5.2	K6.7	S2.0	S2.6	K4.7	B5.6

1	2	3	4	5	6	7	8	9	10	11	12	13	14	15	16	17	18	19	20	21
13. Khuniaon	R46.4	R47.7	R48.8	R45.7	W36.3	W30.5	W15.1	W11.1	RP5.6	RP6.9	RP9.4	K11.0	K4.1	K5.1	WB7.0	WB9.5	01.8	S2.7	K6.9	RP7.0
14. Barhni	R55.8	R53.8	R56.9	R50.1	W27.7	W11.3	W 9.0	RP8.8	RP4.9	RP7.3	RP8.8	K8.3	S2.9	K6.6	K7.0	WB7.8	O2.5	B4.2	B4.7	B6.8
15. Birdpur	R58.7	R69.1	R69.6	R68.8	W27.0	W12.0	WB7.5	WB8.0	RP3.9	WB4.3	W5.7	B4.9	O3.2	B2.2	RP3.7	RP4.3	KP1.3	BG2.1	B3.5	W4.1
16. Naugarh	R56.5	R60.3	R60.8	R61.2	W30.6	W13.2	W 8.0	B11.9	RP4.4	RP5.6	B6.7	WB6.5	O2.7	WB4.8	WB6.5	RP5.6	KP1.2	B4.6	RP6.0	W4.3
17. Jogia	R41.4	R40.7	R39.2	R33.3	W30.5	K12.9	B14.9	B24.6	RP7.8	W10.5	RP11.5	K12.9	K3.9	B10.0	K10.0	RP8.1	O2.7	WB7.1	WB5.8	WB6.9
18. Uska	R44.1	R42.8	R43.8	R40.7	W27.6	W11.8	W10.6	B16.8	RP8.6	K10.8	B10.3	K10.2	K3.8	B9.1	RP9.4	WB8.1	O2.9	WB7.6	K7.5	RP7.7
19. Bansi	R41.9	R33.3	R31.4	R29.2	W40.8	W16.6	B14.1	B25.1	RP7.4	B13.3	K13.3	K12.7	O1.6	K11.5	RP12.5	W9.0	WB1.0	RP7.9	W10.9	RP8.7
Gorakhpur																				
20. Nautanwa	R55.4	R62.2	R64.5	R63.6	W27.6	W22.5	W16.0	W14.0	RP9.5	RP8.7	WB5.1	WB5.6	O3.3	O1.2	RP4.4	RP5.0	KP1.0	WB1.0	B2.2	B3.3
21. Laxmipur	R50.8	R43.0	R54.9	R53.6	W37.7	W36.6	W12.5	WB11.2	RP3.2	RP4.8	WB8.1	W9.8	O2.5	WB2.7	B7.1	B6.2	KP1.0	S2.1	RP4.4	RP5.4
22. Brijmanganj	R48.0	R47.6	R59.1	R57.6	W37.8	W34.5	W10.9	W10.6	RP3.3	RP3.9	WB8.9	WB9.3	O3.0	S2.4	RP6.4	B7.0	S2.7	O1.7	B6.3	RP6.7
23. Pharenda	W39.7	R47.1	R47.5	R47.3	R35.7	W34.5	W14.3	W11.9	RP3.8	RP3.2	B10.6	B11.7	O2.3	S2.9	RP7.7	RP7.9	S2.2	M1.3	WB6.8	WB6.6
24. Nichlaul	R44.2	R49.6	R52.4	R50.6	W30.2	W23.1	W11.6	RP8.7	S9.9	S10.1	S10.0	W8.1	RP8.9	RP9.2	RP5.7	S7.9	O3.1	O2.2	B5.6	B6.8
25. Mithaura	R50.8	R49.8	R54.8	R55.7	W34.1	W30.9	W12.5	RP7.9	RP8.7	RP9.4	B6.2	W7.8	O1.8	S2.9	RP6.0	B6.7	S1.6	O2.3	S5.7	S4.6
26. Maharajganj	R50.1	R51.7	R55.9	R57.2	W38.2	W32.5	W15.0	W11.2	RP4.9	RP6.3	RP6.2	RP8.9	S2.5	S3.4	B6.1	WB5.2	O1.5	O1.2	S5.7	S5.1
27. Ghughli	R45.3	R47.1	R45.4	R46.2	W37.8	W32.1	W16.9	W11.3	S7.1	S9.8	S9.3	S9.3	RP4.4	RP6.0	RP6.4	RP7.9	O1.5	KP0.9	B5.7	B6.9
28. Siswa	R45.6	R46.9	R50.5	R50.4	W35.3	W30.9	W15.5	W11.4	S8.7	S10.2	S8.1	RP7.7	RP4.8	RP5.4	B6.5	B6.5	O1.6	Q1.5	RP5.2	S6.4
Deoria																				
29. Khadda	R35.6	R44.4	R35.4	R36.4	W24.7	S27.3	S27.6	S27.6	S22.2	W15.2	B8.7	B9.7	RP4.0	RP4.2	K7.8	K8.3	O3.5	K2.8	W7.6	W4.8
30. Nebua Naurangia	R36.1	R34.3	R47.3	R47.6	S26.1	S29.3	S12.6	W10.6	W23.7	S24.6	W11.2	S9.7	O4.1	RP3.3	RP9.4	RP9.3	RP2.5	O3.1	B5.0	B5.8
31. Bishunpura	R36.2	R32.2	R59.7	R63.9	S25.6	W26.1	W 5.2	W16.5	W24.1	W23.6	RP3.9	B4.7	O1.9	K5.3	O3.2	KP3.0	RP1.8	RP3.0	WB2.1	WB2.3

Notes: A-1985-1987; B-1975-77; C-1965-67; D-1955-57.
R-Rice, W-Wheat, B-Barley, M-Maize, G-Gram, K-Kodon, RP-Rabi Pulses, KP-Kharif Pulses, S-Sugarcane, O-Oil-seeds, WB-Wheat mixed with barley.

*APPENDIX X**

Calorie output available for ingestion for per hectare of area under food crops for the village of Mangalpur**

Sl. No	Crops	Yields per hectare (kg)	Percentage of total cropped area	Total gross production (kg) Cols: 3X4	Total deduction at the rate of 20% from Col. 5	Net production available for ingestion Col: 5–6	Calorie per kilogram (Aykroyd et al, 1966)	Total calorie as available for ingestion (,000 calories) Col: 7X8
1	2	3	4	5	6	7	8	9
1.	Rice	1,400	47.1	65,940	13,188	52,752	3,450	1,81,994
2.	Kodon	300	2.3	690	138	552	3,070	1,695
3.	Pulses (Kharif)	550	1.7	935	187	748	3,410	2,551
4.	Sugarcane (Gur)	2,500	1.8	4,500	900	3,600	2,170	11,412
5.	Wheat	1,200	20.2	24,240	4,848	19,392	3,460	67,096
6.	Barley	600	2.7	1,620	324	1,296	3,360	4,355
7.	Wheat mixed with gram	1,000	5.1	5,100	1,020	4,080	3,380	13,790
8.	Pulses (Rabi)	650	14.0	9,100	1,820	7,280	3,410	24,825
9.	Oil-seeds	400	2.7	1,080	216	864	5,400	4,646
10.	Vegetables	2,000	0.4	600	120	480	970	466
	Total		98.0					3,12,850

Calorie available for ingestion per hectare = 3,12,850,000 (Col. 9) / 98.0

= 31,92,346

SNU (Standard Nutrition Unit) per hectare = Total calories available / Standard Quantity required

= 31,92,34 / 8,21,250**

= 3.9

* Based on the method used by Singh, J. and given as Appendix II in his book *An Agricultural Atlas of India, op. cit.*, p. 347.

** Taking into consideration the climatic conditions and the population characteristics of the region, 2250 calories per day or 8,21,250 calories per annum may be postulated as Standard Nutrition Unit. This is the quantity which is basically required to keep an average men in good health.

Aykroyd, W.R. *et al.*, (1966), *Nutrition Value of Indian Foods and the Planning of Statisfactory Diets*, New Delhi, Indian Council of Medical Research, pp. 51-86.

INDEX